WAYS OF
SOCIAL CHANGE

Dedicated to the memory of T. R. Young (1928–2003)
Colleague, mentor, friend

WAYS OF SOCIAL CHANGE

MAKING SENSE OF MODERN TIMES

GARTH MASSEY

Los Angeles | London | New Delhi
Singapore | Washington DC

Los Angeles | London | New Delhi
Singapore | Washington DC

FOR INFORMATION:

SAGE Publications, Inc.
2455 Teller Road
Thousand Oaks, California 91320
E-mail: order@sagepub.com

SAGE Publications Ltd.
1 Oliver's Yard
55 City Road
London EC1Y 1SP
United Kingdom

SAGE Publications India Pvt. Ltd.
B 1/I 1 Mohan Cooperative Industrial Area
Mathura Road, New Delhi 110 044
India

SAGE Publications Asia-Pacific Pte. Ltd.
33 Pekin Street #02-01
Far East Square
Singapore 048763

Senior Acquisitions Editor: David Repetto
Associate Editor: Julie Nemer
Assistant Editor: Maggie Stanley
Production Editor: Libby Larson
Copy Editor: Mark Bast
Typesetter: C & M Digitals (P) Ltd.
Proofreader: Susan Schon
Cover Designer: Bryan Fishman
Marketing Manager: Erica DeLuca
Permissions Editor: Karen Ehrmann

Copyright © 2012 by SAGE Publications, Inc.

Printed in the United States of America

Library of Congress Cataloging-in-Publication Data

Massey, Garth.

Ways of social change : making sense of modern times / Garth Massey.

p. cm.
Includes bibliographical references and index.

ISBN 978-1-4129-7987-0 (pbk.)

1. Social change. 2. Technological innovations—Social aspects. 3. Critical thinking. I. Title.

HM831.M3927 2012
303.4—dc23   2011019201

This book is printed on acid-free paper.

11 12 13 14 15 10 9 8 7 6 5 4 3 2 1

Table of Contents

Preface

The year 2011 opened with some of the most significant and stirring events in modern memory: massive peaceful street protests that toppled dictatorships in the Middle East. Tunisia and Egypt saw autocratic leaders abdicate. Bahrain witnessed one-fifth of its people marching in the streets and Pearl Square in its capital. Muammar el-Qaddafi's ironfisted, quixotic rule of Libya appears about to crumble, and in Yemen weeks of protests portend changes of power there. Syria, Jordan, and Iran have also seen massive protests demanding regime change and more democracy. The outcome is uncertain, and change will proceed along various paths, including economic transformations, cultural innovations, shifts in interpersonal relations and public attitudes, and a refiguring of political affairs both within these nations and internationally. Front-page headlines of the region will ebb, but social change has accelerated for tens of millions of people.

Social change is a complex process, even when it is accelerated by tumultuous and visually arresting events. The public upheaval in Middle Eastern nations was prefigured by years of repression, while in much of the world the twenty-first century promised increasing opportunities to match peoples' rising expectations. Young people, especially, refused to be denied what they saw as their birthright. Social networking and the vast array of global culture offered on the World Wide Web no doubt helped spark their imaginations and allowed their dreams to take shape as political dissent. These and other technologies, along with global economic shifts and multinational corporate growth following the end of the Cold War, were part of the mix of factors driving social change. The wars in Iraq and Afghanistan played some yet-undetermined role, as did the seemingly intractable conflict in Israel and Palestine. Seizing state power, as the protesters demanded in the name of their people, made evident the belief that states could and should be vehicles for social change. Making sense of what is happening in the Middle East requires consideration of many forces driving social change, none of which acts alone or in isolation from the others.

The United States has its own quotient of public displays for social change, largely played out in political arenas. Whether conservative or liberal, candidates have gained followings by presenting themselves as agents of change. When it comes to actually effecting change, voters are often disappointed. This isn't the change we expected, they'll say. Or, it's happening too slowly or inequitably or without our consent. Political change *is* often slow, incremental, halting, and inefficient in the kind of democratic political systems governing much of the world today, possibly soon to become the norm in the Middle East.

The prospect of social change seems to be everywhere, but it was not always this way. For most of human history, opportunities to think about, let alone make, a different sort of life were rarely considered. This took a dramatic turn for Europeans half a millennium ago, with the dawning of what is usually called the modern era, and significant changes in the rest of the world were the consequence. Popular uprisings ushered in new religions and new political systems, and still do. Dominant elites created organizations to reconfigure resource use, aggregate wealth, and apply it to new ventures. Wars devastated, and the rebuilding commenced, in a cycle that continues apace. Social movements of every variety proliferated and have become a major force for the enlargement of civil society. Literacy, affluence, and a loosening of restrictions on what could be tried stimulated scientific inquiry and technological development, driven by the prospect of discoveries' conversion into economic commodities. The growth of the national state to carry out a wider range of activities than ever before, by claiming assets and resources and using them collectively, has never been greater. These are the things discussed in this book.

Ways of Social Change, like most books for college classes, had its inception or inspiration from having taught social change both as a topic and as an integral dimension to other subjects over many years. Several social change texts, past and currently in use, are very good at explaining social change. Social change is their topic. What they don't do as well is provide the conceptual tools for students to make sense of social change processes in the world around them. In *Ways of Social Change*, topics are explored to help students make sense of how social change comes about.

Understanding social change, in the final analysis, is a critical-thinking skill needed to unravel the forces shaping the twenty-first century. The world is complicated, and how it changes—in large processes and personal circumstances—is not always obvious. Making sense of modern times requires a historical and comparative perspective, along with the ability to disentangle myth from empirical reality. For social science, the empirical reality of social change can be understood by zeroing in on the forces of change: how they operate, their consequences, and the ways people harness their power as agents of change. This approach is also a device for understanding the resistance to change as well as the resistance to social change being driven by one force rather than another. This is included in the five chapters on drivers of social change.

Ways of Social Change uses a literary device, a biographical approach, to acquaint students with social change as a lived experience. Iris Summers was a simple woman who came of age in a rapidly changing, complex world, trying to remain focused on what was most important to her. The world in which she lived was structured and moved by organizations and powerful forces, some well beyond her recognition, let alone her control. But that is not all of the story. She is the thread that runs through the book, reminding the reader that social change is very real and very personal and that human beings not only experience but make social change.

Two chapters introduce social science research and theoretical thinking, but in no sense are they more than brief introductions. These are not chapters *about* theory and research. Rather, they are about clarifying questions, conceptualization, the logic of inquiry, and the work of pursuing answers to questions using the social scientist's tools of inquiry. They show how research and theory are valuable to making sense of things by subjecting possible answers and reasoning to empirical analysis. These chapters stress research and theory's practical importance for a student who may never again take a social science course, let alone another course in social change. The tools and approaches covered in these chapters reappear throughout the book.

These and the five subject chapters—the drivers of social change—provide a foundation for a range of courses for students in sociology, political science, international studies, business and public administration, or other fields. *Ways of Social Change* is written for the student who is fairly new to the social sciences. It can be an introductory text for undergraduates or for graduate students with academic backgrounds in fields other than the social sciences. It can reinforce what was learned in an initial course in sociology or political science and guide students into more advanced courses.

The most obvious follow-up courses are in social and political change in sociology and political science: globalization, international development, social movements, technology and society, and international organizations. The attention given to the labor process can be continued in sociology of work and industrial sociology courses. It provides background and can be an introduction to studying social inequality and stratification, electoral politics, international relations, formal organizations, ethnic studies, conflict and peace studies, gender studies, and political sociology.

The material in *Ways of Social Change* is not presented as the final word. The voice of the narrator invites a response by the reader. As important as it is to read and digest, having the opportunity to give voice and hear ideas, opinions, and concerns is critical to learning, as every teacher knows. The Topics for Discussion at the end of each chapter extend the chapter's conversation and allow students to think beyond what they have read. They also provide an opportunity for differing points of view and allow students to express their concerns about what they have read. The Activities for Further Study at the end of the chapters offer independent inquiry to the more ambitious or engaged students and can be a starting point for instructors who assign research projects.

I have tried throughout this book to acquaint students with the narratives of contemporary social change. Several case studies recur throughout *Ways of Social Change*, including the modern civil rights movement, the Second World War, and the changing status of women. These are the pivotal events of the last hundred years for people in the United States. They continue to resonate in institutional arrangements, cultural beliefs and practices, political efforts and controversies, and lived experience. While stories about the United States are paramount, there is ample material to whet the student's appetite for more global studies. China, particularly, comes in for close attention. The poorest parts of the world are never far away from any discussion of the affluent, along with the power of the latter to affect the former.

Many of my students deplore the environmental costs of affluent lifestyles characterized by unsustainable consumption, fueled by nonrenewable energy, and requiring international relations backed by the threat or conduct of war. The search for the right paths to international change, helping the world's poorest 40 percent, and empowering people everywhere to pursue their best interests are challenges young people want to undertake. They seem to me eager to be involved in moving in new directions by making a positive contribution to social change. The final chapter provides some guidelines for this. Overall, *Ways of Social Change* takes this desire into the realm of social science where inquiry, clarification, understanding, and involvement can help provide for both informed opinions and a path to effective engagement.

Acknowledgments

F riends and colleagues too numerous to mention have suffered through my not infrequent befuddlement and mental blind spots as this book took shape. Students over many semesters have been coconspirators of inchoate ideas and didactic experiments with barely a murmur of protest. I appreciate that. At the outset of this project I was fortunate to get good advice from a friend who knows the book business, Dan Bowers. Ohio State professor Randy Hodson's encouragement and enthusiasm has meant a great deal to me, as has that of Pine Forge editor, David Repetto. Mark Bast provided both expertise and good suggestions in clarifying what I wanted to say. Longtime friend and University of Wyoming Libraries faculty member Michael Nelson always came through when I was stumped. Former UW grad student Li Qiang and Reed College graduate Charli Krause did great work as researchers, and anonymous reviewers offered incisive comments. My sons were positive sounding boards, readers, enthusiasts, and fact finders. Most important has been my lifelong companion and invariably supportive friend, Sheila Nyhus, who made my working days comfortable, offered smart commentary, and helped me stay focused on the task at hand. To all of them, I owe a huge debt of thanks.

1

The Personal Experience of Social Change

If you could use only six words, how would you describe your life? F. Scott Fitzgerald, author of *The Great Gatsby*, wistfully suggested that he and his wife, Zelda, would write, "For Sale: baby shoes, never worn." One of my students, Joe Hampton, penned, "No plan. Hope it works out." Trying to compose a phrase that captures or summarizes a life is a challenge. Life is long (we hope) and full of twists and turns. As the German philosopher Immanuel Kant wrote, "Out of the crooked timber of humanity, no straight thing was ever made." Our plans, sacrifices, character, perseverance, and common sense help take us where we want to go, but the road was not built by us, nor do we have control over the traffic lights and detours. That is why, in part, the study of society is so important.

We live in a world that could easily go about its business without us, and we leave the world with surprisingly little consequence, especially given all the effort we expend to become who we are. Whenever I arrive at a faraway destination, my first thought is that life there would be absolutely the same if I had missed my flight: taxis whizzing by, church bells ringing, school children rushing down the street, people in shops looking over goods to buy, friends embracing. I think, "I might as well be a ghost," until someone turns and asks if they can help me.

This is the mystery and marvel of studying social change. The human world in all its political, economic, cultural, biological, linguistic, and demographic complexity has been constructed over thousands of years and remains a work in progress. In that time, human beings for several thousands of generations have been born, lived and died, making little or no lasting individual contribution to the social order. They are long forgotten by history. Still, they did have their moment, and the world would be slightly different had they not lived. In their totality, there would be no social world, no culture, no economy, no political system, no war, and no religion if these seemingly insignificant and nameless millions had not lived.

Wait a minute! What about Moses and Abraham, Christ and Mohammed, Buddha, Zoroaster, and Confucius? What about Genghis Kahn, Nefertiti, Christopher Columbus, Julius Caesar, Cleopatra, Hannibal, Napoleon, queens Elizabeth I and Victoria, Adolph Hitler, Joseph Stalin, and Mao Zedong? What about Socrates and Aristotle, Leonardo da Vinci, Emily Dickinson, Charles Darwin, Pablo Picasso, Ludwig Van Beethoven, Madam Curie, Mohandas Gandhi, and Nelson Mandela? Who would we be without Abraham Lincoln, Albert Einstein, and Martin Luther King Jr.? These people and many, many others left their mark; they changed the way we live, think, and see the world. Their visions and efforts altered the course of social life, science, law, religious thought, economics, war, and peace.

True enough, but not alone and not only by dint of their brilliance, egomania, determination, and special gifts does society change. They, too, lived in a particular social world in which their efforts could be successful, even as their actions and ideas disrupted the taken-for-granted order of things. The changes historically attached to their lives can only be understood by taking a longer and wider view of the social world in which they lived. It is the confluence of biography and history, the personal and the public, private efforts and the social milieu, that allows us to understand social change.[1]

A Twentieth-Century Life: Iris Summers

Iris Summers is no one special. She lived a very normal life for her times, but the times changed greatly, and for the most part Iris changed with them. She never talked about social change and would never have considered herself an advocate for change. She was careful with her money (what little she had) and tried to be prepared for life's inevitable setbacks. She believed that some

---

[1]The work of C. Wright Mills guides the perspective that seeks to locate private issues in public history, or what he called "the sociological imagination" (Mills 1959).

things—expressed as aphorisms and maxims—were always true. For example, "A fool and his money are soon parted." Or, "Still waters run deep." And she often quoted maxims when giving advice, like, "Look before you leap." Or, "Never look a gift horse in the mouth." She thought these were always useful, no matter who you were or what was going on in the world. But what happened around and to Iris Summers greatly affected her life and the way she lived. Her life both reflects and makes manifest social change over nearly a century.

She had her children in the decade following the Second World War. These babies joined the twentieth-century Baby Boomers, bought rock 'n' roll records, and screamed in excitement at the sight of Elvis Presley and, later, the Beatles on TV's *Ed Sullivan Show*. They became the '60s Generation who joined the ongoing civil rights movement and fought in Vietnam. Many of them protested the war, experimented with drugs, believed in equal rights for women, and became disillusioned with the American Dream of unlimited abundance and personal consumption. They were followed by Generation X and then the current Millennial Generation, but their large numbers and the social tumult of their formative years made Iris' children feel like they were somehow special and the century's agents of social change.

Sociologist James Davis would disagree: "We tend to think of a conservative, rigid society suddenly modernized by a rebellious post-World War II youth." He finds the opposite: ". . . the strikingly modernized cohorts are those born in the first half of the century . . . [T]he rate of change in the social climate [after WWII] is, if anything, less than that for its predecessor" (Davis 1996: 165). Iris Summers was of that earlier era, and it is her life that tells so much about social change in modern times.

From Farm to Factory

Like many of your grandparents and great grandparents, Iris Summers was born on a farm. Literally, she was born in her parents' bed in the house on the farm, delivered by a woman from a nearby town who had some training and a lot of experience helping women in labor. Birth on the farm—calves, colts, piglets, chicks, pups, and kittens—was an everyday event. The first pictures of Iris were taken in a photo studio, but there are many pictures of her and her family on their farm. When she was a girl, the family had a Kodak Brownie camera, a mass-produced device marketed in 1900 that allowed her generation to be the first in human history to have a visual record of everyday life preserved for posterity.[2]

Just about everyone Iris knew lived on a farm. When she was born, small-scale agriculture was the work of millions of people. Across the nation twice as much energy was used by agriculture as was used in manufacturing. During her childhood, half of all Americans could say they lived or had lived on farms, but that was rapidly changing. Farm foreclosures were common in the 1920s, even before the Great Depression of the 1930s. Small family farms were consolidated into larger farms that spurred rural-to-urban migration and set the stage for the onset of industrial agriculture. By the end of the twentieth century—at the end of Iris' life—agriculture employed fewer than one in fifty American workers.

"In 1992 people were, on average, four-and-a-half times richer than their great-grandparents at the turn of the century," according to Alan Thein (1992: 23). Iris grew up poor. But then, most people were poor by any measure of material consumption used today. Her family bought a car, a black Model T, just like many of their neighbors, when she was in her teens. It was not long after they had gotten a battery-powered radio. Their home had running water in the kitchen, from a catchment built by her father, but there was no hot water for bathing or washing dishes other than what could be heated in a large kettle on the stove. As long as the house stood, it never had a bathroom; an outhouse several yards off the back porch served the purpose. Soon after she left the farm, electricity lines were strung across the countryside and hooked up to the house. This was part of the rural electrification program—a stimulus program of the federal government to get the nation's economy out of the Great Depression and a major reason farming and rural life changed so much in the twentieth century.

In the 1940s, telephone lines were installed, though several families shared a "party line" with her parents. Every household had its specific ring, and people were expected to not listen in on each others' conversations. Many years after Iris left home, the wood-burning stove was replaced by a gas range. The farmhouse was old and poorly built, however, and when farms were being bought up to make way for larger farms in the 1950s, it

---

[2]In 1888, George Eastman coined the name *Kodak* and marketed a camera that took one hundred pictures. When finished, owners sent $10 and the entire camera to the Eastman Kodak Company. They received back the prints and a newly loaded camera. This was popular but only affordable by the more affluent. In 1900, Eastman marketed the Brownie for $1 plus fifteen cents for a six-exposure role of film that could be removed and mailed in for developing. In the first year of sales, a quarter million Brownie cameras were sold. Photography had come to the masses.

was torn down along with many of the neighbors' houses, barns, and equipment sheds. Orchards were leveled and lanes were plowed into fields.

Iris left home when she was a young woman and lived her adult life in small towns. She always described her two years living in a city as "purgatory." She loved to garden and knew how to preserve food, mostly by canning. She could butcher a chicken and make clothes by sewing and knitting, but many of the skills required of her parents to maintain their farm household were of little use to her. She needed a formal education and a degree that qualified her for specific work in a job, one that had not existed a few years before.

When Iris' parents, Herm and Edna, were too old to farm, and the children had left home, they too moved to town. Five years after the 1935 Social Security Act was passed, they began receiving a modest income that continued for the rest of their lives. As poor farmers, there had been little opportunity to save for retirement, and they would otherwise have depended on their grown children for financial help, just the same as had their parents and grandparents. Herm developed "hardening of the arteries," what we would now diagnose as cardiovascular disease, and died soon after leaving the farm. Edna, however, lived to be 93, spending most of her life in a twentieth century that she increasingly didn't understand or really care much about.

Iris married Frank Summers soon after he had a steady job that could support a family. Even though it was the decade of the Great Depression, Frank began working full time when he finished high school and became a skilled laborer. He was soon making a good wage, the criterion for his generation to marry and start a family. All his life he could never understand why young people would marry before they were economically secure, or why those with jobs would remain single. This compact in an affluent society between work, marriage, and family would begin to break down for Frank and Iris' grandchildren in the last decades of the twentieth century.

Unlike later wars in Korea and Vietnam, the draft (conscription into the military) excluded young men working in sectors of the economy deemed important to the war effort, including agriculture and petroleum production. Frank worked in the oil fields, but he and Iris didn't sit out the war. Along with millions of others, they planted a victory garden and purchased war bonds to help underwrite the country's debt from military expenses. Like all Americans, they were subject to rationing and restrictions on what they could purchase: foodstuffs like sugar and cocoa; automobile products including oil and rubber tires; and household items. Much of the country's manufacturing was enlisted to produce military equipment and supplies, not unlike much of the rest of the world, though on a lesser scale than in Britain, Germany, Japan, and other nations prior to and during the Second World War.

Frank Summers, too, was born on a farm and grew up learning the skills and acquiring the habits of mind of rural families. His paternal grandfather was born during slavery and had fought in the Civil War. His grandpa would have been astonished by the mechanization of agriculture during Frank's early youth. The great transformation of American agriculture began about the time Iris and Frank were growing up. Animal traction—plowing, hauling, and harvesting with animals—gave way in those years to tractors and trucks with internal combustion engines. In time, the implements pulled by the tractor would be fitted with their own sources of power and grow to be huge and expensive, capable of working far more land in a single day than Iris and Frank's childhood farms combined. Farms would cease producing food to be eaten by the family. Instead, crops were grown to be sold, and the money was used to buy groceries and everything else the family needed.

Electricity, the cost of land, and agricultural technology and the reduction in human labor that accompanied it dramatically changed the rural landscape, a phenomenon repeated throughout much of the industrialized world in the twentieth century. It was thought that, in time, all countries would produce food in a similar industrial fashion, becoming the model for rural development in poorer countries in Africa, Latin America, and Asia that gained their independence after midcentury. Along with mechanization came increasing hybridization of plants and the development and widespread use of chemicals, many made from petroleum, to stimulate plant growth, kill weeds, and eliminate insect pests. By the end of the twentieth century, the genetic material in seeds was being modified. This technology was a byproduct of one of the most important scientific discoveries of the century, the unmasking of the genetic code underlying the evolution of all life forms.

Iris had six brothers and sisters and fifteen aunts and uncles, all but one who had children. Her husband, Frank, had an equivalent number of siblings, and so did his mother and father. As a result, both Iris and Frank had cousins by the dozens. Though nearly all of them—aunts, uncles, mothers and fathers, brothers, sisters, and cousins—were born and raised on farms, only a couple of cousins were farming by the middle of the twentieth century. Most of the others had joined the urban workforce in industries and services producing the things that distinguished Iris Summers' life from those who had come before her.

Extending the Reach

In terms of material goods, Iris shared the century's prosperity. So much changed: the growth of businesses and thousands of new occupations; cities ringed with endless suburbs; the ability to travel and communicate over long distances; the scope of war and preparations for conflict; and the vast globalization of trade and manufacturing, including the spread of Western popular culture and political systems modeled after those of the United States and European nations. The map of the world displayed the transformation of empires into independent national states, especially in Africa, South and Southeast Asia, and the Middle East. International organizations like the agencies of the United Nations and the World Bank were playing an increasingly important role in coordinating relations among nations. The pursuit of science vastly increased the range of interventions in health, from inoculations to organ transplants. Plate tectonics transformed geology, and the discovery of the double helix opened a panorama to knowledge about how biological life reproduces and changes.

In the first half of the twentieth century, the general theory of relativity and quantum mechanics changed how people thought about not only space and time but also "the way we think about almost everything, not only in physics but in chemistry and biology and philosophy . . . about cause and effect, about past and future, about facts and probabilities" (Dyson 2010: 20). In short, everything from the vast universe to nanoparticles and the unseen subatomic world. Space travel and the photographs it generated opened up the universe to radically new understandings of what Iris called "out there" as well as an appreciation for Mother Earth as a beautiful and fragile environment.

The amount of knowledge and technology in her lifetime grew exponentially. This, and the ideas of human and social improvement impelling much of the work of engineers, scientists, and practitioners, increased the human capacity to affect what to earlier generations had been taken for granted and beyond the reach of human intervention, from weather to mental health, from old age to social inequality. William Gamson sums this up well: "[T]he major thrust of social change in the United States has been to subject more and more forces to the manipulation of conscious decisions." This is true not only in democracies with capitalist economies but across the world (Gamson 1968: 189).[3]

Medical science, civil engineering, theoretical physics, information technology, and sciences whose names and province completely escaped Iris Summers contributed to making the products that became part of everyday

---

[3]The commissioner of the U.S. Patent Office in 1899, Charles H. Duell, is reputed to have offered to close the patent office in the belief that all useful devices and inventions had been patented. There is no evidence that he actually said this, but it was a staple of prideful talk in reference to the abundance of inventions and new technologies of the twentieth century.

life in affluent countries. The landscape of poor countries, too, was changed by the desire of their own people to follow suit, along with multinational corporations' efforts to find raw materials, new markets, and sites for manufacture and assembly.

Wars and civil conflicts throughout the century were often linked to economic and political transformations. Warfare was greatly influenced by changing technologies. Aircraft, armor, communication, and logistics made previously impossible forms of warfare possible. At the same time, nuclear weapons—developed in order to pose such a grave threat that their use would be inconceivable—anchored the Cold War divide following World War II. The growth of national states and the extension of the reach of the state[4] reflected shared ideas about progress, incubated ethnic conflicts, and fueled the powerful privately owned economic organizations—corporations—that increasingly linked people around the world.

Generations of Stability and Change

When growing up, Iris lived near many of her relatives, and her cousins were among her closest friends and playmates. Nearly every Sunday during her childhood relatives gathered at the farm where Iris grew up. Most of what they ate was grown on the farm: fresh fruit and vegetables in season, canned and dried goods otherwise; meat they butchered; and bread from the flour of wheat they grew. Ice cream was churned by hand using ice cut from the river months before and stored under hay deep in the cellar. Sunday afternoons were spent playing with cousins and, for the adults, in conversation or card games. Relatives would stay into the evening, drinking, smoking, and listening to the radio. Some helped with evening chores like milking the cows and feeding the chickens, pigs, and horses. This bucolic life was full

---

[4] "The state" as a driver of social change is the topic of Chapter 8, but to avoid misunderstanding at the outset it should be pointed out that the term, as used by social scientists, is a shorthand reference to the system of power administered by governments. These include local, state, and federal governments that "speak in the name of the state and [are] formally *invested* with state power" (Miliband 1969: 50). Michael Mann, following the ideas of Max Weber, recognizes two dimensions of the state: the institutional and the functional, i.e., what states are and what states do. "The state contains four main elements, being: (1) a *differentiated* set of institutions and personnel embodying (2) *centrality* in the sense that political relations radiate outwards from a center to cover (3) a *territorially demarcated area*, over which it exercises (4) a monopoly of *authoritative binding rule-making* backed up by a monopoly of the means of physical violence" (Mann 1988: 4, italics in the original).

of hardships and hard work, making the periodic abundance and leisure of Sundays seem all the more special.

Daily life was not terribly different from that of her parents and grandparents when they were young. Some of Iris' forbearers didn't approve of card playing, and a few made their own libations—cider, beer, and whiskey—instead of buying them. Rather than a radio, music was provided by local amateurs, including household members. If you went back into her family record far enough, you would find more food of a distinct ethnicity or region of the world, and the people would be speaking a different language: German, Gaelic, and Norwegian rather than English. Everyone had been farmers. Some of the machinery changed, and more work was done using the power of horses and mules. But it was a difficult and precarious life.

Iris' grandparents on her father's side were from Germany before it was Germany. That is, they lived in an autocratic state that was primarily Lutheran and would later be incorporated into Prussia when King Wilhelm and his chancellor, Ludwig Von Bismarck, embarked on German unification. Letters from the old country told of impending war and young men being pressed into the army. The bell and iron fence of a Catholic church were melted down and forged into a cannon, and many of the peasants—especially the Catholics—quietly packed up their farms and fled the region, becoming immigrants in a new land.

Her mother's family had been in the country longer than her father's, though some relatives lived in Canada for three and four generations after first coming to the United States. Her relatives had been horse and mule breeders, masons, and house builders. Others were wheelwrights who built wagons and carts, but all of them had a farm to produce most of what they needed. Iris and her husband had distant relatives with names like *Chandler* (probably candle makers), *Newhouse* (perhaps the family with a new house), *Rakestraw*, *Smith* (blacksmith)[5], *Mason, Miller, Skinner,* and *Waggoner*.

Names tell a story of social change. The practice of adopting a family name often indicates the point at which an increasingly strong state wants to keep track of its citizens, usually to tax them or conscript them into war. Family names change with changing family circumstances. A Norwegian family in 1850 might buy the Skjervem family's farm and henceforth take the name *Skjervem*. Persian, Hindi, or Chinese immigrants,

---

[5] Curiously, the most common family name in the United States is *Smith*. The most common name in Hungary is *Kovacs* (blacksmith), and in Germany is *Schmidt* (blacksmith). I've often wondered if blacksmiths were particularly prolific or if there were just so many of them. Every village needed a blacksmith.

among others, had their name altered when they came to the United States in order for them to sound and be spelled in a way recognizable by people who spoke only English. People whose names identified them with a disfavored nationality or ethnic group changed their names to avoid discrimination and prejudice.

The names of the world's boxing champions tell the story of social change, in particular the story of immigration and upward social mobility. The point at which a prominent ethnic group's names disappear from the list of prizefighters is a rough approximation of when the ethnic group "made it in America." First were the Irish and Scottish (John L. Sullivan, Jack Dempsey, and Bob Fitzsimmons), followed by Jews (Barney Ross and Benny Leonard/Benjamin Liener) and Italians (Rocky Graziano, Carmen Basilio, Rocky Marciano, Jack LaMotta, and Ray Mancini), a century of African American boxers (Jack Johnson, Harry Wills, Joe Lewis, Floyd Patterson, Joe Frazier, Sugar Ray Robinson, Emile Griffith, and Sugar Ray Leonard) and most recently Latino boxers (John Ruiz, Paulie Ayala, Steve Curry, and Tony Lopez). Both of the latter groups have yet to achieve equity in American society and continue to have many fighters in the ring.

In the United States, women have gone through generational cycles of keeping their family names at marriage. Supporters of women's equality who married in the 1960s and 1970s were much more likely to keep their names or hyphenate their own and their husbands' family names than were women marrying in recent decades, the time of third-wave feminism.[6] Following the civil rights movement, ethnic pride among African Americans contributed to an upsurge in creative, often melodic, first names, while conversion to another faith (e.g., from Christianity to Islam or the Nation of Islam) signaled an important social, and sometimes political, trend. No one could miss the point being made by the great prizefighter Cassius Clay when he became Muhammad Ali, when the fiery orator and Black Power activist Malcolm Little became Malcolm X, or when the poet, dramatist, and critic LeRoi Jones became Amiri Baraka.

---

[6]"Waves of feminism" describes the evolution of the women's movement that has realized considerable success in changing laws and social policies (especially suffrage for first-wave feminism) and the passage of equal-opportunity legislation such as Title IX of the Civil Rights Act, as amended in 1972 (for second-wave feminism). Succeeding generations (third-wave feminism) take much of this for granted and see the pursuit of women's equality in a different light by focusing on new issues (Krolokke and Sorensen 2006).

Decades of Social Movements

Iris Summers never marched in the streets or waved a sign, and only rarely did she contribute to a political candidate or cause. She was an observer, like most people, and watched from the sidelines as younger, more confident, and more passionate advocates sought public attention for their causes. Her friends and relatives were not activists and would have found it peculiar if she had become deeply involved or even enthusiastic about political affairs. News came to her in slim local newspapers and short television broadcasts each evening rather than via cable news and across the Internet. She had a positive opinion about social change accomplished through actions of the state, given her experiences as a young woman, but in her last years she was disquieted by the loud voices demanding state action for and against social change.

It wasn't that her generation avoided public controversy by working for social change. It was during her young-adult years that American labor was most successful as a social movement, leading to the creation of the United Auto Workers, the unionization of the country's coal miners and longshoremen, and the establishment of the Congress of Industrial Organization. At the height of the Great Depression, hundreds of people camped out on the Washington Mall in protest of government inaction to stem the financial collapse that put them out of work and shrunk their life's savings. Well before the 1950s and '60s, the seminal years of court and legislative victories, the civil rights movement was active in opposing racial segregation and other forms of discrimination against African Americans. Iris, however, was never a visible part of any of this.

As she got older, the diversity of lifestyle of her children, nieces, and nephews, and their strong opinions and beliefs, helped convince her that the U.S. war in Southeast Asia was a mistake, that unfair treatment experienced by gays and lesbians should not be permitted, that women should have equal rights, and that local, state, and the federal governments should take a more active role in helping the disadvantaged. Her memory of poverty, farm foreclosures, and unemployment gave her a sense that personal problems weren't just for lack of trying. There was racism, sexism, and discrimination; political opportunism and demagoguery; and a lot of things that needed changing. Like most women her age, she held opinions, but she largely kept them to herself.

Public safety campaigns and social movements to change behavior had mixed results with Iris Summers. Antismoking campaigns never fazed her, but she quickly supported the Keep America Beautiful antilitter campaign to stop the way people casually threw trash out their car windows.[7] She was very uncomfortable using seat belts. Though aware of the statistics on seat belt safety and auto collision injuries, Iris believed she would be burned alive if, in a collision, she couldn't unfasten the seat belt in time to escape. She and her husband, Frank, never built a bomb shelter, despite the public announcements in the 1950s and urgings of their neighbors. Nor did she join the antinuclear movement or visibly protest corporate malfeasance by boycotting a company or product. She approved of what many social movements of her life accomplished, but she was largely a bystander.

When her children and younger relatives talked with her about their social and political concerns, she would describe how families used to take care of one another when someone needed help. She could recount how same-sex teachers lived together and, though some people may have suspected a homosexual relationship, this was a private matter and not something needing public discussion. She would tell them, perhaps naively, that in her day the races didn't want to mix, but since young people felt differently, she supposed it was all right.

As a girl, Iris Summers had regularly gone to church, but she stopped attending in her early adult years. With the birth of her children, she decided to return to church and expose them to religion, but she was only partially successful. After their children left home, she and Frank never attended church or gave it much thought. This was somewhat unusual, inasmuch as most elderly people of their generation resumed the practices of religion on a regular basis.[8] Several of Iris' grandchildren, nieces, and nephews became active in fundamentalist congregations that were started in the 1970s and 1980s and considered church attendance and involvement in religious activities an important part of their lives and identities. One very religious granddaughter led the antiabortion movement in her town, but Iris never talked to her about it.

---

[7]Heather Rogers' *Gone Tomorrow: The Hidden Life of Garbage* (2005) describes how this campaign was actually a very successful corporate-sponsored effort to define the problem of everyday pollution as a failure of people to properly dispose of trash, deflecting the public's focus away from the creation of trash by bottling and packaging companies, similar to automobile corporations emphasizing "the nut behind the wheel" discussed in Chapter 7.

[8]Robert Bellah and his colleagues' *Habits of the Heart* (1985) provides a well-researched examination of religious practices and commitment to personal as well as organized religion in the United States. They profile Americans who follow a typical pattern of identifying themselves as "having religion" but deciding for themselves many of the details and practices that best express what they believe, not what the clergy or a congregation dictates.

The Means to Being Modern

Iris Summers thought the best thing to happen in her lifetime was television. It opened a world to her and her family that otherwise they would have never known. Entertainers like Jack Benny and Dinah Shore, performing artists like Van Cliburn, Arthur Miller's dramas, *National Geographic* specials, situation comedies, and the evening news became a major part of her adult experience. She learned things, developed an appreciation for the arts, and became more socially aware because she watched television.

Less well understood or recognized by Iris, she became a member of that great consuming army who discovered through television things they never knew they needed. Though she retained the frugality of her background—and especially her experience with the prolonged worldwide economic depression of the 1930s—she spent much of her life considering, figuring out if she could afford, working in order to have, and purchasing *things*. She was a practical person and not terribly swayed by advertising, or so she thought. Fortunately for her, advertising was far less sophisticated in her middle adult years than by the end of the century.

Hers was an era of rising expectations that saw the U.S. economy grow steadily. The per capita gross domestic product (GDP) doubled between the birth of her first child and high school graduation of her last. Across the globe, economies grew, and following the destruction brought by each world war, Europe's economies also boomed. Dozens of poor countries and newly independent nations were able to link their economies to this growth, and their people experienced a similar shifting of their everyday lives—toward the acquisition of things.

Iris Summers was a lifelong smoker, though men were much more likely than women to smoke for all but the last decades of the twentieth century. She never gave it up or even seriously tried to stop smoking, though it seemed obvious to her that smoking harmed a person's health. She believed the health statistics that told her that life was shortened by smoking, but she didn't really believe in the danger of secondhand smoke. On trips, she cracked her car window, no matter the weather, only in response to her family's complaints. She started smoking not because it was part of the glamour of Hollywood stars like Humphrey Bogart and Greta Garbo but because her friends and husband smoked and seemed to enjoy it. Once addicted, Iris liked to smoke, and her last breath wasn't far removed from her last cigarette.

A Woman in a Changing Society

The expectations for most young women of Iris Summers' generation were fairly simple: marry, bear children, be a faithful and supportive wife, and do whatever it takes to give your children a good chance in life. That meant knowing how to cook, clean, and care for everyday illnesses, having a nice yard, and being a good neighbor. Though never affluent, she knew she was lucky. Her husband, Frank, was a steady provider, was never seriously hurt on the job, and remained her husband until he died. Not so for her children or her siblings. When her sister Ruth's husband was killed in a hunting accident, life's difficulties for a single mother seemed unavoidable and often insurmountable. Ruth's family of four quickly plunged into poverty that was relieved only when Ruth again married and became the mother of her new husband's children. Their "blended family" seemed unusual, especially when compared to what was portrayed in the early days of television, but in fact was not atypical of families throughout history, including the late twentieth century.

As a woman, things had been much the same for Iris Summers as they'd been for her mother and aunts. Times were very different, however, for her daughters and granddaughters. By 1980, it was clear that most American families could not maintain a middle-class lifestyle on only one income. Europe, and especially Eastern and Central Europe, had found this out after the Second World War. In Iris' early years and in other countries, especially in agrarian societies, women worked alongside men and often put in longer hours than their male counterparts. The idea in the Victorian era that women should not be employed outside the home was, in hindsight, a brief interlude and a possibility for only the relatively affluent in capitalist societies. Women were enlisted into the paid workforce during both World Wars I and II. By the 1980s, women of all social classes were increasingly likely to be working in order to help support themselves and their families.

Wives and mothers of Iris' generation were not idle just because they weren't working outside the home. In fact, they were spending an increasing amount of time on housework, chauffeuring children to school, shopping, and other household activities. Of course, single women and women who headed families, and especially nonwhite females, have always worked outside the home in numbers greater than married women, but the numbers increased for wives as well. In 1950, 32 percent of all adult women between age eighteen and sixty-five were in the paid workforce. This increased to 60 percent in 2000.

The women's movement rightfully takes credit for women increasingly seeking careers beyond that of wife and mother.[9] But even as women became

---

[9]Actually knowing the impact of social movements is often difficult to ascertain. Sarah Soule and her colleagues (Soule et al. 1999) have raised some interesting questions about the women's movement as a force for fair employment legislation, with suggestive analysis that women's increased workforce participation was the driving force, not the women's movement itself.

the majority of medical school students and were entering legal and scientific professions in ever larger numbers, the quality of most people's working lives—women included—declined (Sennett 1998). And still the rate of women working outside the home increased. Wage earners on average saw no increase in earnings over the last quarter of the twentieth century, and per capita income declined with the Great Recession of 2007–2010.

As a young woman Iris studied to be a nurse. She was a poor young woman who, like many of her social class, could receive a nursing education in exchange for working in the hospital where she was trained and lived. Work was scarce when she finished her education, however, and she soon married and began traveling with her husband as he went from jobsite to jobsite. Only when her children were well into school did she return to work, and then only part time.

Her daughters and granddaughters had no such luxury. They always worked outside the home, full time and for their entire adult lives. If she had suggested to her granddaughters that they should take some years away from work to raise a family, they would have found this a curious idea and largely impossible. There was a steady growth in educational attainment in Iris' family[10] and great pride in both educational and occupational accomplishments. When President Bill Clinton talked about people who "work hard and play by the rules," Iris felt he was talking about her family. This didn't mean that everything went smoothly, however.

Five of Iris' six children married, and three divorced. That was about average for their generation, and today. Her children moved many times for their jobs or in search of work, to do military service, go to college, or just out of wanderlust. They ultimately settled too far away to make it easy to visit Iris and Frank on weekends, so holidays, vacations, and close relatives' weddings and funerals became the times she occasionally saw her children and grandchildren. When Iris died, most relatives of her generation who were still living traveled fewer than a hundred miles to attend the funeral. Her children took flights in order to be there.

The Changing World of Work

As the years passed, what looked like a good job became more difficult to find. Global competition, deteriorating industrial plants, corporate downsizing and human resource practices, and the declining power of labor

---

[10]This follows the trend for the United States as a whole. Less than 5 percent of college-age youth were in college in the 1920s. This number was 15 percent by 1949, in part due to the GI Bill, and is nearly 60 percent today.

unions combined to create job losses and wage stagnation that changed the way people in the United States lived and the prospects for their children. The labor market changed with the growth of information technology. An aging population's health care needs created new jobs as part of the economy's shift from the production of things to the provision of services.

Like many others from rural backgrounds, Iris had no experience with labor unions. Her husband, Frank, alternated between ambivalence and skepticism when the idea of joining a union came up. Only a small fraction of American workers were members of unions in 1900, nearly all of whom were skilled craftsmen. When things became very difficult for working people in the Great Depression of the 1930s, the federal government responded to labor unrest by passing legislation that made it easier to form unions, negotiate agreements with owners, and so improve workers' pay, health, and retirement benefits. By 1956, nearly two in five workers in the United States belonged to unions, and the conditions for nonunion workers improved correspondingly. By midcentury, U.S. workers were the best paid in the world. Corporate America clawed back, however, especially with the passage of the Davis-Bacon Act in 1948 and incremental pro-business legislation through the Reagan years of the 1980s and beyond. In 1980, one in five workers were union members, but by 2009 less than one worker in eight (fewer than fifteen million workers) belonged to a union.[11]

More importantly, the decline in unions was the result of a shifting labor market. Just as agricultural work diminished in the first half of the century, industrial work shrank in the second half. Many of the jobs that were the heart of the American labor movement went overseas, and in their place were service jobs, from retail sales and other low-skilled jobs to positions requiring college and professional degrees, few of which were unionized.

Women, too, were joining the labor force in ever-larger numbers, more often than not in service jobs and in nonunionized manufacturing and assembly plants. By century's end, women began moving into all branches of work, their faces gradually becoming more common in the newspapers' business section. Their place in science and medicine is no longer remarkable. Women are now more than 50 percent of those working full time in managerial and professional positions in the United States, though they are paid, on average, 72 percent of what males earn (BLS 2009).

---

[11]In 2010, the portion of the total U.S. workforce that belonged to unions (11.9 percent) was the lowest in more than seventy years. Private-sector workers in unions fell to 7.1 million, or 6.9 percent. For the first time, more than half of all union members were public employees, 7.6 million or 36.2 percent: teachers, police officers, and so forth (BLS 2011).

The Personal Challenge of Social Change

The Second World War was a watershed in Iris Summers' life, but in ways that Iris did not always recognize or understand. It was not the only major event that helped forge her worldview, but the changes in global power and economic might that shifted to the United States vastly changed the world in which she lived. While the war's end was a cause for great celebration, in some ways the United States remained at war. The federal budget deficit, created to pay for the war, declined, but the cost of being a global military power—which the United States became—and the Cold War with the Soviet Union required huge outlays of tax dollars. It seemed, however, that the country could afford it.

Much has been written about the "Age of Conformity" following the Second World War. The movie *The Best Years of Our Lives* swept the Academy Awards in 1946 by telling a realistic and painful story of returning GIs, but quickly Hollywood began producing hero stories that glossed over the complex realities of war and wartime. Best-selling novels like *Man in the Grey Flannel Suit* and books like William Whyte's *Organization Man* seemed to describe the norms of conformity, complacency, and material satisfaction, while Ayn Rand found a large audience with her novels of heroic individualists in the shadows of sinister governments. Against the grain, books like *Peyton Place* and early rock 'n' roll found large audiences and challenged the myth of conformity, setting off a national debate about public morality, freedom of expression, and the cultural bonds that defined the nation. It became increasingly difficult to locate "the mainstream" and assume a common outlook and purpose in a nation of 250 million people.

Perhaps because several of her children came of age in the 1960s, Iris Summers was greatly aware of the events of that decade. The war in Vietnam and Cambodia changed many of the ways Iris looked at the world. It was the first televised war. Walter Cronkite, the iconic newsman who presented the *CBS Evening News*, provided the "body count" on a nightly basis, reminding viewers that hundreds of young men and women were dying every week. The Pentagon compared these numbers to the reported number of enemy killed as a measure of the war's success. This interplay of military violence and nearly instant communication was a dramatic new reality, changing how wars were to be carried out and publicly understood in the decades ahead.

Iris Summers was at a loss when the modern civil rights movement took off in the mid-1950s. She had grown up in a white world. She had never had a neighbor or friend who was African American, and few of her husband's workmates were ethnic minorities. She had never liked hearing racial slurs and jokes, believing they were hateful and meant to hurt and demean people. But she didn't comprehend the deep sense of injustice African Americans felt. She had difficulty understanding the commitment with which civil rights activists pursued the cause of racial justice. Watching the marches on TV and hearing the speeches of leaders like Dr. Martin Luther King and Shirley Chisholm, she would ask her children, "Why do they keep saying, 'I want to be free'? This is a free country, isn't it?"

Probably like millions of nonminority people, Iris' thinking about race and civil rights followed a crooked path, and small things had a great impact on her. The 1956 movie *Imitation of Life*, with its parallel stories of a mother and daughter—one white, one black—touched her deeply. She loved Scott Joplin's music and watching Sammy Davis Jr. tap dance. She thought Jackie Robinson was a truly brave individual. And when she watched Bull Connor and his police force, along with chanting and cursing crowds of Whites, taunt and attack civil rights marchers, she believed she knew whose side she was on. Four little girls blown up in a Birmingham, Alabama, church and the murders of civil rights workers like Viola Luizzo and Medgar Evers touched her heart. She began that long road so many of her generation had to travel, rethinking not only the rights of African Americans but the everyday relations among people in an increasingly diverse society.

When Iris was older, she and Frank lived on social security and a small pension from Frank's last job. They had always been frugal, and it suited their lifelong habits to buy little more than what they needed. Her children described this as a consequence of their having lived through the Great Depression, empirically verified in Glenn Elder's *Children of the Great Depression*. While they were growing up, there were many arguments in the household about money. It seemed to her children that they lived poorer than they really were. In order to have what they wanted, her children had to get jobs. Much of the post–World War II affluence that came to the United States, and in the next decades to Europe and Japan, affected Iris' approach to life very little. Certainly, looking back at her childhood, she had left behind what Clair Brown (1994) describes as "the bleakness of everyday material life" of most working-class and farm families. Brown's comparisons of material consumption and household expenses in the early and later years of the twentieth century were true for Iris, Frank, and their children. Like tens of millions of others, they benefited greatly from a rapidly growing economy. By the time their children became adults, the economy had cooled, however, and the promise of doing better than one's parents continues to pose a challenge to Iris' grandchildren.

From the analysis of decades of survey data, Norval Glenn (1987) identified "growing individualism" as the "master trend" in postwar society, made possible by greater affluence and promoted through corporate advertising. Iris was not immune to this, though her personal ambitions remained much as they had been in her formative years. She was fortunate to have the choice of calling her children on the phone, getting in a car and visiting sisters and cousins, and taking time to arrange generations of family photographs in albums. She never quite became the rugged individual so often celebrated in popular culture. Having, helping, and being helped by family and friends was always most important to her.

It could be that, in many ways, the times passed Iris Summers by. When she tried to get a credit card, she was turned down because she had no credit rating, having always paid cash for her purchases. That included their car and the last house she and Frank bought. She paid for things with cash and checks, always knew the exact balance of her checking account, and never used an ATM. While others her age got computers and communicated through e-mail, Iris would rather talk on the telephone (on a land line, not a cell phone) and write letters. She never used a digital camera, let alone a mobile device, to take pictures. She had never heard of social networking, had never Googled information, and the idea of putting speaker buds in her ears to hear music had no appeal.

In the last quarter of the century, as the environmental movement grew with public awareness of the ways human beings were threatening the environment, Iris' tendency to reuse things most people dispose of, her ability to weave rugs from strips of rags, her distaste for "wasting water" while brushing teeth and for leaving lights on when nobody was around caused her to seem prophetic rather than ancient. Her modest childhood and the deprivation of the Great Depression made her something of a forerunner to green living. Frank was the same way. He composted kitchen scraps and kept his car running to its final demise. He planted grass that required very little watering and preferred to cut it with a push mower.

If everyone lived that way, the economy of the United States and the world would look very different than it does, and many peoples' idea of happiness would never be realized. But we also know that a lifestyle fueled by nonrenewable resources and financed with borrowed money can also be very problematic. Did social change pass Iris by? Perhaps she missed out on the fads and fashions as well as the revolution in information technology, but the changes in the economic, political, and social fabric were unavoidable. Her resistance to social change was largely on a personal level. Her experience of structural changes was inevitable, and as a consequence, her choices and options, too, grew with the social changes of her lifetime.

Not Every Person's Story: Capturing
Social Change in Personal Experience

Though similar to millions of others, the story of Iris Summers is only one person's story. She is not an "average American" in the sense of fitting the profile of a statistical average or numerical mode of personal characteristics. For example, by a wide margin Americans lived in cities and suburbs during the last half of the twentieth century, but Iris lived, with the exception of a couple of years, in small towns. Hers is a story of the rural Midwest, not the South, the Southwest, or the West Coast. It is not an immigrant's story, the story of an African American, or one of an upwardly mobile white man. She didn't fight in a war or suffer a debilitating injury.

Digital technology and the consumer products it spawned largely passed by Iris Summers' everyday life. She didn't have a passport and never traveled to a foreign country, didn't know a language other than English, nor was she particularly interested in international cuisine. Iris may have experienced discrimination because of her gender, social class, and rural background, but these could not compare to the experiences of those whose life chances were severely restricted and, sometimes, whose personal safety was threatened or violated because of their race, ethnicity, sexual orientation, or nationality. She was only an occasional observer of these others, and in this way she was like a great many of her fellow citizens.

In Iris Summers' life can be seen the influences of the historical periods through which she lived. Obvious as well are the consequences of passing through the stages of life. Social change is experienced differently depending upon whether you are young, in your middle years, or elderly. Her peers— the millions of people born early in the century—made up a generation with unique features that placed their own demands on the world around them. They, more than any other generation, shifted from rural lives to industrial and postindustrial work. They used the power of the state to help them succeed, through a public commitment to education, guaranteed loans for home ownership, federally insured bank deposits, and military industries that fueled regional and local economies.

After getting started as adults in the midst of the worst economic crisis of the century, her generation was asked to support and fight the century's most destructive war. The sustained economic expansion from 1947 to 1970 benefited them at the height of their careers and filled their lives with an abundance of material goods and opportunities. Theirs was the first generation to have a substantial economic safety net of social security to keep most of them out of poverty as senior citizens. When they grew old, their health care was largely paid for through a public health insurance program. They saw home-town banking shift to powerful Wall Street financial institutions and the stock market go from trading millions of shares daily to billions of shares on a regular basis.

The American labor movement rose and declined in Iris Summers' lifetime, while the women's movement, the civil rights movement, the environmental movement, the public interest movement, the gay rights movement, and other efforts of millions of people and thousands of organizations to create social change emerged and moved the world in new, uncharted directions. Her generation developed a more tolerant attitude than that of their parents and grandparents, and they increasingly lived in a more ethnically diverse and culturally contested environment.

The forces of social change experienced by Iris Summers and those around her are the same ones that impel social change today, not only in the United States but across the globe: social movements, corporate activity, state initiatives, war, and new technology.

Defining and Understanding Social Change

People seek to, and successfully do, redefine themselves. A young person joins the military, hoping to get the structure, respect, and skills that will help her in pursuing a steady, productive life. People marry, divorce, and remarry, establishing relationships and shifting affections that define who they are. Individuals of all ages, but especially youth, decide to live differently. They change their clothes, adopt a new hairstyle, get a tattoo, take up a pastime, and buy things to redefine who they are and how others see them. They try on new identities or publicly assert an identity they claim is theirs.

For most people, what they take to be social change feels very personal and local. A child's family moves: suddenly there is a new house, new school, and new friends. During the life course, new social circumstances challenge and provide opportunities, doors open while others close behind, and age qualifications are met. Being old enough to go to school, date, leave school, drink alcohol, join the military, drive, vote, retire, and qualify for Medicare and Social Security initiate significant personal changes in an individual's social circumstances. Personal experiences, however, are a part of something bigger than themselves, something social.

Think for a moment about social change as the sum total of many peoples' personal changes. Millions of youth come of age during a massive war that needs soldiers and conscripts hundreds of thousands of young men into the military. Millions of people lose their jobs or leave school and can't find another job, facing a daunting economic recession or depression that doesn't allow most of them to fully use their education and talent. Babies are born during the years of civil rights legislation, stripping away legal barriers to their equal opportunity. Theirs is a vastly different social milieu from that of their parents. They grow up assuming they will vote, pursue an education, get a job, buy a home, and live in a community without overt discrimination toward their race, ethnicity, sexual orientation, or gender.

You may have been born into a household full of computers. Information, entertainment, and communication involve sitting in front of a flat screen or walking around with a powerful computer in your hand that takes videos and photos, tells you where you are, talks to you, and links you to friends 24/7. It feels like you, your change. And it is, but it is also happening to millions of others. That's social change.

Most definitions of social change enumerate the things that are social and the things that change. For example, Wilbert Moore (1972) defines social change as a "significant alteration in social structure" and proceeds to discuss the things that are social structure: groups, organizations, and so forth. Moore's definition is extended by Harper and Leicht, in their popular textbook *Exploring Social Change*, as "the significant alteration in social structure and cultural patterns through time."

> [S]*ocial structure* [is] a persistent network of social relationships where interaction between persons or groups has become routine and repetitive . . . persistent social roles, groups, organizations, institutions, and societies . . . Culture is the shared way of living and thinking that includes symbols and language, . . . knowledge, beliefs, and values, . . . norms, . . . and techniques ranging from common folk recipes to sophisticated technologies and material objects. (Harper and Leicht 2007: 5, italics in original)

Relationships, group norms, beliefs, technologies, and the other things enumerated extend beyond the individual. The idea that social change is an experience differing from what others—at an earlier time—have experienced is incomplete if it does not examine the social processes and networks of relationships surrounding the experiences. They are widely shared, indicating a degree of patterning through the influence of the things that impel both personal and social change in one direction rather than another.

Robert Nisbet (1969: 169) defines social change as "a succession of differences in time within a persistent identity." What Nisbet means can be seen by comparing Iris Summers and her children's marriages. Marriage is a "persistent identity," something Iris and Frank had in common with their children. One "succession of differences" is the greater preponderance of the children's marriages ending in divorce. What changed? Marital life or, if you like, fidelity to a marriage. Two or three generations earlier, when women died in high numbers during their childbearing years, men remarried and fathered more children. This changed when maternal health and life expectancy for women improved. Married couples could look forward to many more years together and many more years of possible discord resulting in divorce.

What is causing these personal changes that manifest themselves in the lives of large numbers of people? That is what we want to uncover and in so doing understand the ways of social change.

A Very Brief History of Human Societies (With Apologies to Mel Brooks)

As a scientific—rather than a speculative—endeavor, understanding social change requires a multidisciplinary approach, combining an interest in culture, social structure, political process, anthropology, and economics, as well as the physical and biological sciences. Foremost, recognizing social change is a venture into history, reconstructing the human record in which can be seen the forces and processes that have configured social life and are likely to impel the future. These are the drivers of social change examined in upcoming chapters. If we hope to work effectively to solve problems of the present and chart a positive future, it is worth taking a moment to examine, however briefly, the key elements of the historical record.

Before the Last Ice Age

Most of what we know about societies, culture, and human accomplishment is in the record of human history following the last Ice Age ten thousand years ago. What we know about earlier human social life before that is very fragmentary. We do know that the species that became today's humans changed biologically and migrated to populate the habitable portions of the planet. The archeological and anthropological record shows that life changed very, very slowly, but human accomplishments were far from insignificant. Groups fought and probably merged with one other, developed symbolic communication, and established religious practices, group norms, and a sense of identity. Human beings developed pivotal technologies that enhanced their capacity to survive, reproduce, and lay the foundation for what came next.

Isolated family groupings left behind their debris but no written records for many thousands of years. The Paleolithic drawings in the caves of

Lascaux and Trois-Frères in southwestern France are fascinating in both their artistry and the scenes and animals depicted. From them can be inferred a bit about how life must have been lived and how people may have seen themselves and their world almost twenty thousand years ago. Archeological studies of human remains tell something about the religion, adornments, skills and crafts, and occasionally the pleasures of life. Pollen and DNA analysis provide many fascinating facts about the health and nutrition of people before the last Ice Age, but we have only vague clues about their changing social organization and cultural forms.

The corpses found in bogs and glaciers, as well as other more contemporary human remains, fill in some blanks about technologies, practices, and skills, and from these are drawn links between our ancestors' and contemporary cultures. The "discovery" of isolated human groups in the remotest parts of the world by adventurers, journalists, and researchers in the past three hundred years encourages tentative inferences to be drawn about the social life of early humans, especially when the comparisons can match fairly well the archeological records of earlier peoples.

The overall picture painted of preliterate Homo sapiens is one of inventive adaptive strategies largely governed by environmental conditions. There was very slow population growth, with gradual migration and eventually much diversity in practices and beliefs. People spent their time in small groups foraging and hunting, with a minimal division of tasks that was primarily along gender and age lines. We know little about the spiritual beliefs they held but infer that these emerged from their efforts to eke out a living, reproduce their own kind, and define themselves within the natural and celestial worlds.

World Population Growth

It is difficult not to be startled at the most basic facts about human population growth. The number of people living on the planet was about one million (the population of a small city) in 10,000 BCE and had probably been about that number for millennia. Ten thousand years later, after the rise of the great kingdoms in what is now Iran and Egypt, the world's population had grown to 200–300 million people, about the number of people now living in Japan. In the next centuries, the world's population continued to grow, and it doubled by 1000 CE. It doubled again by the time the United States Constitution was ratified. By 1800, the total number of people living on earth reached one thousand million, i.e., a billion people.

It took another 130 years for the world's population to reach two billion. That was in 1930, about the time Iris Summers was in high school. In her lifetime, the world's population nearly tripled.[12] It grew to three billion in 1960, four billion in 1975, and five billion in 1989. It was six billion in 2000, will be seven billion people in 2012, and more than eight billion by 2030. The world's population grew rapidly first in Europe, North America, and Japan. Since 1950, growth has been much faster in the rest of Asia, Latin America, and Africa, posing an enormous challenge to the economies and environments in those regions.[13] Fortunately, global population is expected to stabilize by 2050. Population and its dynamics, including the factors influencing population growth and the consequences of this growth, are central to understanding the history of human societies.

Urbanization

The growth of cities challenged public order and safety, health and sanitation, and human adaptability, but once the process of urbanization began, it continued unabated. In 1500 CE there were only twenty-four places with a population of at least 100,000 people; four were in China and none were in Europe. In 1800, three of the world's ten largest cities were in China (Beijing was the largest, with 1.1 million people) and three were in Japan. This was dramatically different in 1900. By then, 13 percent of the world's population was urban, largely as a consequence of industrialization. The world's largest cities were London with 6.5 million people, New York City with 4.2 million, and Paris with 3.3 million. They were followed by Berlin, Chicago, Vienna, Tokyo, St. Petersburg (Russia), Manchester (England), and Philadelphia. In the United States, 30 percent of the population was urban in 1900, up from 10 percent in 1860.

In the past sixty years, nearly all population growth worldwide has taken place in cities, due both to natural increases and rural-to-urban migration. Worldwide, three out of ten people were living in urban areas in 1950. Fifty-five years later, five out of ten were living in urban areas.[14] More than three-quarters of the people in wealthier industrialized nations now live in cities,

---

[12]Wes Jackson (2010, 7) observes that no one living before 1930 "lived during a doubling of the human population."

[13]Alene Gelbard and her colleagues (1999) provided a clear and fascinating overview, as part of an invaluable series of the Population Reference Bureau in Washington, D.C.

[14]An urban area has no agreed-upon definition, but the U.S. Census Bureau's definition is similar to most: an area with 50,000 people or more.

and by 2030 this will be true in poorer countries as well. Only in recent times has this dramatically new urban way of life become the norm for billions of people across the globe.

New Forms of Production and the Development of Capitalism

Why has the human population grown at such a rapidly increasing pace? The answer probably lies with food. After the last Ice Age, many human groups began to cultivate their environment rather than only foraging and hunting. Horticulture, or the application of basic farming techniques, provided a more predictable food supply and an opportunity for at least a partially sedentary life. Animals were domesticated, and seeds were selectively culled from plants that best met the needs of people. Living a portion or all of the year in one place, people honed the skills—particularly pottery and metallurgy—to create goods for settled living.

Over the next several thousands of years, careful seed selection and breeding of animals made agriculture more productive. The increased food supply not only supported more people but made possible and necessary many new social forms. A surplus of food meant that not everyone needed to produce food. A very small portion of the people could specialize in religious practices, metal, wood and pottery crafts, and soldiering. It opened up opportunities for some to accumulate the small surplus and use it as a source of power toward others. Statecraft, writing, long-distance commerce, and the arts of war were developed. They ushered in a radically different way of living in those areas of the world where population growth was most rapid.[15]

First in the fertile valleys of present-day Iran and then in areas surrounding the Mediterranean Sea, empires of conquest and colonization grew, spreading languages, belief systems, technology, and forms of governance of the dominant groups. Trade and commerce as well as paths of conquest brought into contact diverse cultures and social formations that borrowed from one another, altering long-held ways of seeing, believing, and doing.

Contrary to what you may have learned in grade school, the end of the Roman Empire around 400 CE did not usher in a Dark Ages lasting several hundred years. A vibrant and immensely creative world outside the European continent—in the Middle East, Africa, Asia, and in Central and South America—reflected human ingenuity, new needs, and the will to power.

---

[15]Gerhard Lenski's *Power and Privilege* (1986) is a fascinating examination of the process of accumulation and inequality, as are sections of Jared Diamond's *Guns, Germs, and Steel* (1997).

Empires and dynasties emerged, new technologies were cultivated, literature and philosophy flourished, handcrafts and architecture gave a distinctive local look to material culture, and new ways of making war continued apace. Life in what is now Europe changed as well. Most people were farmers, subsisting on the work of their families and living in a local world only occasionally breached by a traveler, a government official, or a religious pilgrim. These were enough to alter in small ways the pattern of living and the sense of possibilities for another way, and little by little social life changed.

Plunder, Mercantile, and Industrial Capitalism. Most social historians designate the fifteenth century as the point when social change speeded up and modern times began. This was due to two "master processes . . . [i.e.] the development of capitalism and the formation of powerful, connected national states" (Tilly 1984: 15). Prior to the fifteenth century, those with enough power to coerce others had relied on tribute (taxes) to support themselves, pay for building projects, and fund military engagements. The empires of the Moguls, the Romans, the Muslims, the Ottomans, and others did little to alter the way of life of those under their control. Their design was to extract a portion of the surplus created by their subjects through a network of governors and tax officials exercising "tributary power" (Wolf 1982: 85).[16] By propagating myths and belief systems that expressed a claim to supernatural or superior status, backed up by the ability to coerce through violence, empires became rich and powerful. Only later would states seek to transform economies in new ways that allowed them to grow, amass wealth, and engage in even greater imperial adventures.

One of the social sciences' gifted early practitioners, Max Weber (1864–1918), examined early capitalism as practiced wherever "the possibilities of exchange, money economy, and money financing have been present," including the Chinese and Roman empires (Weber 1964: 279). Of greater interest to him, however, was the first modern version of mercantile capitalism, initially practiced in parts of Europe around 1500 and characterized by rational behavior in finance, ownership, trading, and marketing. It organized production for profit.

Going beyond simply taking possession of what others created, mercantile capitalism increased the value of goods (wealth) by relocating them closer to markets. The Silk Road across Asia and its tributaries had for centuries connected sources and markets for merchants in Asia and Europe. By the fifteenth century, many other trade routes had developed and became the major focus of a burgeoning capacity to create wealth. Global trade and

---

[16]Another early way of extracting wealth included, in Weber's (1964: 280) phrase, "political capitalism" where "political events and processes [including the explicit use of violence and state power] . . . open up opportunities for profit."

mercantile capitalism continued to develop as sources of wealth extraction well into the eighteenth century.

It is sometimes difficult to distinguish mercantile capitalism from plunder capitalism, especially when those who resisted trading with European merchants and states were met with armed violence. The British East India Trading Company (as well as the French, Portuguese, Swedish, and Danish East India companies), in the process of amassing fortunes, killed or subjugated indigenous people who tried to maintain their sovereignty and way of life against the traders' encroachments. As will be seen in Chapter 7, at first trading companies had their own armed mercenaries, but they were increasingly aided by their respective governments, and in time many became state corporations.

Among the most important mercantile systems was the triangle trade between England, Africa, and the New World. The death of slaves from the brutal conditions of labor on sugar cane plantations in the Caribbean and South America required the continual importation of new slaves from West Africa. Ships loaded with cloth, hand tools, and weapons were sent from England to trade for slaves. Human beings, captured and sold into slavery, were deposited in the New World where they became plantation labor. The ships returned to England with sugar that sold for a high price. Much of the sugar was made into gin and consumed by the growing, impoverished urban poor whose families had been small farmers.

The rural areas of England became the sites of production that funded a growing class of merchants and traders. They, in turn, became the backbone of a more flexible, economically aggressive political system: representative democracy. Serfdom became less common in the twelfth century and "by the fifteenth, serfdom was almost universally abolished in the west" (Anderson 1971: 143).[17] Rural families were contractually obligated but legally independent. By 1700, the declining rural standard of living, in part because of the consolidation of land and creation of sheep pastures that accompanied the rise of the woolen industry, made rural families eager for any source of income. Some peasants turned their homes into workshops (hence the term "cottage industry") that processed and spun thread and wove woolen cloth under contracts with merchants. The capital accumulation from this arrangement in time allowed merchants to shift the work to new urban factories. These could be powered by water and, with increasing technological sophistication, steam, reducing further the need for human labor.[18] Thousands of the redundant poor were shipped to England's new colonies, including Australia.

---

[17]Serfdom did not officially end in England until Queen Elizabeth's proclamation in 1574 and in France until 1781.

[18]Eric Wolf's *Europe and the People Without History* (1982) tells this story well.

Industrial capitalism and the first factories began the process of proletarianization, relocating peoples' place of work to large-scale production sites where, for a specified amount of labor time, they received a wage. Many factories retained the shop-oriented work teams headed by master craftsmen who did their own hiring, owned their tools, and possessed the knowledge of production. By the late nineteenth century and the advent of assembly line production, this, too, was changing. Today, most people who make things or provide a service accept the arrangements of workplace hierarchy and control and would be offended to be called proletarians rather than employees and middle-class consumers.

Eric Wolf (1982: 267) sums up the transformations in capitalism, culminating in an industrial revolution, by emphasizing not only the accumulation of capital but technology as the critical factor driving social change. "Technology and labor power were subjected to the calculus of creating surplus value. The result was to speed up the pace of technological change." To this could be added the need for the state to become more responsive to a significant portion of new and would-be elites and for the use of state power to help corporations acquire and control resources and markets around the world.

The problem of funds to run the state and pursue its ambitions required an ever-growing economy. This gave an advantage to mercantile capitalists who needed the state's legal and military protection in order to pursue global trade. When industrial capitalism began to create a wealthy class of private owners, legal protections in the form of corporate laws, agreements on levels of taxation, and the state's provision of protection against worker unrest and foreign confiscation were brokered that benefited economic elites in exchange for keeping the state well funded (Tilly 1992: 195–197).

Over the course of the nineteenth century, nearly all of the world's people encountered colonial incursions, first by European nations and followed by those of the United States, Japan, and China. Their ability to do this rested heavily on superior technologies, especially technologies of war. Under colonial rule, regions of the world were transformed into satellites of the colonial powers in order to provide raw materials and, in some cases, finished goods that were the exclusive property of the colonial power.[19] By 1886, Europe's Berlin Conference had divided the entire African continent into colonial holdings, divisions that became *de facto* national borders in the two decades of independence movements following the Second World War.

---

[19]The exception was Latin America, whose nations were formed in the nineteenth century but remained closely tied to one or another European nation and functioned much as colonies did elsewhere.

Dominance of the National State

Charles Tilly's (1984) second "master process" is the growth of national states as the dominant form of political organization. Empires that preceded nations were largely content to claim large areas of land encompassing a variety of peoples speaking different languages, practicing different religions, and involved in various ways of farming, animal husbandry, and fishing. The empire's interest was in the physical security for its capital and major cities, uncontested control of trade in its domain, and the extraction of taxes from its subjects.

The national state—unlike empires—was much more active in remaking the life of the people over which it held control. Because it was deeply involved in the expensive activity of making war, the national state had to solve several economic and political problems. It could purchase an army, but this became prohibitively expensive. It could conscript an army, but resistance would in time erode the legitimacy of the state. A better solution was to emphasize a common national identity through the adoption of a single language and support for a state religion. Creating a national, "imagined community," in Benedict Anderson's (1983) phrase, of personal attachments and social unity allowed the state to enlist loyalty and patriotism of its subjects in making war. Even this approach, however, was not without costs.

A bargain had to be struck with the citizens of the national state that gave them at least the sense of justice and equality of rights as well as a voice in the country's affairs, i.e., political democracy. Subject people were in turn expected to respect the laws, pay taxes, and help fight the wars of the national state. The late-nineteenth and early twentieth centuries ushered in an expansion of social democracy by the national state that included a broader agenda of social welfare and improvements in peoples' quality of life. This expansion of the state was loudly debated throughout the twentieth century and remains among the most contentious divides in the ideologies of democratic nations today.

One of the ways to influence the state's agenda was the mass social movement and the occasional general strike. As will be seen in Chapter 5, the movements to abolish slavery, win legal equality for women, provide material protection for the nation's most vulnerable citizens—beginning with widows and children (Skocpol 1992)—establish workers' unions that owners were legally obligated to recognize, remove obstacles to full citizen participation without regard to race, and the myriad social movements following in the wake of the modern civil rights movement became a major force for social change in the twentieth century.

Science & Tech

The power of scientific research and the technological solutions it provided contributed greatly to the economic foundation of industrial capitalism and the capabilities of the national state. Science, discovery, and technology were supported by growing affluence of an emerging middle class. The nineteenth-century state's provision of general public education and specialized education in institutions of higher learning created a better-educated populace for a rapidly growing economy. Most importantly, scientific progress led to a dramatic shift in how the world was viewed, the capabilities of human beings to affect their environment, and the social relations and obligations among people. From these shifts grew the modern social sciences and efforts to understand and influence the process of social change.

Iris Summers' Time and Place in Global Context

A commonplace observation is that the nineteenth century was the European century, and the twentieth century was the American century. Some suggest that the twenty-first century may belong to another nation or continent, e.g., China or Asia. Such statements are simplifications if the intention is to imply that whatever happened of significance happened in only one part of the world. More often the implication is that power—economic, political, and cultural—is concentrated in one part of the world and then shifts elsewhere. Europeans would be hard pressed to accede the twentieth century to the United States, and the historical record bears them out. By the same token, Latin America, South and East Asia, Russia, and Africa were hardly standing still the past two hundred years. There was plenty of power and influence being exerted throughout the world in pursuit of territory, economic advantage, natural resources, security, and the intention to dominate, exploit, and change.

Many of the major forces of global social transformation in your life have their origin or were significantly developed in the centuries before you were born. Global trade intensified in the nineteenth century and was the source of several wars between alliances of European nations. Industrialization emerged first in England early in the nineteenth century and soon took hold throughout Europe, the United States, and Canada. This shaped the lives of much of the population, in the kind of work people did, where they lived, family life, and the kinds of social inequalities that justified privilege and poverty. Nineteenth-century colonization and client states were central to both economic expansion throughout Europe and the transformation of nonindustrial societies. The consolidation of territory and nation building, along with the creation of coercive judicial and military institutions, gave industrialized nations unrivaled authority. The corporate form allowed private companies, backed by state power, to form armies that forced others to participate in European economies. Invention, the diffusion of technology, and the impetus to engineer new possibilities for commerce and war accelerated change in material, political, economic, and cultural life worldwide.

More than a hundred years later many of these same forces are driving social change. The concentration of economic power and capital in multinational corporations and the global pursuit of earth resources, foodstuffs, a pliable workforce, and eager consumers now go by the name of economic globalization. Global military expenditures are at an all-time high, despite the decline in the Cold War great-power rivalry between the United States and now-defunct Soviet Union, and nuclear weapons remain available to obliterate much of the earth. The causes of human rights and social justice, as well as threats to resources and economic livelihoods, mobilize tens of thousands of people who increasingly rely on Internet-based social networks and cell phone communication to share their grievances, develop a compelling ideology, coordinate protest marches and public displays, and challenge corporate and state authority.

Some people fear that international organizations and regional coordination will subvert national sovereignty. To date, this has not happened. The national state remains the most significant political actor on the world stage. National self-interest more often than not dominates international conferences and the workings of international organizations like the United Nations and the European Union.

The capability of information to influence social change, especially amid a social movement for political change, is nothing new, whether the information is rumor or fact. What is new is the speed and immediacy of digital information, including images and sound, and the almost instantaneous dissemination of it around the world. For example, amid the Iranian protesters' street marches in 2009, a young woman, Neda Agha-Soltan, was shot and killed. Her death was captured by two cell phone users, one of whom sent the two-megabyte file to an Iranian exile living in the Netherlands. She then posted the video on Facebook, and it quickly proliferated on YouTube where it was picked up by CNN and broadcast worldwide (Stelter and Stone 2009).

Information technologies now link people globally in surprising and significant ways. They are another manifestation of globalization, akin to the spread of popular music, movies, celebrity news, fashion, and fads. These culture forms often enter traditional domains and provoke a challenge to values, aspirations, and relationships that may split generations, spark political movements, or unlock opportunities for mass marketing. They provoke both resistance to social change and an insistence that social change be locally, rather than globally, controlled.

A More Crowded Continent, a More Crowded World

When Iris Summers was a baby, nearly a hundred years ago, the U.S. population was ninety-two million. It now stands at well over three hundred million. Global population has gone from 1.79 billion to 6.8 billion people. Thousands of small towns declined, disappeared, or were absorbed into expanding cities, while small cities became massive metropolises. Fewer babies and toddlers died, and people began living longer. They not only increased in number, but their consumption of natural resources, especially energy and water, went up dramatically.

Globally the same thing happened. Population growth came later in less-industrialized parts of the world and in regions that had been colonial hold-ings, such as the Philippines, the British Raj (what is now India, Pakistan, Burma, and Bangladesh), Kenya, and Algeria. Families of six children or more remained the norm up to the end of the twentieth century in many parts of the world. A steady rural-to-urban migration created massive popu-lation centers that often could not accommodate millions of people living there. Crowded cities became overcrowded, with vast slums, squatter com-munities, and shantytowns ringing the urban core. When birth rates began declining, however, they did so quickly.

Changes in population can have many consequences. For instance, a grow-ing or declining population may affect the supply of labor relative to demand, shifting wages in favor of or to the detriment of wage earners. An influx of people from other nations may spark prejudice and nativist movements that lead to violence and repressive laws. Infanticide of young females and abortion of female fetuses, as happens today in India, China, and elsewhere, alters the sex ratio and may result in outmigration of young males and more men remain-ing bachelors. Population growth can create competition for land, especially when environmental degradation reduces the amount of good land for farming and grazing. This can lead to violence and occasionally genocidal practices as happened in Rwanda, Burundi, and the Darfur region of Sudan.

Each year 140 million people are born. This looks like a frightening num-ber, but it was 173 million fifteen years ago, and the number of people dying annually, currently about 57 million, is increasing. As a consequence, global population growth is slowing. At today's rates of declining growth, popula-tion could stabilize at less than nine billion people by 2050. According to the U.N.'s Food and Agriculture Organization, because of growing economies there will need to be a 70 percent increase in food production and a doubling of food availability in developing countries. The use of nonrenewable resources is not likely to decline. Nor will the pollution created by industry, the burning of fossil fuels, and especially the demand for water. Where the energy and water to produce this food will come from, and whether it will be done in a sustainable way, will pose the greatest challenges and dilemmas—and the most difficult choices—of the twenty-first century.

DO POPULATION DYNAMICS DRIVE SOCIAL CHANGE?

The significance of population changes—including migration—lies in the meanings people attach to these changes (Are they worrisome or welcome?) and their response. Is a change in population itself a driver of social change? Many demographers, i.e., people who study population trends and population geography, treat "population pressure" as a force influencing social change. Paul Ehrlich, author of *The Population Bomb*, and others who find global population growth alarming see it as cataclysmic in the not-too-distant future, somewhat akin to the well-known prediction of impending starvation made nearly two hundred years ago by the clergyman Thomas Malthus.[20]

Malthus' prediction has not been realized, though massive famines have happened. China's Great Leap Forward contributed to millions dying of starvation from 1958 to 1960. Recognizing how uncontrolled population growth would thwart improved standards of living, China's one-child policy was instituted and has reduced what would have been the current population by at least two hundred million people. Globally, growth rates are dropping precipitously. In European countries, Russia, Japan, and the United States, the fertility rate is below the replacement level of 2.1 births per woman. Annual world population growth is expected to decline from 1.3 percent to barely .3 percent by 2050 (U.N. 2010).

Is this social change, or is it a trend that seeks an explanation? Social scientists are keen to know what social changes are causing this dramatic reduction in fertility. There surely are many contributing

---

[20]Malthus urged "positive measures," and particularly abstinence, on the part of the lower classes in order to delay the inevitable day that population would outstrip the food available to sustain the world's population. He doubted this would actually happen, however, and a day would come when there was not enough food to support the population, resulting in mass famine.

factors, including increased availability of birth control, an improved standard of living, government policies penalizing large or favoring small families, and improvements in the literacy and status of women. Numbers of people per se, even when changing quickly, are facts to be explained by the forces that impel social change.

In the late 1800s, cities in the United States adopted programs and regulations that led to improved housing and reduced overcrowding. As discussed in Chapter 8, public health and sanitation measures included landfills for trash and chlorination of drinking water to reduce waterborne diseases like cholera. State and local health departments adopted vaccination programs, beginning with smallpox in 1900, that greatly reduced communicable diseases. By midcentury, childhood vaccinations were common. The most deadly diseases in 1900 were pneumonia and tuberculosis; these have been treated with antibiotics since the 1940s, dramatically reducing their death toll.

In 1918, half a million Americans died in a worldwide flu pandemic. With the exception of HIV/AIDS, there has been nothing comparable since. Americans today can expect to live thirty years longer than in 1900 when life expectancy at birth was 47.3 years. Children under the age of five accounted for 30 percent of all deaths in 1900, and now they are less than 2 percent. These are features of change to be explained. What drove these changes?

The forces impelling demographic changes like these include scientific discoveries, government efforts, and the development of the pharmaceutical industry. It is more complex than this, but the study of social change in most cases treats population dynamics as part of the social change process and a subject to be explained rather than a driver of social change.

The More Things Change . . .

In her *Shifts in the Social Contract* (1996), Beth Rubin writes that social change is happening very rapidly, especially since the mid-twentieth century. In a popular textbook that emphasizes the evolutionary perspective, Patrick Nolan and Gerhard Lenski conclude that "rapid social and cultural change has been the exception rather than the rule *until recently*" (Nolan and Lenski 2009: 52, italics added).

pendulum change

Social change proceeds neither at a constant speed nor in a straight line. Social change has been variously described as being linear, cyclic, and like a pendulum. One of the more surprising features in twentieth-century social history, revealed in the life of Iris Summers, is the return by the end of the century to many of the features of one hundred years earlier. Certainly much has changed, as her life story reveals, but there is also a pendulum swing that makes some aspects of her early life more similar to that of young people today than young people at midcentury.

When Iris and Frank Summers married, and again today, the average age of marriage is several years later than in the 1950s. A hundred years ago as well as today, more people are living in multigenerational households, with adult children residing with their parents, even after becoming parents themselves, than was the case at midcentury. There is considerable personal uncertainty about financial security, and the ethos of material accumulation and consumption is tempered by the ideas of conservation and sustainability. In Iris Summers' experience, this came from growing up with less, while today it stems from economic uncertainty and a concern for the well-being of the environment, the earth that future generations will inherit.

The most important swing of the pendulum in the United States—back and forth over the century—was the gap between rich and poor. It was wide at the start of the century and remained wide into the Great Depression. The industrial buildup for World War II, postwar economic prosperity, government taxation policies, strong labor unions, and a rising minimum wage "lifted all boats" over the next decades, expanding the middle class and narrowing the gap between rich and poor.

After 1970, however, that trend began to shift. An economic slowdown in the 1970s was followed by minor recessions in each succeeding decade and the Great Recession of 2007–2010. Changes in the federal tax code favored the wealthy and shifted federal revenue away from unearned income.[21] The declining strength of labor unions, a stagnant

---

[21]Household income increased between 1976 and 2006 for most people because of the increased number of household earners (especially wives entering the labor force and adult children living at home), despite the nearly stagnant average for wages. In this thirty-year period, the poorest fifth of all households had an 11 percent increase in household income; the next fifth had an 18 percent increase; the third fifth had a 21 percent increase, and the forth quintile had a 32 percent increase. The top 20 percent of households saw their income rise by 55 percent. Households at the very top, the top 1 percent with 2006 incomes averaging $1.2 million, had a 256 percent increase in income between 1976 and 2006 (Hacker and Pierson 2010: 23).

minimum wage, and global competition that induced manufacturers to close shop in the United States and set up production overseas all increased the division between rich and poor. Today, income and wealth inequality in the United States is greater than in any other industrialized country and is greater than it has been since the 1920s when Iris Summers was a girl.

Claude Fischer and Michael Hout's *Century of Difference* (2006) concludes that education plays the most crucial role in influencing people's life chances today. This contrasts to the huge influence on life chances of nationality and immigrant status in 1900. Their findings parallel the life of Iris Summers. Differences between rural and urban life narrowed throughout the twentieth century, largely because fewer people were living and working in the kind of rural isolation of their great grandparents. Roads and highways, automobiles, telephones, radio and television, and the Internet came to link virtually everyone.

The children of farmers, many of whom rented their land at the turn of the century, became factory and craft workers, and their children became service employees. Job opportunities became less a matter of where a person came from (their nationality or ethnicity) and more a matter of what they learned. Despite this change, overall chances of moving up (and down) in the social hierarchy or class structure changed little. They are about what they were a century ago.

Drivers of Social Change

The story of Iris Summers' life and times highlights the things that most strongly affect social change. These are the topics of Chapters 4 through 8. Robert Merton, a major twentieth-century sociologist, called these "social mechanisms," i.e., "social processes having designated consequences" (Merton 1968: 43–44). They can also be thought of as drivers of social change or major forces to which social change can be attributed.

The five drivers or mechanisms are often themselves the outcome of changing circumstances, and so it is a mistake to imagine that social life is unchanging until one of these mechanisms kicks in. Equally erroneous is to think that they act independently, that social change is the result of one or the other of the mechanisms. They often work in tandem or sequentially. Studying them in each chapter is only a means to thinking about and understanding the way each influences the speed, direction, and scope of social change. Understanding them helps unscramble situations of social change that might otherwise seem totally inexplicable.

Chapters 2 and 3 address two questions prior to studying the mechanisms or drivers of social change: How do we recognize social change? How do we understand social change? In a more formal presentation, these would be chapters on methods and theory, two staples of social science that are the foundation of its accomplishments. Entire careers of academic professionals are devoted to one or another of these topics, but most students new to the study of social change would rather skip them. That would be a mistake. It doesn't take long to recognize how befuddled and frustrating inquiry becomes without some guidance in working through a maze of information and ideas. Everyone needs what Charles Hampden-Turner (1970) calls the toolbox of research methods and theory. These chapters provide a few basic tools.

Chapter 4 discusses social change due to new knowledge and ways of applying knowledge to the solution of problems. This is the general concept of technology and includes science, discovery, invention, and new applications of existing techniques. Much social change associated with technology occurs through processes of diffusion from the point of origin, adoption, and creative adaptations in order to solve problems, to say nothing of the problems new technologies create. Technology is nothing new, but for the past half millennium it has proliferated and proved to be a very powerful and transformative force.

Chapter 5 explores the ways that people challenge authority and mobilize for social change. Social movements grow out of grievances that impel people to join together in diagnosing a problem and proposing a solution toward which the social movement devotes resources. One of the most fascinating aspects of social movements is the framing process, i.e., putting forth a perspective that recommends a course of action and motivates participant involvement. Why so many social movements focus on enlisting the state and, increasingly, on the abuses of corporations, is part of the story of twentieth-century social movements, from suffrage to unionization, civil rights, and the environmental movement.

Chapter 6 takes a sobering view of transformation through war and revolution. The third driver of social change is international war and other coercive conflicts carried out by nations, groups within nations, and stateless groups. Ethnic groupings and nationalist aspirations have been central to civil conflicts for much of the past century and promise to be prominent in the decades ahead. The chapter concentrates on the way societies and regions are changed in preparation for war, during wartime, and in the aftermath of war. Revolutionary wars are motivated by hopes for social change, but even defensive wars that seek to hold back social change rarely succeed in maintaining the status quo.

This chapter incorporates much of what has been examined up to this point and what comes after. War's conduct is highly contingent on the technologies of warfare. The pursuit of nationhood or ethnic autonomy usually begins as a social movement. And wars shift the field for economic actors, altering the competition and—in the case of the "resource wars" in Africa—are promulgated by individuals and corporations in pursuit of wealth. Diplomacy and international organizations provide mechanisms and frameworks for the prevention of war and the pursuit of peace that have altered modern warfare and hold out hope for the diminution of war as a driver of social change.

Chapter 7 looks at the making, selling, buying, and consuming of the things that pervade our lives. The fourth driver of social change is the actions of large corporations in pursuit of economic gain. This change is rarely planned, but its cumulative effect has transformed economies, cultures, personal identities, and the relationships of everyday life. It sometimes seems that the power to create goods and services, as well as our sense of what we need, rivals the power of nature. The work we do, the health of our communities, our sense of well-being, and avenues to realize our life's dreams are tied to corporate decisions. The worldwide pursuit of resources and markets influences foreign policy and military actions that change the world order. The influence and power of global corporations is a story critical to any understanding of social change.

Chapter 8 examines social change through the use of common resources and collective power. The fifth driver of social change is the state, the repository of collective power. Whether authoritarian and acting in the interests of a few, or democratically working on behalf of the people while protecting the rights of the minority, what modern states seek to accomplish affects everyone. This chapter examines how this is done and why, by telling the stories of several state actions in the United States and then focusing on modern China's efforts to become a major economic power.

Chapter 9 takes up what many students want when they originally venture into the topic of social change. It offers some ideas on how to make social change happen. This final chapter stresses the importance of awareness, engagement, leadership, and taking responsibility for one's actions. Rather than trying to predict the future—a valiant but dubious effort—this chapter focuses on making the future.

Topics for Discussion and Activities for Further Study

Topics for Discussion

1. Iris Summers is not a person of color, not a male, an immigrant, lesbian, or a person with disabilities or special needs. How might her life story have been different had she been any of these? She lived only a few years in a city, never went to jail, and spent her adult life as a wife and mother. What does this leave out of the picture of social change in Iris Summers' lifetime?

2. Has contemporary social change caused the world to get larger, or has it gotten smaller? More or less diverse? Examine both sides to these questions, especially the one that seems less obvious to you. What are the things that make the world larger, smaller, more diverse, and less diverse?

3. Charles Tilly's two "master narratives" cover a lot of ground in describing 500 years of social change in Europe and the rest of the world influenced by Europe. What might be left out? There was a radical change in religious thought, from Luther to liberation theology. Could this be a third master narrative?

4. Some students are from rural areas and some from urban areas. Because the shift from rural to urban was so critical in contemporary social change, explore the differences—some of which are described in this chapter—and their significance. Do you think the rural/urban divide is as important or obvious as it once was?

5. As a thought experiment, imagine a tremendous decline in global population. Would social change go backward, returning to bygone days? What would be the main challenges of a smaller population in a world that is trying to accommodate nearly seven billion people today? Or is it?

Activities for Further Study

1. Take a few minutes and read a good newspaper. Make a list of the social changes described and discussed in one day's reporting. What data are cited? Why are these changes newsworthy?

2. Collect words, especially old (archaic) words that tell you about social change. Most obvious are words for technology and equipment. Who talks about typewriters or scythes these days? What about food? With the typical family eating 40 percent of their meals outside their home, has the language of cooking and meals changed? How about warfare? What do "interdiction" and "counterinsurgency" say about war in the twenty-first century? What social changes are reflected in the words you collect?

3. Have a conversation with someone in your family, a close friend, or someone you're comfortable talking to who is more than sixty years old. Ask him or her to describe the three or four most important social changes in his or her lifetime. Discuss these with him or her: Why are they important? What were their causes? and How have they affected the person you are talking to? Present this to the class.

4. Why and how would anyone resist social change? Think of something that appears to be changing in the world around you that you might resist. Rather than just opting out or saying no, think about the source of this change and how you (and others) might oppose it or change its course. What concrete steps could you take? Who might join you? What would be your alternative course?

5. Go to a library that has a collection of popular magazines going back at least fifty, and preferably seventy-five, years or more. Read enough in them to imagine yourself actually living in the months the magazines were published. How would you be different? What thoughts would you not have? What would you not be doing? How would your view of life and the future be different from what they are now?

2

Recognizing Social Change

Barely twenty years ago, a large and powerful nation collapsed. Its economy imploded, the government was disassembled, and people were told overnight that they were no longer citizens of the country. Because it was such an important nation, its collapse sent shockwaves throughout the world. Other nations' economies shrank overnight, generating unemployment and poverty on a massive scale. Trading partners found themselves without a major buyer or seller. Powerful, well-established global diplomatic and military activities ceased. Alliances and agreements, ideologies, and the global rhetoric of friend and foe quickly became irrelevant or evaporated. Life was no longer as it had once been.

This account may sound like an overly dramatic novel or an action movie, but the collapse of the Soviet Union was very real. Few shots were fired. There was little immediate property damage or loss of life. But the change was as great as what had happened seventy years earlier in the Russian Revolution that created the Soviet Union (Kotkin 2009).

It was a cataclysmic social change whose reverberations continue to be felt worldwide. Nearly two hundred million people were made citizens of other countries, 150 million of whom became citizens of Russia. The Soviet Union's economy shrank to 53 percent of its previous size between 1989 and 1997, throwing millions of people into unemployment and poverty. Billions of dollars in assets were transferred to private hands as the state-run economy was taken over by those who previously had managed enterprises in the name of the people. Often called a kleptocracy, they and members of the former political elite became billionaires and multimillionaires. Crime, including homicide, and suicide increased significantly. Birthrates fell by a quarter as death rates shot up. The average life expectancy declined from 68.5 years to 63.2. Much of this was the consequence of "alcohol poisoning," that is, drinking oneself to death, especially among males, whose life expectancy fell to 56.5 years.

Today Russia, Kazakhstan, Latvia, Ukraine, Georgia, and the nine other countries that were republics of the former USSR continue to change as their governments exert national sovereignty and find their place in the world of nations. Like a ripple effect, countries closely allied with the Soviet Union were set adrift in the 1990s, not always finding a peaceful resolution to their upended situations. Poland, already in the process of social changes brought on by social movements and street demonstrations led by *Solidarność* (solidarity) in the 1980s, went through "shock therapy" that shrank its economy, sent its pensioners into poverty, and caused widespread unemployment. Hungary, Romania, and Bulgaria did much the same thing. Czechoslovakia underwent a "velvet revolution" and split into two countries, again nearly bloodlessly. Conflict broke out within the new borders of Russia, where ethnic minorities sought greater autonomy and were suppressed by Russian forces. The most prominent of these was in Chechnya. Possibly 200,000 civilians were killed in two wars where resistance and terrorist violence were met with massive Russian military retaliation and state terrorism.

The demise of the communist leadership in the USSR signaled the death knell of Yugoslavia's own communist party that had controlled the country since 1945. The country was already on the brink of collapse by 1988 with a rapidly failing economy and hyperinflation. Earlier nationalist movements were revived, and three republics declared independence from Yugoslavia. The government of Serbia, one of Yugoslavia's republics, used the army to try to hold the country together, but the conflict escalated into ethnic violence and civil wars that killed possibly 200,000 people, many of them victims of what came to be called "ethnic cleansing." Six nations now exist in what was once Yugoslavia.

Most social change is not this dramatic. Fortunately, people rarely experience social change by looking out the window and seeing tanks roll down the street, thousands of marchers waving guns and chanting threatening slogans, artillery bursts punctuating the night, plumes of smoke rising over government buildings, and jets bombing and strafing outlying hills where rebels are camped. No, for most of us and for most times social change is barely perceptible.

There is significant social change going on around us all the time, but it goes on less dramatically, often out of the public eye. In their *Hallowing Out the Middle*, Patrick Carr and Maria Kefalas (2009) examine the decline of rural communities, not a new phenomenon but one that has reached crisis proportions in many parts of the United States. The family farm's demise, widely publicized in the early 1980s, seemed to condemn rural communities, only to get a brief reprieve when manufacturing plants relocated there in search of a low-wage workforce. Global competition has now caused many of these plants to shut their doors, again threatening large swaths of small towns. Increasingly, poverty and social problems like high divorce rates, illegal drug (especially methamphetamine) use, and nonmarital childbearing are among the highest in the nation. The outmigration of the best educated and most ambitious young people has left thousands of small communities bereft of the human capital needed to reverse their declines. This "class-structured migration pattern reinforces a level of uneven development [between rural and urban America] not seen since the Civil War." As Carr and Kefalas observe, "perhaps the rural crisis has developed so slowly that symptoms of decline have been easier to ignore" (Carr and Kefalas 2009: 8).

Another example is incarceration and imprisonment in the United States. Today there are approximately five times as many individuals in prisons or jails as there were in 1980, just over thirty years ago. Is this due to a tremendous upsurge in crime? Does the criminal justice system more effectively apprehend and successfully prosecute wrongdoers? Probably neither of these to any great degree. The more likely reasons are more onerous sentencing guidelines, stricter penalties for violating parole, and prosecution of crimes involving even small amounts of drugs. Many people believe court leniency allows criminals to avoid jail, and the state has responded with legislation that incarcerates more people for longer periods of time.

The implications for social change due to rising levels of incarceration are great, if for no other reason than the public spending burden this imposes. Money for prisons (tens of thousands of tax dollars per prisoner each year) means less money for schools, parks, libraries, public transportation, and infrastructure. Gradually, the costs mount. Growing expenses elude most people until state and local budget deficits threaten school and library closures and the early release of prisoners from overcrowded jails. Then people recognize that something is happening!

Ways of Recognizing Social Change

Lapses of the calendar, comparing two points in time separated by historic events, is a familiar way to recognize and appreciate social change. Rip Van Winkle, Washington Irving's fictional character, went to sleep and woke up twenty years later. His wife had died, and his daughter was a young mother. Unaware of the American Revolution in his somnambulistic absence, he dangerously swore loyalty to King George III. He had missed the tumultuous war and the creation of a new nation. A popular German movie, *Goodbye, Lenin*, has a great deal of fun with a Rip Van Winkle–type story. A woman goes into a coma and wakes up after the collapse of the Soviet Union. Her country, the German Democratic Republic/East Germany, has been absorbed into the Federal Republic of Germany/West Germany in the shifting political and economic landscape of the Soviet Union's collapse. Her son frantically tries to shield her from the abandonment of Lenin's ideals for a communist society, to much comic effect.

During the early days of space flights, there was talk and no small amount of head scratching about the idea that a person could be sent into outer space and return hundreds of years later. He or she would barely have aged due to the effect of speed on their experience of time (remember Einstein's formula $E = MC^2$?). How would space explorers cope with a world in which their nuclear family had died generations before? New technologies befuddle them. Norms of behavior strike them as exotic. Social and political arrangements seem alien. This speculative motif helps emphasize social change by omitting the incremental changes between one point and another, one year and years ahead, one lifetime and another. In this way social change is much easier to recognize.

One of the most popular forms of literature takes the reader back in time, evoking an empathetic response to what life must have been like "back then." Movies, plays, music, and paintings do the same. For millennia, songs, stories, and poems were the principle device for relaying past events, including those that bound people to a tribe, a nation, or an empire larger than themselves, transcending their own mortality. In providing a shared legacy, the central narratives of heroism, victimization, and being a chosen people could be adapted to altered conditions and so remain salient guideposts for collective efforts and identities. The refrain was about permanence in the face of change.

Archeology and architectural studies also provide a strong message of change in people's lives. A small house with low doors, few if any windows, an outside kitchen, no bathroom, and a sleeping area shared with animals (possibly dogs, but also sheep, cows, and chickens) tells of times past when social conditions were much different. Perhaps extended families banded together, multiple generations occupied the same domicile, or powerful men had several wives with families living in a shared compound. The ruins of an ancient walled city or fortified market town, Hadrian's Wall or the Great Wall of China, evoke images of bandits, invaders, and warfare. That's the way things were, "before everything since" (Watson 2010: 130).

The majority of people leave no record of their meager lives. On the other hand, architectural grandeur and servants' quarters, crafted tiles and mosaics, luxurious gardens, ornate furniture and items of art, baths, and opulent venues for religious events provide a glimpse of a past system of inequality. Thousands of workers and peasants created the wealth enjoyed by only a few. They tell us how far we have come (or how little some things have changed).

What are other ways of recognizing social change? We may quiz our grandparent or great-aunt and uncle. We may read a biography or historical novel. We may take a break from studying and thumb through an old copy of *Life, Look*, the French magazine *Photo*, or the fashions in old issues of *Vogue, Esquire,* or *Harper's Bazaar.* The photographs reveal changes in material culture—cars and clothing, especially—and the advertisements evoke laughter both for what is being sold and how they seek to appeal to the reader. Go back far enough, and blatant sexism is obvious. There are few if any people of color in the advertisements. Carefully peruse the photographs: no computers or mobile handheld devices, most of the men are wearing hats outside, and ball caps are only worn at a game. Women who are described as beautiful may be much more plump than most of today's celebrities and models. If curiosity is sparked by these photographs, the next step is to seek out information about social change by putting these photographs in social context to make sense of them.[1]

Most of the time, we barely recognize social change. Revolutions, invasions, cataclysms, and other traumatic events that transform the taken-for-granted social and political world overnight are rare. In lieu of these, a special effort is required to perceive and appreciate social change. As James Davis (1996) points out, even significant changes like birth rates and mortality are sometimes imperceptible. Year-to-year changes are slight and can only be captured by a very deliberate process of data gathering and analysis. Only when viewed over several years do many small social changes become apparent.

An example is the educational attainment of U.S. women in the last half of the twentieth century. In 1960, only one in twenty graduates of a U.S. medical school was female. By 2000, two generations later, more than half of all medical students and nearly half of all law students were women.

---

[1]Less mainstream photographers like Lewis Hine, Walker Evans, Robert Frank, Lee Friedlander, Diane Arbus, and Bill Owens intentionally captured social change and the trauma of dislocation. Photography has itself influenced the older art of painting. Georgia O'Keefe's images of New Mexico's deserts and flora are magnified, enlarged, and cropped as if they were photographs, possibly owing to her close friendship with the great photographer Alfred Steiglitz.

Similarly, among women born between 1926 and 1935, about one in five worked outside the home full time year-round; for women born thirty years later, between 1956 and 1965, this figure was almost two out of three.

SCIENCE AS A SPECIAL APPROACH TO INQUIRY

Michelangelo told his inquirers that the figures he sculpted (e.g., David, the Pietá) were always within the stone. It was his task to find and release them. It is not unusual to think of science as discovery. In this sense, secrets lie within nature, to be uncovered, studied, appreciated, and destined for the storehouse of scientific knowledge. This view implicitly assumes some hidden code or an order to nature—rather than random spontaneity—to be found in particular events and applied across similar events. This is what Einstein meant when he said that God doesn't play the dice with the universe.

An inductive approach guides most original scientific research. Evidence from the natural world, i.e., empirical information, is gathered under controlled conditions, i.e., very consciously and systematically. Interpreting or making sense of it begins with the process of connecting the dots from A to B and knowing that A leads to B, or A is associated with B, under certain, but not all, conditions. Understanding *why* A leads to B is the key to creating knowledge that goes beyond a particular experiment, survey, observation, or investigation. Being able to make more general statements, proposing a conceptual scheme, or developing a theory is the other side of empirical research. It allows for deduction, i.e., making sense of what is found in other empirical cases by reference to prior research. This is discussed in Chapter 3.

How does scientific induction work? Take the example from anthropology. Human groups that lived for centuries in forests worshiped a pantheon of spirits and deities associated with the forest. Pastoral herding groups worshiped no such deities. Rather, many of them conceived of powers greater than themselves in heavenly bodies and celestial constellations, so much a part of their night panorama unimpeded by a forest canopy. The recognition of the connection of a people's environment to their belief systems encouraged anthropologists to examine and study the links between religious beliefs and practices and the material basis of their lives, e.g., the topography or physical world in

(Continued)

(Continued)

which different peoples live (Harris 1979). From a series of particular studies, researchers could apply principles or a theory that links environment and belief systems when making new inquiries across time and space. They make sense of new events in terms of the understandings they developed from prior study.

As a special kind of inquiry, however, science does not seek confirmation. The new cases are studied with the intention of *rejecting* accepted explanations, even ones that have been derived from excellent research. When empirical inquiry seeks—but fails—to reject an explanation, there is greater confidence that the explanation is the correct one. The failure to disconfirm strengthens our knowledge. In many cases this disconfirmation effort leads to some modifications or improvements in the explanation. Thus, scientific knowledge grows—not only as a body of facts but as a system of explanations about which we have more and more confidence. It is a never-ending process, however, and one that keeps absolute certainty at arm's length.

How Is Research Done?

Because it is a lot of work, it may sound strange to say that research is fun. But, it is. What could be more fun than finding out things nobody else knows? It is challenging, often frustrating, and it is intriguing. Life in all its complexity is also full of mysteries, only a few of which are amenable to social research. But research is one way to solve disputes, clear up ambiguity and gain clarity about a lot of things, and figure out what to do. If you are interested in knowing things, research is empowering. If you are someone who asks a lot of questions or if you are a problem solver, research can become your passion. This chapter intends to encourage your efforts at finding out what no one else knows by discovering new knowledge and contributing in whatever small way you might to recognizing social change.

Many students struggle with research because they have a great deal of trouble figuring out what they really want to know. In essence, they don't really know what their question is. Once the question is made clear, it is much easier to determine what information is needed to answer the question, how this information can be gathered, what to do with the information in order to arrive at a reasonable answer, and how to evaluate the strength of the conclusions. This is like five fingers (or four fingers and a thumb) on a hand:

- The thumb is the question;
- the pointer finger is the information needed to answer the question;
- the longest finger is how to gather this information;
- the ring finger is what to do with the information, often called analysis; and
- the pinky is the evaluation of how well the information provides an answer to the question.

Research may lead to policy. If we know something to be the case, what policies could be undertaken to implement an action that addresses a need, problem, issue, or concern? This is the full hand, or both hands, making something happen.

Asking Good Questions

Research should originate with a strong interest on the part of the researcher. For a moment and for the next few pages, imagine that you are a bicycle enthusiast. You look around and see your professor riding a bike across campus. Your sister is now an assistant chef in a new restaurant and sometimes bikes to work. Members of a bicycle advocacy group in town regularly attend city council meetings and want the city to be made safer for bikes. The radio and television updates on commuter traffic, accidents, snarls, and tie-ups send a shudder down your spine. Stuck in traffic? You think, I'm so glad I'm not there! You know that few people regularly bike to work or for shopping but think that many more would if some changes were made. Its benefits are obvious to you. More people should do it. So, why don't they?

Having more people using bicycles to shop and commute is both social change and a response to changes. Biking is less expensive than using a car, and in tight economic times that's important. It lowers the cost to towns and cities for maintaining streets, providing parking spaces, and so forth. Bicycles emit no greenhouse gasses to operate and, except for their manufacture and maintenance, don't contribute to global climate change. They don't add to noise pollution. Biking is good exercise. Increasing numbers of overweight people have unwanted implications for the health care delivery system as well as shortened lives, restricted opportunities, and increased public costs. You and a lot of other people would like this to change.

It is tempting to jump over the simple descriptive questions and ask more complex questions or questions of policy. For example, Why don't more people bike to work? or How can we get more people to ride their bikes safely? These are good questions, but not good questions to start with. Start simple, and get the facts right. What is really going on?

Descriptive and Analytic Questions. The simplest social-science research question is a descriptive question that asks for information. What is happening? Who is doing it? Where are they located? For how long? This is where historical and comparative information can be very useful. Even if the time and place are different from your own, the facts of other cases can help to clarify your own.

Analytic questions, more complex than descriptive questions, pose questions of why and how. As explained in the boxed discussion "Concepts and Variable Language," social science poses analytic questions in variable language, asking about the variation in one thing in relation to variation in another thing or several things. For example, if all your classmates ran a race, they would finish with varying times. You would try to explain why there is variation in running speed by linking it to variation in a variety of things: leg length, athletic experience, quality of footwear, motivation, and so forth. You might find that the longer one's legs, the faster he or she runs. The stronger the desire to win, the faster he or she runs, and so forth. The logic of research is to explain something by linking its variation or change to the variation or change of something else.

CONCEPTS AND VARIABLE LANGUAGE

Concepts. When talking with someone, words are used that have meanings shared by the speaker and listener. The sound made when referring to a cat—*kat* or *chat* or *gato*, depending on your language—evokes in the listener an image of a feline creature. If adjectives—*gray, striped, fat, stealthy, green-eyed, old*—are added, the image the speaker is trying to convey and the image the listener acquires is more particular. The speaker and listener will share a general idea of what a gray, striped, fat, stealthy, green-eyed, old cat is. Concepts in the social sciences are a special category of words and meanings.

Russell Schutt (2009: 101) calls a concept "a mental image that summarizes a set of similar observations, feeling, or ideas." Social scientists have provided thousands of new concepts to our vocabulary. They are the things social scientists talk about, study, and use to describe why people do what they do. Just as words and their meaning are the building blocks of sentences, paragraphs, chapters, and books, concepts are the building blocks and provide the language of research and theory.

Constants, Variable Concepts, and Variable Language. Some concepts provide context and are often called constants. "In wartime" is a constant. So are "in organizations that provide employee benefits" and "in single-parent households." These constants help to narrow the range of cases to which a research question applies.

In order to formulate research questions that can establish associations or relationships, variable concepts are required. Paul Lazarsfeld (1970: 67) explains variable concepts as dividing "the imagery [of a concept] into components. The concept is specified by aspects, dimensions, etc." For example, peoples' attitudes about things can be thought of as having several dimensions: favorability, strength, and so forth. A researcher may ask people how strongly they believe in God in order to tap the variability of the concept of "religiosity" or how supportive they are of military intervention to stop humanitarian crises in order to measure the political attitude of "interventionist".[2]

Now you can ask a question using variable language. The simple descriptive question, Do people bike to work? is better asked this way: How often do people bike to work? Biking to work is now seen as a variable concept; it varies across a population. Some people do it all the time, some do it occasionally, and some do it not at all. When you ask a question this way, you are speaking with variable language.

When you ask an analytic question, you link two or more variable concepts. Your research poses relationships between variable concepts that provide answers to why people do what they do. You might ask, What is the relationship between peoples' ages (a variable) and how often they bike to work (also a variable)? In this way, you can begin to explain things.

---

[2]In his essay "The Bearing of Sociological Theory on Empirical Research," Robert Merton (1968: 139–155) provides a discussion of concepts that guided a generation of social scientists. Russell Schutt's (2009, Chapter 4) discussion of variable concepts, and especially operationalizing and measuring variables, i.e., deciding on what actions, attitudes, or characteristics to actually get information about, is very useful. Hodson's (1999b) creation of, and research using, the concept of "management citizenship" is an excellent illustration of how good social science involves concept formation, measurement, and empirical testing.

From Questions to Hypotheses

Searching the Literature. Others may have already tried to answer your initial, descriptive questions. You can benefit from their work. Think of them as your colleagues, and be sure to give them credit. Finding out what happened is often a matter of researching existing sources: e.g., newspapers and magazine articles, websites, historical writing and other nonfiction books, public records and official statistics, polls, and published survey data. A good researcher sorts through fact and fabrication, consulting several sources before concluding what facts are correct.

To continue with our example of biking to work, you may do an online literature search and find several articles on the topic. Reading these, you clarify your own ideas and concepts, measurements, plans for data gathering, and modes of analysis. Some authors intend for their work to provide an empirical foundation for policies. For instance, Tilahun and his colleagues (2007) found that people are more willing to commute by bicycle if towns and cities designate streets and trails for bikes only, even if it takes longer to get to work. That's just the kind of research you want to use as a model and possibly to build on or go beyond with your own research efforts.

Independent and Dependent Variables. Variable concepts are of two types: dependent and independent variables. A good question distinguishes clearly the thing to be explained, the *dependent* variable, from the things that make it happen, the *independent* variables. The propensity or tendency of people to bike to work is, for the moment, the dependent variable you want to explain. It can be measured (operationalized) by finding out how often people bike to work. The contributing factors or reasons associated with this vary across the population. Because you are not trying to explain these things, they are independent variables.

You are now poised to compare things that vary with what you want to explain, in this case the dependent variable, the propensity to bike to work. Some things that might cause variation in the propensity to bike to work are the following:

- Distance from home to work
- Age of the person
- Season of the year or weather conditions
- Difficulty and cost of parking a car near one's work
- Need to carry a large amount of goods to work
- Perceived safety of biking
- Number of friends and co-workers who bike to work

Note that all of these are variable concepts, and information about them can be acquired by asking people questions. What other factors can you think of? The need to take children to school on the way to work? Whether or not the person even has a car?

Direct and Inverse Relationships. When concepts (e.g., independent and dependent variables) vary in the same direction, we say there is a direct relationship. For example, there may be a higher likelihood for someone biking to work if several friends and co-workers bike to work. When the concepts vary in opposite directions, an inverse relationship exists. For example, as distance from home to work increases, the propensity for biking to work decreases. These relationships may or may not be correct. We don't know until we gather information and carefully analyze it.

Hypothesized Relationships. To be honest, most research begins with some idea of what the answer might be. This isn't a bad thing, as long as the research is done properly and in such a way that the researcher's prior ideas *can* be proved wrong. We make statements that can be accepted or rejected with the information we are gathering. The questions are written as testable statements—hypotheses—that propose a relationship and the type of relationship, direct or inverse.

The questions look like statements when phrased as testable hypotheses:

- Hypothesis 1: The age of a person is inversely related to propensity to bike to work.
- Hypothesis 2: The degree to which weather is inclement is inversely related to propensity to bike to work.
- Hypothesis 3: The perception that biking is safe is directly related to propensity to bike to work.
- Hypotheses 4–7: State them for yourself. It's good practice. Like riding a bike: once you do it, you'll never forget how.

Your job is to gather information and see if it conforms to these statements or not. Note that none of them points in the direction of causality. Obviously, some statements imply a causal order. You can't change your age by biking, so age "causes" differences in the likelihood a person will bike to work. The same with weather. Does biking to work improve a person's feelings that it's safe to do, or does the perception "cause" the behavior? Determining causality is not always possible or even necessary for policy formation, but it is a research challenge.

Explanation and Causality. Explanation rests, to a great degree, on isolating or enumerating the things that cause events to happen. But woe to the person who claims to have found *the* cause. The search for a "prime mover" or single cause raises not only the hackles of other scholars but threatens to explain nothing by trying to explain everything. Statements such as "War is caused by human nature" are as meaningless as "The source of good health is clean living." Such phrases may be good for bumper stickers, but they are not good social science.

In studying social change, questions of cause and effect are always bounded by the historical social context. Especially when the question is about the reasons social change went in one direction rather than another, at one speed rather than another, and happened more or less broadly, causal imputation must be understood within the changing social context of what is being studied. For example, the significant changes in social interaction (e.g., patterns of intimacy between new lovers) made possible by electronic communication (e.g., when the telephone became a fixture of everyday life) can only be understood in the historical context: before and with the increasing availability of the telephone, increasing affluence, and changing norms about sex and gender. [3]

Causality must satisfy several well-recognized requirements: (1) an association, relationship, or correlation between two (or more) variables; (2) temporal sequencing, i.e., the independent variable occurs before the dependent variable; (3) a sensible idea of why there would be an association between them; and (4) nonspuriousness or independence between the variables, i.e., that their apparent relationship is not due to their both being "caused" by a third thing (Schutt 2009: 205–214).

That causality requires an association and temporal sequencing is fairly straightforward. An increase in spousal abuse has followed recent economic recessions. As the economy goes bad, police and social service agencies report more marital discord and incidences of violence. Marital discord doesn't cause the economy to sour. The causal arrow is pretty obvious.

The third requirement is to make explicit why one thing leads to another. In the earlier example of romance and telephones, the inquisitive researcher keeps a keen eye on creative uses for the phone, well beyond what was imagined by its inventors, in order to know why this happened. Perhaps telephones gave a romantic couple a new venue for private conversations but also allowed them to check up on each other with greater ease. Whatever the

---

[3]Tamara Hareven (1991) discusses the "grand processes" that have influenced changes in the family form for more than two centuries and the complexity of interactions that make it impossible to think about social change of families apart from the changing historical context.

reasons, it is up to the researcher to draw inferences that make sense, i.e., to find a causal mechanism.

The fourth criterion, nonspuriousness, is often demonstrated with the example of ice cream consumption and rape. There is a strong direct statistical correlation between increases in rape and increases in ice cream consumption. This association is so implausible, however, that the idea of causality is rejected out of hand. The two, of course, are spuriously related; both increase in warmer weather and both decrease in cooler weather. A third variable, average atmospheric temperature, contributes to their changes and the consistent statistical relationship between the two. A change in weather seems obvious for ice cream consumption. For rape, it may be due to the likelihood that windows are left open more often at night, that people spend more time in public when the weather is warmer, and so forth.[4]

Finally, an awareness of sufficiency (A is sufficient to produce B) and necessity (for B to occur, A is necessary) reminds us that B may not always happen, even though A is present, and A isn't the only thing that is needed in order to produce B. Robert Putnam, author of *Bowling Alone*, reminds us that "social change is . . . inevitably uneven. Life is not lived in a single dimension" (Putnam 2000: 26). It is important to try to pinpoint causal factors that are necessary and/or sufficient for something to occur. But one can never tell the whole story or tell the story completely. To some degree, any imputation of causality will always be incomplete.

The Ecological Fallacy. One of the more interesting mistakes in imputing causality is the ecological fallacy. Social scientists study data about groups as well as individuals. Their concepts describe what groups do, group characteristics, and group processes as well as those of individuals. The most common ecological fallacy is committed when findings about aggregates of individuals (groups) are inferred to the individuals themselves.

For example, I might find that factories in southern states report more days of absentee workers than factories in the Midwest. To infer from this that southern workers are more likely to stay home from work might well be wrong. The rates of southern absentee workers as a group (i.e., per factory) might be caused by a small percentage of workers who are always absent on

---

[4]This example should in no way suggest that rape is not a very serious problem deserving of rigorous research and effective policies. Understanding the causes of rape range from the psychological predisposition of perpetrators to widely shared attitudes about the victims of rape, the community characteristics that inhibit or facilitate rape, the role of police and the courts, and the relationship of rape to other forms of violence, such as robbery and warfare.

Fridays, while most never miss work. These few workers "skew" the results of their respective factories. Workers in the Midwest might be more modest about regularly missing work than the few southern workers who skip out on Fridays but more likely to miss work than the majority of southern workers. The moral of the story (barring the use of sophisticated statistical techniques): If your data are about individuals, apply your conclusions to individuals. If your data are about groups, apply your conclusions to groups.

TRACING AND UNTANGLING CAUSALITY

Iris Summers' sister, Martha, wanted all of her children to get a college education. She and her husband, Jake, saved part of their paychecks every month, and the children saved from their summer and part-time jobs. Everyone talked about how good grades help secure scholarships. Martha and Jake met regularly with their children's teachers. Housekeeping obligations and recreation were put off if there was homework to do.

A lot of Martha and Jake's savings went for tuition, books, room and board, and other expenses. At one point when money was very short, Jake got a loan from his uncle so their oldest daughter, Doris, could finish the semester. The children knew that their parents supported and had high expectations for them. They often thought of this when they were considering whether or not to go to the library or get started on a research paper. They learned from an early age how to organize their time, putting a priority on education and hard work. This helped them later.

On the other hand, several high school teachers also encouraged the children to go to college, and counselors helped select classes that prepared them for college work. Several college professors motivated and advised the children as well. The city where the family lived had several community colleges and a public four-year school to choose from. Urban mass transportation could be taken at a special student rate. The college where Tom went also offered low-cost health insurance. When he became ill, he would have had to quit school to pay the medical bills if he hadn't had the insurance.

Like all families, nothing was perfect, and there were setbacks. For one, Martha and Jake divorced while the two younger children were in high school. Only Tom received a good scholarship. Loans were required as well as part-time work, and Doris took nearly nine years to graduate. She met her future husband, had a child, quit school to support the family, and then returned to school when her husband found a good job. And what about Sandy? She was a good student but left college and joined the navy.

From a strictly empirical point of view, it is difficult to sort out the various causal factors. It is clear, however, that without Martha and Jake's financial support, encouragement, and help in developing good work habits, their children's college success would have been very unlikely. The urban environment and the benefits of actually being in college—subsidized mass transit, low-cost health insurance, scholarships—made a difference, but only after the children had enrolled in college in the first place.

We know there is a necessary causal link between having the money to pay tuition and gaining access to a college education. We know there are causal connections between childhood socialization (a good self-concept, work habits, ambition) and life experiences. Studies have shown that "cultural capital" and educational attainment have a strong association, i.e., between things outside of school—books and reading, family travel, respect for education—and how well children do in school.

These facts are particular to this family, but the links between them can be explained (some social scientists would call these "path-dependent") in terms of what is known more widely, i.e., that they are true for other families as well. Statistical analyses of large random-sample surveys have established many of these links. In this example can be seen both "internal" and "external" factors (things they did and things outside the family) leading to the outcome. There are also indirect and unanticipated consequences, as well as the possibility that some actions had not only unanticipated but also negative effects. For example, the family's financial anxiety was a source of stress between Martha and Jake, contributing to their divorce. In the final analysis, logical connections and empirical support justify our drawing some conclusions about causality. What the family did, in combination with the social environment, strongly contributed to—but cannot entirely explain—the children's educational attainment.

Gathering Information

The third part of the research task is gathering information about the concepts in the hypotheses. Returning to the biking example to test your hypotheses, you will want information about several things: peoples' ages, distance between their home and work, difficulty and cost of parking at or near where they work, whether they need to transport large amounts of material (tools, books and papers, building materials, etc.) to work, their perception of the safety of biking on the streets between their home and work, and the number of friends and co-workers who bike to work. You will want information about the weather and seasons in your town or city, and you probably want to know how the individuals in your study evaluate the weather conditions.

There are many ways to gather information. First, try to find out if the information has already been gathered. Maybe some exists and is waiting to be analyzed. Ask around, go online, read similar studies, and see if other researchers have data you could use. You may turn up something, saving yourself time and money.

To do this project, you need to get information from people who bike to work regularly, of course, but you also want information from people who bike to work occasionally and people who do not bike to work. You are gathering information about bikers and nonbikers, but you are actually gathering information *about biking*. Nonbikers have something to tell you about this.

The most straightforward thing to do is to give out a questionnaire to, or interview, a random sample of people, in this example both bikers and non-bikers. It is expensive to do this, however, and it may be difficult to ensure randomness by ascertaining how representative people in your study are of the entire population. Not gathering information from a random sample limits the applicability of a study's findings, but even a nonrandom sample can provide strong evidence and useful answers. This is explained in the boxed discussion.

SAMPLING AND DRAWING INFERENCES

A sample is the individuals or groups (e.g., businesses, communities, gangs) about whom information is actually gathered. If the characteristics of the sample match those of the larger population (e.g., the town or city in which you live), you can generalize about what you find from the sample to the larger population. Reputable pollsters such as Gallup, Harris, Roper, Pew, and Zogby have nearly perfected the selection of a sample of 1,500 people to represent the adult population of

340 million Americans.[5] There will always be some sampling error, i.e., the degree of accuracy of results from the sample. You may read about a poll that says something like, "Fifty-five percent of college students intend to vote in the next election. These results have a sampling error of plus or minus 3 percent." The "true" figure is between 51 and 58 percent. You could only know for sure if you literally asked every student if he or she intended to vote, rather than rely on a sample of the population of college students.

There are less representative ways to gather information—from nonrandom samples—that can still be very useful. A nonrandom survey should include the full range of people you would include in a perfectly random survey, in order to give you a full range of information. You might want to use "snowball sampling" by talking to a diversity of people until you are comfortable that you are not getting new information. Your ability to make sense of what people tell you about themselves, their circumstances, their beliefs, and so forth allows you to draw inferences that can, with caution and qualifications, be generalized or inferred to a larger population.

For example, you might find that about half of all young adults who you interview bike to work sometimes, and they rarely mention bike safety. On the other hand, you might find that nearly all people over the age of forty—those who bike and those who don't—mention bike safety as a concern, but those who bike feel it is safe and those who don't bike feel it is unsafe. You could conclude that there is an age difference in concern about bike safety (a direct relationship), and uncertainty about being safe is a greater influence on the likelihood of not biking to work for older people (an inverse relationship). As you study your data, other connections may appear. These may be less than definitive, but they are still useful enough to begin crafting policies, such as how and where to focus public safety awareness campaigns.

---

[5]Many people today think there is far too much polling and that it is an intrusion into their privacy. Charles Tilly, however, calls the period in Europe from 1870 to 1920 the "golden age of official statistics and social surveys" (Tilly 1984: 13). In the United States, the first polls were done by Roper in 1935, and they have gradually improved to where their reliability today is usually quite high. They tend to concentrate on attitudes and opinion but also provide some useful behavioral data, e.g., rates of smoking, newspaper readership, or cell phone use. Journals such as *Public Opinion Quarterly* and *International Journal of Public Opinion* are devoted primarily to studies using polling data.

Types of Information. Social change is recorded in words, images, and numbers. Personal diaries and journals, ship's logs, e-mail messages, autobiographies and nonfiction literature, and magazines and newspapers preserve written records of information that tell a story of events, phenomena, and processes. In her study of the war tribunal that sentenced more than a thousand Japanese officers, soldiers, and civilians to death following World War II, Kazuko Tsurumi studied the diaries, letters, poems, and essays of those convicted. "They were written on . . . writing paper, wrapping paper, toilet paper, handkerchiefs, a piece of sheet, and the margins of books"(Tsurumi 1970: 139). Photographs, movies and documentary films, drawings and paintings, and videos capture images and provide data. Useful data exist in records of possessions and transactions, population and longevity, wedding dates and times of death, speed, size, and volume, and anything else that can be measured. Together, words, images, and numbers are the information or data for the study of social change and, though they seem obvious, require some explanation.

Qualitative Information. Among the many sources of information used to study social change are the official and unofficial records of both famous and ordinary people. Thomas Jefferson, Frank Sinatra, Golda Meir, Thomas Edison, and Virginia Wolff had a good idea that someday people would be interested in their correspondence, their personal journals, and other items that were part of their life's work. Ordinary people, too, produce information but more humbly doubt that anyone will be interested in it, and they are sometimes right. When Iris Summers died, her children put boxes of newspaper and magazine clippings, banking records, sales receipts and product warranties, holiday brochures, and recipes for food never made into recycling bins. Phonograph records, clothing, bedding, furniture, books, and glassware went to Goodwill and the Salvation Army. But letters and photographs found a special place in their homes and were occasionally looked through and appreciated for the past life they recorded.

Personal narratives, diaries, photographs, scrapbooks, letters, and the like are qualitative information that help researchers reconstruct important historical events as well as more mundane, day-to-day life. W. I. Thomas and Florian Znaniecki's *The Polish Peasant in Poland and America* ([1918] 1958) was perhaps the first major sociological study to systematically scrutinize the letters that passed between immigrants and their families in Poland, and it told a story of both profound personal change and a historical watershed. The ordinary, individual lives of these turn-of-the-century immigrants recount social change that adds to the understanding of immigration in other times and places, whenever large numbers of people move from one society to another due to war, natural catastrophes, or in pursuit of opportunities.

People who are prominent in periods of social change leave behind personal records that become the staple of historical biographies. The correspondence of Margaret Sanger, one of the founders of the family planning movement in the United States, archived at Smith College, provides not only a look at one historical moment but an insider's account of a social movement that empowered women and men by challenging patriarchy, medical practitioners, and laws on public and private morality. The personal documents of inventors, political leaders, explorers, entrepreneurs, and generals contribute to the analysis of social change and the forces that impel change, especially when their accounts reveal the social context of their efforts and accomplishments.[6]

During the Great Depression, photographers like Gordon Parks, Dorothea Lange, and Walker Evans were employed by the federal government's Farm Security Administration to travel across the United States recording the images of African Americans (Parks), migrants to California (Lange), and southern Whites (Evans). Their photographs could not be statistically analyzed but are valuable data nonetheless. They tell a gripping story of tens of thousands of people, not only the individuals photographed, struggling and coping with poverty and lost opportunities.

At the same time, George Orwell (the pen name of Eric Blair, author of *Animal Farm* and *1984*) was hired by the Left Book Club "to write a documentary report on conditions among the unemployed in the north of England" (Orwell 1958: vii). He visited the mining communities hard-hit by the Great Depression and published his experiences, conversations, and observations in *Road to Wigan Pier*. The often unflattering accounts of the people Orwell met and the places he stayed are written in a blunt language that nonetheless evokes tremendous empathy for people in situations of economic distress. Nonfiction and documentary reports, when written with an eye for significant details, provide a rich source of information for understanding social change.

Many great qualitative research projects have few if any numbers. Careful research that accurately hears and records what people say and what people do, and uses empirical information to understand their circumstances, thinking, motivations, attitudes, emotions, and values, makes a valuable contribution to the understanding of why people do what they do.

---

[6]A useful guide is *The Uses of Personal Documents in History, Anthropology, and Sociology* (1945), edited by Louis Gottschalk, Clyde Kluckholm, and Robert Angell.

Floyd Burroughs, cotton sharecropper, Hale County, Alabama. Walker Evans, photographer, 1936. Migrant mother. Dorothea Lange, photographer, 1936. American gothic: portrait of government cleaning woman Ella Watson. Gordon Parks, photographer, 1942. Farm Security Administration Photographic Archives, Washington, D.C.

Quantitative Information. Record keeping by government agencies, schools, hospitals, and other public institutions, military units, businesses, and non-profit groups generates quantitative information. Many organizations specializing in data gathering and analysis, including social-research institutes, focus on what is happening right now. Over time, these data, too, become an invaluable source for the analysis of trends, cycles, and historical patterns. Often using statistical operations more elaborate than the original researchers used and guided by questions about specific factors influencing social change, these numerical data spawn a great deal of social-science knowledge. Karl Marx spent years analyzing the "Blue Books" in the British Museum in London that contained thousands of pages of trade and manufacturing information of early industrial capitalism. The result was his monumental *Capital* (1967; *Das Kapital*) about the transformation of capitalism as an economic system and the changes in social life it was bringing about.

The U.S. decennial census is authorized in Article 1, Section 2 of the Constitution of the United States to ensure proper proportional electoral representation. It does much more than merely count the people, and it has increasingly become one of the most important sources of information about the composition and conditions of the U.S. population (Anderson and Feinberg 1999). Comparing the information gathered in the census from decade to decade, as well as its Current Population Reports where the analysis focuses on select groups in the nation's population, is among the most important ways we can describe social change in one country.[7]

The distinction between qualitative and quantitative research is not absolute. When compiling information from informal interviews, a researcher may record the number of times people mention something, use a term, or make an erroneous observation. Content analysis of newspaper and magazine articles involves counting the number of times certain words are used, topics are mentioned, or a perspective is expressed. Putting numbers to qualitative material blurs the distinction between what is qualitative and what is quantitative.

MEASURES OF CENTRAL TENDENCY AND ASSOCIATION

Quantitative analysis of numerical data uses statistical techniques to calculate, and statistical measures to display, numbers. The most common and simplest way to do this is by comparing measures of central tendency: averages/means; the largest category/mode; and the center

(Continued)

---

[7]Another very useful source of quantitative information is the *Statistical Abstracts of the United States*, published annually and available on the World Wide Web. It compiles data from dozens of agencies and bureaus in the federal government and often presents these for a period of time, making it a handy reference for seeing trends and changes.

(Continued)

point/median that equally divides the cases. For example, you might find that among non-bike-to-work people, the distance from home to work is, on average, 7.3 miles, but the most-often-mentioned distance is four miles from work, and half of the nonbikers live less than four from work. If you compare these to the mean, median, and mode of bikers, viewed as a table, chart, or graph, you could possibly draw an inference about one of your hypotheses.

Because explanations involve the relationship between variables, statistical techniques have been developed to measure associations. These statistical measures involve some sense of probability, i.e., they provide an indication that the relationship is not due simply to chance. Talking to only one person or a few people is a little bit like a lucky streak at the gambling table. Even with chance occurrences, someone can have an improbably lucky streak. Over time, however, the laws of probability will catch up, the streak will end, and the true chance of winning will be apparent. Similarly, talking with a lot of people provides a better measure of what's happening than talking to only one person or a handful of people. That's why statistics that measure associations are best when there are more cases in the study rather than fewer.

A useful measure of association is a correlation. When graphing the relationship between two variables, this statistic shows how strongly the two are associated. For instance, if you graph all of the answers obtained from eighty people about (1) the distance from their home to work and (2) the frequency of their biking to work, you might calculate a .75 correlation between the two variables. A perfect correlation would be 1.0, and no relationship would be 0.0. A .75 correlation is a fairly strong relationship. How strong can be shown by another statistic, designated by the Greek letter Rho (ρ), to indicate whether this relationship, based on probability, is greater than chance and how much greater. There are many statistical-association calculations used in the social sciences, not just correlations, that help to draw inferences from quantitative data about the rate, scope of, and factors influencing social change.

Analyzing Information

Ethnographic, Historical, and Path-Dependent Analysis. Social scientists, especially anthropologists and sociologists, do ethnographic and field research that takes them to the site of what they are studying. They interact with the people and may participate in the events that interest them, though how involved they become varies greatly. Most qualitative researchers say that they are analyzing their information all the time. They are trying to make sense of what is going on, formulating concepts, and putting forth explanations that are tested day-to-day as their fieldwork proceeds.[8]

More often than not, qualitative analysis provides a more intuitive but multidimensional understanding of the people being studied by allowing for more personal details and the particular features of a situation. The result is a portrait or a narrative that may be full of insights based on empirical information. It is, however, difficult to apply conclusions from the qualitative study of individuals and groups to individuals and groups who were not studied.[9]

Knowing what happened and when it happened—the most basic historical knowledge—is a second kind of qualitative research. Historical studies often are based on a minimum of quantitative information, relying instead on physical artifacts and narrative accounts such as journals, diaries, newspaper articles, and so forth. Accuracy of information can be the most critical analytic problem. Cross-checking facts by having more than one source, or having a sound reason to be confident in a single source, increases the likelihood that historical data are correct.

Social scientists seeking to reconstruct historical events such as a revolution or economic collapse often talk about "path-dependent sequences" where a series of events and situations narrow the range of outcomes. The outcomes themselves become new events and situations in a sequence of cause and effect. For example, Geoffrey Herrera's *Technology and International Transformation* (2006) uses path-dependent analysis to explain the sequence of political and social factors that explain the development of the atomic bomb. At several junctures in the development of nuclear weapons, he shows why the path taken did not unfold differently, given the particular

---

[8]Many field researchers have only loose guidelines for analyzing their notes and other information they gather, but there are steps that can be taken to help sort through qualitative data. An excellent discussion of this is in Carol Bailey's *A Guide to Field Research* (1996).

[9]Many students are initially attracted to the social sciences by reading books based on qualitative research, like Elliot Liebow's *Tally's Corner* and Barbara Ehrenreich's *Nickled and Dimed*. Hodson (1999a) provides guidelines for the systematic analysis of such research, thus increasing its usefulness and reliability as qualitative data and the ability to generalize beyond a particular setting or case study.

circumstances and the constellation of political and technological forces at play. This is a form of causal analysis using qualitative empirical information.

Recognizing Social Change in Three Study Designs. Recognizing social change has greatly benefited from developments in quantitative analysis. Some of this work uses powerful statistical techniques. The ability to do these statistics—and less complex statistics as well—depends on how data are arranged so they can reveal trends, patterns, and influences. Three ways of gathering and arranging data are most useful in the study of social change.

The Cross-Sectional Study. Most social research is about a particular time and place. Because research is expensive and people want to know what is going on right now, even very comprehensive studies usually gather information at only one point in time. They provide a snapshot. If special care is taken to organize the data in particular ways, however, even a one-time study can provide a picture of social change.

Imagine a photo of the elderly Iris Summers' family reunion showing several generations: infants and toddlers, children and adolescents, young adults and middle-age parents, and the elderly and the ancient. In the photo are two adolescents with piercings and tattoos. There is a middle-aged man with a tattoo on his bicep; it looks like something from his days in the navy. All of the elderly people are white, and all of the elderly couples are heterosexual. Two of the young couples are interracial, and a lesbian couple in their thirties are holding hands. There are numerous aunts and uncles, great aunts, and great uncles. The middle-aged couples have fewer children. Cousin Valerie has two children, but she doesn't have and never has had a husband.

You could infer social change from this photo. If you had several photos of family reunions taken the same year you could feel more confident in the inferences about social change that might be drawn. In the same way, the data from a one-time, cross-sectional study of adults can be divided by age groups, and the information about each age group can be assembled and compared. Caution, however, is advised. There is much to be careful about.

In analyzing a cross-sectional survey, it might be found that the oldest people on average married in their mid-to-late 20s. Those born between 1930 and 1950 usually married as soon as they left high school. The next age group married on average in their mid-20s, and more of them never married. Similarly, each succeeding group might have had more divorces and fewer children. Such vital statistics can be very good indicators of social change, but they are not ideal. In particular, this could be a small number of people and may not be representative of a larger population. A good random sample survey done at one time with thousands of people, however, can help avoid this problem and offer strong indications of social change.

There may be other reasons to be cautious of a cross-sectional study. Imagine interviewing a very large number of veterans of U.S. foreign wars. They represent many wars and conflicts, including the most recent wars in Iraq and Afghanistan, United Nations peacekeeping missions in Bosnia and the Korean War, the Vietnam War, the Persian Gulf War, and World War II. Data gathered might include each person's personal information as well as their record of military service and combat. Much of this can be accurately reported.

Answers requiring distant recall may be less reliable. Memories of experiences both pleasant and unpleasant change over time. Feelings, attitudes, opinions, and values once held may be colored by experiences in the intervening years. The maturation process—the "age effect" discussed shortly—alters ways of thinking and views about one's self. When the only way to study a topic of social change is through memories, it can be done as well as possible. A bit of skepticism and a lot of fact checking, however, are required.

THE PROBLEM OF RECALL

"Changes in the social environment and in the self inevitably produce transformations of perspective, and it is characteristic of such transformations that the person finds it difficult or impossible to remember his former actions, outlook, or feelings...He cannot give an accurate account of the past, for the concepts in which he thinks about it have changed and with them his perceptions and memories" (Becker and Geer [1957] 1969: 330).

Because social change research is interested in the past as well as the present, it is not unusual to design research that asks people about the things they remember. Despite the problem of inaccurate memories, if there is no alternative to their recollections, social scientists are happy to have this type of information. In many cases it is the only history that can be assembled for people who left few records or whose lives were upended and who lost everything they once had.

For example, with the exception of Frederick Douglass and a small number of other former slaves, there are few written records that capture firsthand the slaves' experience of slavery. In light of this, the

(Continued)

(Continued)

Federal Writers' Project, an effort of the Works Progress Administration to provide employment to writers and artists during the Great Depression, gathered oral accounts of over 2,300 former slaves—of the one hundred thousand then still alive—between1936 and 1938. If no one had deliberately sought them out, and in many cases tape recorded their recollections, much of the personal record of slavery in the United States would have been lost. Similarly, the Shoah Project has recorded the recollections of 52,000 Jewish Holocaust survivors, i.e., Jewish persons who were incarcerated but escaped death or who hid to avoid capture during World War II.[10]

Normally, the time between a past event and the present is inversely related to an account's reliability. Recorded statements at the time of the event are most preferred. E. P. Thompson (1963) was fortunate to be able to use the court-recorded statements of individuals, many of them workers resisting early industrial practices, a century and a half ago, at the time of their sentencing. Through this record we can understand the circumstances and feelings of people displaced and impoverished by the onset of industrial capitalism. Collections of personal recollections—for example, Studs Terkel's *Hard Times* (1970) in which people recount life during the Great Depression—enrich our appreciation of past times, though they may have the feel of a novel rather than objective social science research.

---

[10]Both of these archives are available by going to their websites. You can access the "Slave Narrative" material at the Library of Congress "American Life Histories" website: http://memory.loc.gov/ammem/snhtml/snhome.html. The Shoah Project is accessible through the University of Minnesota Visual History Archives at http://www.lib.umn.edu/vha. See also George Rawick's *The American Slave: A Composite Autobiography* (1972–79). *These Were Our Lives* (Federal Writers' Project 1939) is one of many books chronicling the lives of the people interviewed. Ralph Ellison was part of the WPA Federal Writers' Project that employed more than six thousand unemployed writers during the Great Depression to collect the stories of average Americans. Ellison's *Invisible Man* is deeply informed by the interviews he did, as are the stories of other novelists who were employed to do this project.

The Longitudinal Panel Study. Ideally, social change research is more like a film or video than a photograph. An individual's biography, such as that of Iris Summers, covers a long period of time and provides a narrative of social change. Because it is only one person, however, it can only be illustrative or anecdotal, no matter how richly it links personal experience to history.[11]

The best social change research would randomly select hundreds or even thousands of individuals across a wide spectrum of society and follow their life course over the next several decades. Typical situations could be distinguished from exceptional cases, the differing experiences of women and men, rural and urban, immigrant and native could be charted, and the effects of historical events and the aging process could be more readily understood as these unfold for particular birth cohorts, i.e., people born in the same year range. This is essentially what a longitudinal panel study does. To do this with more than a handful of people, however, can be extremely difficult and expensive.

A unique, early longitudinal research project, published by Glen Elder as *Children of the Great Depression* (1974), is a sociological classic that tries to answer the basic question so many Baby Boomers ask: How did the Great Depression make Mom and Dad the kind of people they were? In 1931, the University of California's Institute of Human Development launched a study of children in Oakland, California, to "identify economic changes in family life," including the "physical, intellectual, and social development of boys and girls" (Elder 1974: 5) who were growing up in the Great Depression of the 1930s. Children selected for the study were born in 1920 and 1921. During their teen years, as much as a third of the labor force nationwide was unemployed and looking for work.

Data about the 162 children, ages eleven to seventeen, selected for the study were gathered at several points between 1932 and 1938, including information obtained from schools and teachers. Information about parents and family members was part of the study, providing a rich picture of the family life of the children. In the 1950s, three follow-up studies of 145 of the children included lengthy interviews, personality inventories, and psychiatric assessments. By looking at intracohort (generational unit) differences and emphasizing the dynamics of family life, Elder was able to grasp some of the ways social change comes about, not just through calamitous economic conditions but through the process of young people adjusting and adapting to new circumstances and personal changes. In so doing, the children developed identities, attitudes, and values that persisted throughout their lives.

---

[11]Isabel Wilkerson's *The Warmth of Other Suns* (2010) tells the stories of three people in her history of the Great Migration of six million African Americans from the South between 1914 and 1970 to escape Jim Crow and find better economic opportunities. In her profile of Mae Ida Brandon Gladney that occupies much of the book, Wilkerson tells an amazing story of the United States in the twentieth century.

Few things interest social scientists as much as the transition from youth to adulthood and, in particular, the movement from school to work. To study this, a series of national longitudinal panel studies have been conducted, interviewing the same people every two years over a forty-year period. These have yielded valuable information about a wide range of topics.

The National Longitudinal Surveys of Young Men and Older Men began in 1966 to study career choices, earnings and wage differentials, late working-life conditions, and retirement plans, as well as life-course changes and marital and family histories of a random sample of individuals selected as the panel. This was followed in 1968 with the National Longitudinal Surveys of Young Women and Mature Women. More than five thousand men and women were selected for each of these surveys and were contacted for follow-up interviews every two years for the life of the project. The surveys of women chart their increasing workforce participation as well as their experiences with wage discrimination, career tracking, and exposure to the glass ceiling. They show the various ways women negotiated responsibilities and roles at home and at work, with few guideposts and path-breaking role models.

Interest in labor force activity, as well as a wide range of sociological questions, motivated the National Longitudinal Surveys of Youth (NLSY) in 1979 and 1997. The first study selected nearly thirteen thousand young people, and the second included approximately nine thousand young people for study. The same people were interviewed every two years. The 1997 NLSY also collected information from parents about family history that could help researchers understand the vicissitudes of life changes across the age cohort born between 1980 and 1984. Longitudinal panel studies provide an in-depth portrait of people's lives in a changing society, making for compelling, empirically sound ways to understand social change.

The Multiple or Repeated Cross-Sectional Study. Imagine trying to find a thousand or five thousand people every two years in order to ask them some questions. You have their names and addresses, but at least 20 percent of them have moved, many have married and changed their names, some are overseas, the whereabouts of a few are a mystery even to their families, and others have died. This makes longitudinal panel studies difficult and very expensive to carry out.

An alternative is to do identical studies at regular intervals with a random sample of people who, in each study, represent the entire population. When the study is a reliable representation of the population, it is more affordable, but it requires a long-term commitment to gather information several times. Research institutes, state and local governments, and university researchers do this, but people are most familiar with this research design in the reports of pollsters.

Pollsters chart changes in attitudes, outlooks, opinions, and some behaviors. They report these as changes in the population over time. These often are reported as comparisons by gender, age group, ethnicity or race, region, and whatever other characteristics are thought relevant, but they regularly show how opinions, attitudes, and outlooks change over time. Polls rarely ask questions of the same people at time A and time B, but the representativeness of their samples allows them to draw inferences about changes in the groups they poll.[12]

Drawing Conclusions From Empirical Data

Earlier in the chapter the question was asked, What explains the varying propensity to bike to work, including not biking at all? There were several hypotheses put forward. If the analysis of observations, conversations, and possibly a survey provide the information needed to test your hypotheses, you have gone a long way in answering your initial question.

- If you fairly consistently find that no one living beyond ten miles from work uses his or her bike to get to work, that is important information.
- If you find that people of all ages bike to work, but that most bikers seem to be younger people, you have important information.
- If you find that biking to work falls off by half or more when the weather turns cold and wet, you have important information.
- If you find that most people who bike to work say they would hate to pay for parking, while nonbikers think that parking costs are reasonable and affordable, you have important information.
- If you find that people who have a significant amount of stuff to take back and forth to work (many books and papers, heavy tools, etc.) rarely bike to work, while most bikers have no more things than would fit in a small backpack or panniers, you have important information.
- If you find that young people who bike to work don't think it is dangerous, older people who bike think it is dangerous but believe they can make their ride safely, and those who don't bike think it is unavoidably dangerous, you have important information.
- If you find that most people who bike to work have at least three colleagues or friends who also bike to work, but nonbikers can rarely name more than one person they know who bikes to work, you have important information.

---

[12]One of the most important cross-sectional studies for social scientists is the General Social Survey (GSS). Conducted with a random sample of people since 1972, and with many questions repeated annually, the GSS provides a fairly detailed picture of the U.S. population, including its attitudes, circumstances, and behaviors. The GSS is an accessible and readily analyzed data set that students can use to study social change in the contemporary United States.

You now know several things that may inhibit or encourage people to bike to work. You found that a combination of physical factors (biking distance, things to carry, weather/season), social factors (friends and colleagues who bike), and personal factors (age, attitude about parking and safety) were important. You know quite a lot, though you only know those factors about which you gathered information.

Perhaps your conversations and observations during the research process revealed other important things, and you adjusted your data gathering to test other hypotheses. Perhaps you noticed that people biking to work are less likely to own a car. Perhaps you were told that protecting or saving the environment was a high priority for bikers, and bikers often referred to it as a form of exercise, but nonbikers rarely mentioned these things as being important to them. If you have sufficient information, you can add these findings to your conclusions.

However many reasons you find and however many earlier suspicions you discard, there will always be more possibilities. This is the nature of scientific inquiry. Answers to questions are only as good as the best available research. This bothers some people who want absolute truth. Science cannot provide this. It can only reveal the best that is known at any given time. Think of a world record. How good is it? Only as good as the best performance of the current record holder. The record may fall soon or may last for decades. It's the same with science. An answer may be very good, but it is rarely or never the final answer.

RESEARCH ETHICS AND A CAUTIONARY TALE

The ethics of social research are important. Research that does harm should always be avoided, and projects that cause injury or illness are unacceptable. As well, there are limits to which research can intrude into people's private lives, especially when this exposes them to retribution or public ridicule. People studied should be told what the research is about, give their consent to participate, and be able to withdraw at any time. Researchers must be honest with subjects and resist the temptation to see themselves as undercover reporters or investigators. There is, however, the need to balance some forms of intrusion, avoidance of the whole truth, and discomfort against the potential or likely benefits of the research.

Few things are more private than sex between consenting adults. It is one of the more difficult things to study for both scientific and political reasons. In the following account, ethical and moral objections were raised in the political domain, with little effort to balance potential benefits against the ethical concerns normally considered in doing science. Sex research is very important. It is also objectionable to many people. Should it be?

In 2007, more than 40 percent of all new births in the United States were to unmarried women, up from 34 percent just five years earlier, according to the National Center for Health Statistics (Stein and St. George 2009). Is this indicative of a moral crisis, or was there a sea change in economic circumstances that made it more difficult for sexual partners to marry, even when a child is born? By 2009, the marriage rate in the United States was at an all-time low.[13] Are these out-of-wedlock births planned or the consequence of promiscuity? Or, is this increase due to more complex changes in sexual and marital practices?

In Iris Summers' lifetime, there were two "sexual revolutions," the first in the 1920s and the second in the late 1960s and early 1970s. Is a third going on today? Are women, more economically independent than ever, stepping over another barrier (marriage?) to motherhood, freer than ever before to be mothers without marriage? Other matters of sexual practices have important implications for everyone. The spread of socially transmitted diseases (STDs), a continuing HIV/AIDS worldwide pandemic, and the need to track genetic maladies seem to make obvious the value of having accurate information about changing sexual practices.

In 1955, the first large-scale survey research of sexual practices in the United States was carried out (Reynolds 1996: 53). Public funds paid for the famous and groundbreaking, if sometimes flawed, Kinsey studies of sexual practices after 1957.[14] For the next thirty years, nearly all social survey research on sexual behavior was funded directly or indirectly by the federal government. Because public funding is involved, privacy

---

[13]In September 2010, the U.S. Census Bureau reported that only 52 percent of adults (age eighteen and older) were married, the lowest proportion in one hundred years of record keeping. Among adults ages twenty-five to thirty-four, the portion who had never married increased from 35 percent in 1999 to 46 percent in 2009 (Eckholm 2010).

[14]Alfred Kinsey's research at Indiana University was funded from 1941 to 1957 by a combination of support from the Rockefeller Foundation and the National Research Council. The Kinsey Institute for Sex Research gathered information from 18,000 individuals, mostly through face-to-face interviews, during Kinsey's lifetime.

(Continued)

(Continued)

issues and the idea that some research is intrusive and voyeuristic can easily become fodder for political debates and posturing.

Edward Laumann and his colleagues' experience with government funding provides a cautionary tale about the political benefit of opposing and often ridiculing research in the name of protecting the public's morals and/or tax dollars. In the midst of the HIV/AIDS epidemic in the 1980s, the National Institute of Child Health and Development solicited research on "social and behavioral aspects of health and fertility-related behavior," that is, human sexual behavior. After receiving initial funding to design a study, Laumann and his associates were asked by five federal agencies to apply for funding to gather information on AIDS, including its transmission.

When a scientific publication made reference to their intended research, Senator Jesse Helms of South Carolina and other conservative members of Congress attacked the project. Right-wing publications and religious radio broadcasts raised an outcry, and Congress passed legislation banning federal funds for a national survey of sexuality. The fear was that the findings would "normalize" widespread practices such as marital infidelity, masturbation, homosexuality, and other "religiously or morally unacceptable modes of conduct." Laumann and his colleagues concluded that "orchestrated ignorance about basic human behavior as salient as sexuality has never resulted in wise public debate" (Laumann et al. 1994: 36).

After some delay, they acquired funding from private foundations and, in 1992, began surveying more than 3,400 individuals. This resulted in several publications, including *The Social Organization of Sexuality* and *Sex in America.* Their work is seminal in accurately portraying changes and stability in the sexual behavior of adults, with important policy implications.[15]

---

[15]Not until 2009, more than fifteen years after the Laumann study, was another major sexual behavior survey carried out, the National Survey of Sexual Health and Behavior. It was funded by Church & Dwight, manufacturer of Trojan condoms.

Social Policy and Social Change

Throughout this chapter the question has been asked, How can we get more people to bike to work? Only when the reasons for people biking to work are better understood can an effective policy be developed. Despite what your research may have found, some things are not amenable to policies. For instance, you can't change the weather. You probably can't change the distance between home and work, but you can make it easier to drive some distance to work and then park the car and ride a bike the rest of the way. A free or rental bike station at park-and-ride sites is an idea being adopted in many U.S. and European cities. Bike trails and bike-designated streets improve bike safety in inclement weather.

The hassle and costs of parking don't seem to dissuade nonbiking commuters from using their cars, but what if things became even worse? What if parking fees were increased by 100 percent, or parking lots were turned into parks? Would this be a good policy? Because bikers are less likely to own a car, what about increasing car registration fees by 500 percent to encourage people to get rid of their cars? How about banning cars from the streets between seven and eight in the morning and five and six in the evenings to increase bike safety? Would this be good policy?

Okay, what about less radical policies? Businesses with bike lockups encourage employees to bike to work. What else can firms do? Build changing rooms with showers and lockers where people can freshen up after biking to work, on the company time? Some have done this, too. Not much can be done about the need people have to carry loads of things to and from work. Or can it? Maybe new bike designs (technology) are an answer. In some cities and towns, large amounts of goods are delivered on bikes, similar to what has been going on in Asia for decades. Panniers and racks can hold a lot of things, but there may be problems with stealing when the bike is parked. Bike parking spaces can be made more theft proof.

And what about perceptions of danger and attitudes about parking? It may be that, though older bikers also perceive biking as dangerous, they are willing to rationalize this while older nonbikers aren't. Can anything be done to change this? Is it a perception problem or a safety problem? It is important to make streets safe for bikes and for drivers to be courteous to bikers using the streets. Public relations campaigns, civic meetings, and public education can help. Maybe police should issue more tickets to lawless bicycle riders. The fines could pay for new bike paths.

Social policy is often made with little systematic, empirical understanding of the situation, including how policies have worked elsewhere and what doesn't work. Little research goes into mandatory sentencing laws such as three-strikes-and-you're-out legislation that incarcerates serial perpetrators of crimes. These laws were largely driven by anger, fear, and an inflated perception of the chances of being victimized. They have probably helped reduce crime by keeping would-be violators off the street. But at what cost? Some states now spend more money on incarceration in prisons and jails than on higher education. Other policies could well have been less expensive and more effective.

On the other hand, some social policy has considerable scientific backing but is not enacted because opponents resist changes in the status quo. Global warming is a case in point.[16] It is perhaps a dream to think that social policy always will be based on good research and that social policy backed by good research will always be adopted. It is, however, worthy of serious consideration.

Generations and Social Change

What generation are you? Are you an early or late Baby Boomer? A member of Generation X? The Millennial Generation? It depends of course on when and where you were born. Someone born in Cuba around 1910 was part of the Revolutionary Generation. If you were born in Japan in the late 1970s or 1980s, you would probably identify yourself in the Hi Panda Generation, named for the omnipresent and hugely popular bubble-headed, big-eyed doll that morphed into thousands of variations. Though not without problems that bother some social scientists, referring to generations is one of the most popular ways of talking about and recognizing social change. Critics of the concept of generations prefer more empirically precise concepts such as birth and age cohorts. These are valuable refinements of the concept of generations; the ways cohort analysis recognize social change are much the same for those who study generations.

Understanding social change by studying generations invites questions about causality. Do changes in the social, economic, and political environments in which a generation grows up and lives account for their differences from other generations? Is one generation different in ways (e.g., larger, less susceptible to illness, better educated) that explain its different attitudes and behavior? To what degree are generational differences simply a matter of age differences? These questions—often called period, cohort, and age effects—arise in the study of generations, cohorts, and social change.

---

[16]The National Academy of Science, the U.S. Geological Survey Global Change Science projects, the MIT Center for Global Change Science, the U.S. Global Change Research Program, and the Intergovernmental Panel on Climate Change, winner of the 2007 Nobel Peace Prize, have all concluded that climate change is real and most likely caused by human activity, particularly the burning of fossil fuels. See also Oreskes' (2007) "The Scientific Consensus on Climate Change: How Do We Know We're Not Wrong?"

The Concept of Generations

Research in social change often uses the concept of generations, and it has become perhaps the most-often-used concept in studies of contemporary social change. Duane Alwin and Ryan McCammon (2004: 27) define generations as "groups of people who share a distinctive culture and/or a self-conscious identity by virtue of their having experienced the same historical events at roughly the same time in their lives." Importantly, members of a generation itself must identify, at least in part, with their own uniqueness and see themselves as set apart from those who came before and those who followed.

As a sociological concept, it provides a glimpse of history made personal, or what Norman Ryder (1965: 857) calls a "macro-biography." At the edges of every generation there is some overlap, and far from everyone in a generation fits the description. In some sense "generation" is merely a way of saying that, with time, people who share formative experiences have more in common with each other than with those who don't.

Not all parts of a person's biography are of equal value, in terms of identity and experience. The concept of generations favors young adulthood. Hughes and O'Rand (2002: 2) observe that "events in early adulthood are usually considered especially important." Norval Glenn (2005: 36) echoes this view that "influences typically have greater effects among young adults than among older ones." Mirroring Iris Summers' experience of the Great Depression and World War II, there is a great deal of research supporting the idea of a receptive and formative adolescence and young adulthood, after which beliefs, values, and ways of seeing the world are more or less fixed, though attitudes and opinions do continue to change throughout life (Alwin and McCammon 2004: 36–39; Schuman and Scott 1989).[17]

Generations in the Past Century

Writers like Ernest Hemmingway, F. Scott Fitzgerald, and John Dos Passos left the United States following the First World War and lived as young adults in France and throughout Europe. They were called the Lost Generation. For young people in the United States, this came to describe their own rebelliousness (the 1920s saw the first sexual revolution) and their adult opportunities thwarted by the Great Depression. In France, a similar

---

[17]In very simple terms, beliefs are strongly held ideas about what is and what is not. Values involve desired end states and ways to accomplish goals (e.g., nonviolently or through discussion and consensus). Attitudes are "latent predispositions to respond or behave in particular ways toward objects" (Alwin and Scott 1996: 75).

generation was called the *Génération du Feur*, the disillusioned young people who came of age during the First World War and struggled with their wartime experiences as well as a Europe in tumult during the interwar years.

Those persons born in the late 1940s and early 1950s in China are the Red Guard Generation, the Third Generation, the *shiluo de yidai*—lost generation, or the *sikao de yidai*—the thinking generation (Rosen 2000: xiii–xix). They came of age during the tumult of the Cultural Revolution (1966–76). As discussed in Chapter 8, millions were required to leave school and work ("sent down") in rural areas, only to return home after a decade to witness the end of Mao's rule.

The Lost Generation was followed by the Depression/War Babies/Silent Generation. Economic distress in the years preceding World War II, and the effort to "catch up" by millions of GIs who were away from home and had delayed the onset of career and family, resulted in a population growth "boom" that spread throughout the economy and culture. Their large families became the Baby Boom Generation, born between 1946 and 1964. The size of this generation and the growing affluence of most families provided a visibility and buying power of youth to fuel popular culture in an era of expanding prosperity in the United States.[18]

Baby Boomers. "The self-aware, or self-absorbed, feel less self-fulfilled, and thus are racked with self-pity," is how a *New York Times* writer described the Baby Boom generation as the first of them reached age sixty-five (Barry 2011: A1). This largest-ever generation often seemed larger than life in the tumultuous counterculture revolution and antiwar era from the mid-1960s into the 1970s when the early Baby Boomers were reaching adulthood. The first children to be the concerted focus of corporate advertisers, they grew up thinking they were different, and many have never lost their sense of being agents of social change.

The times made the Baby Boomer generation as much as the generation made the times. The U.S. economy doubled in per capita size, and the growing prosperity was shared surprisingly equally, in contrast to the

---

[18]Generations, unlike birth cohorts, are not based on a fixed number of years during which a particular group is born. Most discussions designate a generation as spanning from ten to twenty years. An exception is Hughes and O'Rand (2002, 4) who designate six generations born between 1906 and 1964, each about ten years in length: Young Progressives, Jazz Age Babies, Depression Kids, War Babies, Early Boomers, Late Boomers. Others designate four generations by combining Hughes and O'Rand's Depression Kids with War Babies (born between 1926 and 1945) and Early and Late Boomers (born between 1946 and 1964).

growth in inequality that began in the 1970s and has continued over the next four decades. The United States, though challenged by the Soviet Union for geopolitical influence, was far and away the world's largest economy. The dollar was the benchmark for global currencies, English became the language of science and business worldwide, and U.S. military might was felt throughout the world. A third of the early Boomer males served in the military during the Vietnam War (Hughes and O'Rand 2002: 9–10). Baby Boomers were the best-educated generation in U.S. history. Married women's employment dramatically increased, as did the use of birth control that sparked the second sexual revolution.

Five changes helped shape the experiences of the Baby Boomer generation: (1) the shift from an economy that primarily made things to one that relied on information and provided services; (2) the reduced role of government in distributing resources; (3) a more open door to immigration and consequently a more diverse national population; (4) the sheer size of the Baby Boomer population; and (5) shifts in cultural values, especially regarding gender roles (Hughes and O'Rand 2002). To this can be added a sixth formative influence, participation in social movement activity, the topic of Chapter 5.

Stephanie Coontz (1992: 263), author of *The Way We Never Were*, describes the postwar United States as a "golden age" for the increase in the material standard of living, including wages, housing values, and job security. These things particularly benefited the Depression/War Babies Generation. Baby Boomers reached young adulthood at the *end* of a period of widely shared rising prosperity, just as wage gains began to wane. Job benefits secured when unions were strong declined with accelerating speed over the 1980s and 1990s. The world was changing as they entered adulthood and faced new challenges. Late Boomers as adults are "more unequal in wealth, net worth, and home equity, such that . . . the American dream of upward mobility may have been denied to a significant share of the baby boom" (Hughes and O'Rand 2002: 11).

Gen Xers. Between the Baby Boomers and the Millennials is Generation X. This generation is much debated, as can be seen in contrasting titles of books about the generation: Jean Twenge's *Generation Me* (2006) and Jeff Gordinier's *X Saves the World* (2008). During this generation's formative years, the United States, Britain, and European nations embraced more conservative politics and political leaders. The most popular college majors shifted from the humanities and fine arts to business and information technology. "Helping others who are in difficulty" as a major goal in life gave way to "being well off financially," as revealed in poll after poll.[19] The preppy look replaced hippy attire, haircuts became shorter, young women again shaved under their arms, and new brides took their husband's family name.

The label *Generation X*, applied with a broad brush, connotes uncertainty and lack of social consciousness, hence its other name, the Me Generation. Some writers have characterized the entire generation as cynical, sarcastic, unable to commit to anything, and plagued by hopelessness and frustration (Shanahan and Macmillan 2008: 126). The generation has often been unfavorably, and unfairly, contrasted to the high-minded idealists of the 1960s. Gen X young people have been compared to what some writers called the Silent Generation of the 1950s (Wilder 1953).

The Silent Generation (a subset of the Depression/War Babies Generation) was similarly characterized in unflattering terms. A closer look at this group reveals a much more complex picture. It is often described as conformist, epitomized in the sociological classics *Organization Man* and *The Lonely Crowd* and in the title of a popular 1950s novel, *The Man in the Grey Flannel Suit*. The Silent Generation, however, spawned the Beats, iconoclastic poets and writers like Allen Ginsberg and Jack Kerouac who paved the way for the 1960s counterculture.

That generation broke the color barrier in music. White youth, through their buying power, helped talented and innovative black singers and musicians become mainstream celebrities. A generation that grew up in the conspiratorial anticommunist atmosphere of the McCarthy era took up the challenge of a handsome young president, John Kennedy, to become the first Peace Corps volunteers. They were the youthful activists in the antinuclear weapons movement and civil rights movement. Perhaps in time Generation X will be seen in a different light as well.

Millenials. Most of you reading this book are part of the Millennial Generation born between 1986 and 2005, a generation twice removed from the Baby Boomers. At the start of your generation, unemployment was low for several

---

[19]When given the choice to answer affirmatively to several life goals, both of these were picked by nearly three-quarters of all first-year college students in 2009. When prioritizing them, however, helping others was gradually picked less often, and financial success was picked more often through the 1980s and 1990s. A very good source of information on student attitudes is the UCLA Higher Education Research Institute's annual surveys of more than a quarter of a million first-year and fourth-year students at approximately four hundred colleges and universities. The survey results are published regularly in the *Chronicle of Higher Education*. See, for example, Bartlett (2002).

years. The United States was not fighting a major war anywhere. In fact, the United States seemed poised to be the sole global superpower and could actively promote democracy and capitalism everywhere. That began to change by the turn of the century, and, as Frank Furstenberg and his colleagues (2004) have written, growing up became harder to do for Millennials.

For Iris Summers' generation, World War II had enormous significance. The touchstone event for Baby Boomers is the day President Kennedy was assassinated or Neil Armstrong's walk on the moon. For Gen Xers, it could be the Space Shuttle Columbia disaster or Kurt Cobain's suicide. If you are a Millenial, your touchstone memory is probably the 9/11 attacks on the World Trade Center and the Pentagon. You are familiar with the epidemic of ADHD (attention deficit hyperactivity disorder) and bipolar disorder and have read or heard a lot of concern about the pharmaceutical solutions to your behavioral problems. You are much less likely than earlier generations to identify yourselves with any religion; more than one in three of you have no religious identification (Putnam and Campbell 2010: 99–105).

The number of adult children living with their parents (the "boomerang children") has been growing and accelerated in the Great Recession of 2007–10. In 2007, one out of eight parents reported that grown children had recently moved back home, and more than half of all young men and women ages 18–24 were living with their parents. Aging parents' reliance on support from their adult children shifted 180 degrees in the twentieth century, to greater support given to young adults by their aging parents (Zernike 2007). This social change reflects changes in economic circumstances and even the meaning of young adulthood. In Europe, and increasingly in Japan and the United States, marriage may actually be of declining importance for Millennials who often practice serial monogamy outside of marriage and increasingly are becoming parents without the benefit of marriage.

Gen Xers had to learn to use the computer, often in school. Millennials grew up with the personal computer and became computer savvy by learning to use the Internet in ever-more creative ways. Powerful hand-held mobile computers (smartphones or iPhones) for social networking and orientation are now indispensable and ubiquitous for all but a few Millennials. Music sharing obviates their owning a CD player. New computer technologies, inexpensive and powerful, gave this generation the means to become recording artists and filmmakers free of corporate backing. Most war babies recognized three genres of music: classical, country, and popular. Young people's music today has too many genres to count.

Popular movies are likely to portray a very average guy (but one worth rooting for) who gets the girl but remains an average guy, just like most of us. Today, Millenials as well as Gen Xers are more socially tolerant than those older than them.[20] They have been described as the first generation to think and act like America actually looks in terms of relationships among ethnic and nationality groups. For many, a postracial society looks possible, and the civil rights movement seems like ancient history.

Birth Cohorts and Social Change

The concept of generations provides a way to describe social change at the personal level. The concept annoys some social scientists, however, who prefer the less evocative but more precise concept of birth cohort. Birth cohorts are people born within a specified time period. A birth cohort can be those born within a certain number of days, months, or years. Most social research compares five- or ten-year cohorts, i.e., people born within a five- or ten-year period, with starting years ending in 0 or 5 (e.g., 2000, 2005, 2010). This is somewhat arbitrary from a historical point of view, but it does lend precision to data analysis.

Cohorts are statistical constructs of a sort, but the persons so designated can experience the same historical events during their formative years, a key feature of generations (Schuman and Scott 1989: 378). For example, by comparing individuals—mostly males—who completed their education and began working full time prior to the outbreak of World War II (those born between 1915 and 1919 who were ages 22 to 26 in 1941) to men ten years younger (born between 1925 and 1929 who were ages 22 to 26 in 1951), Elder (1974) shows how a birth cohort can be characterized as a generation. Many of the first group experienced an interruption in their adult work and education careers when they were drafted or volunteered to serve in World War II. The latter spent the war years growing up, before going on to college or to full-time work. For the former, many of whom were married at an early age, their long absence during the war contributed to the astonishingly high number of divorces in 1946. The latter were largely spared this experience, though their divorce rate was much higher than their parents' generation. As Elder shows, the characteristics of a generation can be seen in what are otherwise simply statistical categories, i.e., birth cohorts.

Grouping data into five-year and ten-year periods provides a way to study social change by selecting several variables or measures and charting

---

[20]An example is young peoples' attitudes toward same-sex marriage. Support for same-sex marriage increased nationally by 20 percent between 1996 and 2010. In 2004, only three states had a majority of the population supporting same-sex marriage. In 2010, a majority supported it in seventeen states (Gelman, Lax, and Phillips 2010), with the strongest support coming from younger adults.

them over a period of time for several successive birth cohorts. This can be seen by comparing two figures. Figure 2.1 shows the percentage of working-age men and women in the civilian labor force from 1950 to 2005 and projected to 2050. What it shows is important in understanding social change: women increased and men decreased their labor force participation throughout the second half of the twentieth century. The reasons for, and the consequences of, this are profoundly important in understanding social change in the United States, just as they are for any other country.

Figure 2.1   Civilian labor force participation rates by sex

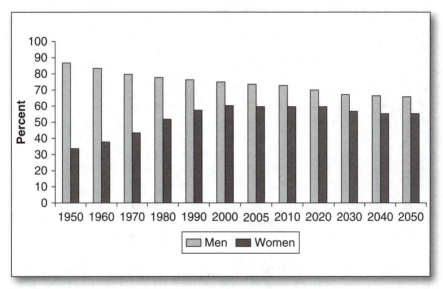

Source: Monthly Labor Review 2007. http://www.bls.gov/opub/ted/2007/jan/wk2/arto3.htm.

Figure 2.2, women's workforce participation at different points in time, provides a clue about the reasons for the rising workforce participation of women. Hodson and Sullivan (2008) constructed trend lines as older cohorts were replaced by younger ones. Look at the lines and think about the work and family experiences of different cohorts, what Hodson and Sullivan call the diminishing M. Then read the footnote.

When comparing birth cohorts, we see, or at least infer, that three things are happening. First, we see that which is measured, in this example the changes in labor force participation. In Figure 2.1, we similarly see an increasing proportion over time of women working outside the home. We see it in more detail in Figure 2.2 by recognizing how it increased for successive birth cohorts. Second, we infer something is happening in the society, and questions come to mind. What is going on to cause women to increasingly be employed in such higher numbers? What social processes, structural changes, economic demands, cultural attitudes, and technological changes are influencing this trend? We want more information.

Figure 2.2   Labor force participation of U.S. women by birth cohort[21]

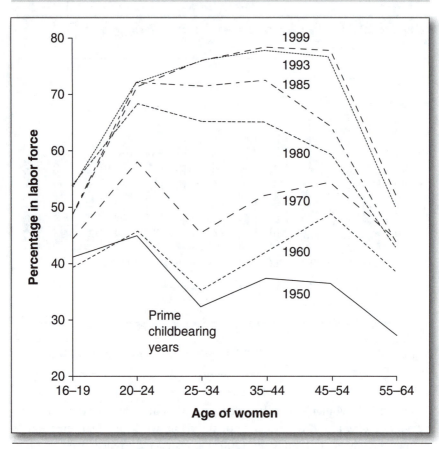

Source: Hodson and Sullivan 2008: 114.

[21]Women born later in the twentieth century were more likely to work outside the home, to start work at a younger age, and stay in the labor force longer. Most importantly, the trend throughout the century is women's decreasing tendency at ages 25 to 45 to suspend their work careers when they had children at home. For a more detailed explanation see Leibowitz and Klerman (1995).

The third thing we ask about is the agency of women themselves. People create social change by their intentional actions. Women see opportunities and risks, and they act accordingly. They adopt or craft identities that include a set of actions, motivations, and priorities. Did women increasingly see themselves as something more than wives and mothers? Amid the structural changes going on, they made choices about their time and effort, however socially constrained.

The effect of the birth control pill can also be seen in Figure 2.2. The oral contraceptive was widely available to married women by the early 1970s and within a short time later to all women. This was a desirable option for women and couples that had several significant consequences, not the least of which was the expectation that a sexual encounter or relationship would not result in pregnancy. For young women it created a markedly different situation from what had been their mother's (Goldin 1990; Leibowitz and Klerman 1995). The comparison of their work careers to other birth cohorts would forever be informed by the availability of the pill and the ways it allowed women—both married and unmarried—to make decisions and organize their lives with a much reduced uncertainty that sexual intimacy would be likely to end in pregnancy. They became agents of social change.

Cohort, Age, and Period Effects on Social Change

Cohort Effects

In the simplest sense, the human condition is to be born, mature, and die. This is a form of societal population turnover, what Norman Ryder calls "demographic metabolism." The comparison of the lives of successive waves of people is an important way to study social change, especially because "the continual emergence of new participants . . . compensates the society for limited individual flexibility" (Ryder 1965: 845). Because cohorts age and die, to be replaced by new and different cohorts that have new experiences, cohort replacement becomes a process of social change, though it says little about the direction, scope, or speed of change.

Studying and carefully comparing birth cohorts helps make sense of the direction taken by social change. In some cases social change is a consequence of the particular historical experiences of a significant segment of the birth cohort. For example, persons born after 1975 were often taught reading and writing by phonics, learned how to do visual math, and read textbooks containing more discussion of women and minorities. These things not only distinguish the birth cohort but have implications for the cohort's educational attainment and experiences with others beyond the classroom, which in the aggregate amounts to social change.

Social change may also result from a feature of the birth cohort that strongly influences its experiences, such as a cohort that is much larger than the cohort preceding it. When Baby Boomers entered the job market, their large numbers made for much greater job competition, and their education was of less career value because of this. The job attainment and earnings of Baby Boomers, especially the later Boomers, has never recovered from the consequence of this "oversupply of labor" in the U.S. labor market.[22]

Conversely, the death toll of Soviet citizens during World War II was enormous, and births dropped during the war due to extreme wartime hardships. The result was far fewer people—those born between 1940 and 1945—available to assume positions of authority in Soviet society thirty years later. Added to the lower birth rates during the preceding Great Depression, within a generation this shortage of older leaders thrust younger, possibly unprepared individuals into managerial and leadership positions. Could this have hastened the crises that befell the Soviet Union in the 1980s and contributed to its collapse in 1989?

Age Effects

If you need a reminder of aging, take a look at a recent photo of Mick Jagger or Keith Richards. Aging happens to all of us, and it has many individual consequences: we get bigger before we get smaller; our shape changes; we either lose our hair, see it turn gray, or both; it takes longer to heal from an injury; we wish with increasing fervency that we'd never started smoking; and we vow to exercise with greater fidelity. If fortunate, we are more likely to have money saved, our car insurance costs less, and we are more apt to vote. When changes associated with aging accumulate for many people, we speak of age effects on social change.

A simple example is often used to explain age effects, or what are sometimes called life-cycle effects (Alwin and Scott 1996). A well-documented

---

[22]Annette Bernhardt and her colleagues' *Divergent Paths* (2001) shows in quantitative detail the declining fortunes of later Baby Boomers, especially in their diminished likelihood that they would live more affluent lives than their parents. Richard Easterlin's (1987) explanation of their reduced life chances as the result of demographic pressure (the generation's large size) and economic opportunities, what has come to be called the Easterling Effect, essentially mirrors Bernhardt's.

pattern of church attendance begins with low attendance among young adults. When they marry and have school-age children, their attendance goes up. Once the children are grown, they may again be less inclined to attend church, but as they grow older people often return to church and become regular attendees. Does aging actually *explain* variations in church attendance? As a biological process, no. What aging does is approximate a life situation that makes something like church attendance more or less likely. As the age profile of a society changes, e.g., as more of the population constitutes older adults, social change happens. In this case it's simple arithmetic.

Not only growing older but adopting a new social status can result in age effects: being a graduate, getting married, becoming a parent, owning a home, entering retirement. All of these, and more, typically mark a change of status and may result in a change of attitude toward certain topics. When more young people go to war, when they delay or decide against marriage and parenting, are unable to purchase a home, and when an economic recession forces the delay in millions of older peoples' retirements, aggregate society-wide status changes are altered, attitudes shift, and social change happens.

Many rural communities are, in the demographic sense, getting older. Their young people have difficulty finding jobs in the community, many go away to college and never return, and the community has fewer young families. As older citizens become a larger portion of the community, business activity changes to meet their tastes and needs. Medical facilities and assisted-living arrangements become more prominent. Older persons provide less public revenue from sales taxes, and many vote against property taxes in light of their fixed retirement incomes. Younger people find the town less interesting and exciting, providing a further motivation to move away. Without an available young workforce, the prospect for business startups diminishes. Unless it can attract new residents with an inexpensive real estate market or a plan to become an attractive retirement community, the outlook for the community is not good (Carr and Kefalas 2009). This is social change in rural areas.

Period Effects

Implicit in the idea of both generations and cohort effects is the notion that some events are more significant for teens and young adults than for the rest of the population. Sometimes called generational imprinting, these events are formative in the life experiences and life-long perspective of the teens and young adults in ways that are not as strong for children and older adults. This is not to say that major events and changes of circumstances don't affect the entire population, regardless of age, but they affect young people differently and more formatively. These are called period effects.

A war affects the entire society. Younger adults do the fighting and die. Older citizens pay most of the taxes to fund the war. They are the ones who have children eligible to fight, some of whom will be killed or maimed. No one escapes the effects of a massive, sustained war, but the effects vary by age, and for those in young adulthood, the effect will have a more lasting impact.[23] Similarly, few escape the effects of a significant economic depression. Terrible famine such as China's of 1959–61 and Ethiopia's of 1972–73 were most devastating for the poor, but nearly everyone was affected. For those just starting out as adults, however, the effect was jarring and influential in ways and to a degree not felt by dependent children and more established and older adults.

Everyone experiences, directly and indirectly, a technology that has society-wide implications: the mass-produced automobile, radio and television, home and office computers, or cell phones and smartphones linked to clouds of wireless computer networks. But for older people like Iris Summers, digital technology has had a more peripheral role in their lives. Not so for you! The discovery of polio and smallpox vaccinations, the development of antibiotics, and the ability to map genes and trace DNA have left few, if anyone, unaffected, but different age groups are affected in different ways. While the Watergate scandal of the Nixon presidency accelerated a growing distrust of government and altered political journalism and news coverage, it also helped spawn a generation of young political skeptics and motivated thousands of young people to pursue investigative journalism.

Recognizing social change involves uncovering the ways events of a historical period are experienced and interpreted by young adults such that the times became part of who they are and for a lifetime inform their understanding of themselves and the world around them. These are period effects that have a significant impact on the direction, pace, and scope of political, economic, social, and cultural change.

---

[23]Norman Ryder observed that "period effects are not universally felt . . . period influences typically have greater effects among young adults than among older ones" (Ryder 1965: 452).

Topics for Discussion and Activities for Further Study

Topics for Discussion

1. Why are the trends discussed in the chapter—women in higher education and women in the paid labor force—important? The most obvious consequences are changes for the individuals themselves and those persons close to them. Social change is less obvious. List several social changes for each trend, including changes for men and in gender relations.

2. This chapter talked a lot about biking to work. You could just ask people why they do or don't bike to work. Why would this not be a good idea? Here's a hint. People in the United States marry because they believe they love each other. If you ask people why they are married, that's what they would probably say. Would you feel like you understood why people marry if that's all the research you did?

3. What technology in later generations could be comparable to the birth control pill for Baby Boomers? Was "agency" unleashed by this technology, the way the pill gave women and couples new choices and career opportunities? What might be the downside of these technologies?

4. What period effects seem most important for the younger members of the class, say those under twenty-five? Period effects are often not apparent for years or decades later. Why do you think the period effects you are discussing will have a lasting imprint?

5. Many writers have suggested that a new stage in the life cycle is emerging: young adulthood. Read some discussions of this, and try to find three or four reasons (structural conditions in society today) that support the idea of a new phase in life. What does this imply about "growing up" and adulthood in a complex, affluent society? What are the implications for social change?

Activities for Further Study

1. Take a trip to the nearest landfill where trash is dumped on a regular basis. What can you see about the way people live, but also the past they are discarding? What does this debris and detritus tell you about social, and particularly material, change?

2. If your family has a photo album with pictures going back several generations, take some time to look at these for clues about social change. What objects are and are not in the pictures? How do the subjects' postures and positioning relative to one another differ from what might be photographed today? On what occasions are many of the photos taken? Is this different?

3. There are many sources of information that record social change. Here are a few of them available on the web. Go to one and find some data over at least a thirty-year period. What do the data show? What social changes could you possibly infer from these data? Why should you be cautious about attributing the changes seen in the data to social change? What questions do the data raise? Can you find information elsewhere to answer these questions?

- Vital Statistics Reports at the Center for Health Statistics (http://www.cdc .gov/nchs/products/nvsr.htm)
- Current Population Reports (http://www.census.gov/main/www/cprs .html)
- Consumer Expenditure Surveys at the Bureau of Labor Statistics, Department of Labor (http://www.bls.gov/cex/)
- Gallup News Service (http://www.gallup.com/home.aspx)
- U.S. government data site (www.data.gov)
- Pew Center for People and the Press (http://people-press.org/)
- Statistical Abstract of the United States (http://www.census.gov/compendia/ statab/)

4. Formulate a research problem that requires you to ask questions in variable language, moving from descriptive to analytic questions. Imagine a causal model that would explain your dependent variable. Do this for a social trend that requires you to consider the changing social context in which the concept occurs. State at least one question as a hypothesis, and give at least one "constant" that qualifies or conditions the hypothesis.

5. Look through some social science research journals and select an article that sounds interesting. What is different about this article, compared to reading an article in a more popular, mass-circulation magazine? Make a list of at least five ways they differ.

6. If your library subscribes to *Contexts*, a journal for students published by the national professional sociology association, find an article that addresses some aspect of social change that interests you. Read it and summarize it. Is there anything about how the research was done that is left out of the article? What else might you want to know?

3

Understanding and Explaining Social Change

T he young woman is leaning forward, observed in profile looking casually at a computer monitor. The photograph shows her sitting with one knee up, her foot perched on the edge of the chair. She has a pencil in her hand, or is it an unlit cigarette? She could be a student, maybe twenty years old, with short dark hair, stylish glasses, bracelets, wearing a light-color crewneck top and midcalf slacks. You glance at the caption, which begins, "Chinese blogger . . ."

What do you know about this young woman? What can you conjecture? What makes her interesting? Why is this photo newsworthy? The answers have something to do with social change, but in what way? Is the newspaper article about digital technology or social networking? Is she working in the global economy? Is she reading a discussion about the need for political change in China? Could she be blogging about new fashions in New York or Shanghai? Is an international pop star coming to her city, evoking a reaction from her? Is she wondering who will win the Nobel Peace Prize? China, a country with four times the number of people as the United States, has the second-largest economy in the world and consumes more energy than any other country. Modern technology is part of the lifestyle of hundreds of millions of Chinese. The urban young are very much connected to the global-communications network as well as the world of popular culture. China has the world's greatest number of people active in social networking. In ways both personal and fundamental, Confucian culture is being challenged as much as is the Communist Party. Women found a central place in China's political and economic life decades ago and, despite being outside the small circle of China's top leaders, remain there today. Do these observations matter in order to appreciate the photograph of this young woman? She could well be emblematic of some larger trends, some forces of change that have catapulted millions like her into a way of life unimagined by earlier generations. How can we understand what is happening?

The place to begin lies in our critical ability to make sense of the world around us. When thinking about social change, experiences and what we know about life guide our thinking. This is not enough, however. It is necessary to go beyond the ways we usually think about things. The social science mantra is that things are not always as they appear. And many things necessary to understanding are not apparent at all. They must be sought out and assembled or refashioned in such a way that greater clarity, comprehension, and more reliable answers can be found. The answers begin in culture and social structure.

Social science concepts are often about groups, institutions, organizations, and social structure, i.e., the patterns of relationships and recurrent interaction that give social life some predictability. Focusing on social structure is not to lose sight of individuals who, like yourself, are implicated in social processes, institutions, and organizations "larger" than individuals. The challenge is to locate people in a changing social context, with its shifting cultural, political, and economic dimensions. This requires skills not normally acquired in everyday life. These are the skills or the habits of mind C. Wright Mills (1959) called "the sociological imagination." Less mysteriously, in this chapter they are described as ways to understand social change.

The Ubiquity of Change

Is social change inevitable? Of course it is. Imagine attempts to hold back social change or to return to an earlier period of time. These efforts can appear to be modestly effective in the short run, when it is possible to isolate a society, enforce repression of any sign of change, and banish anyone who seeks change. But even these situations have brief lives. The most basic, sustainable human habitation invariably alters the environment, requiring social responses that are buttressed by new norms of behavior, status systems, and beliefs. If successful in meeting environmental challenges, life goes on. If the pattern of habitation is unsustainable or challenges meet with failure, both human life and the environment of plants and animals can suffer extinction.

In the few stable ecosystems supporting isolated societies, untouched by outside influences—whether these are armed invaders, disease, or new material culture—social change can be very slow. It is possible to speak of each generation repeating the life of those before them, of sons and daughters essentially replicating the life of their parents and grandparents. But social change continues nonetheless. Even the most isolated family group in the most impenetrable environment sees contrails overhead and satellites blinking in the night sky, now added to the celestial bodies as a source of wonder and mystical reasoning. Affluent societies' communication pathways, pollution, and quest for far-flung natural resources, food, labor, and markets penetrate every corner of the globe. New states seek to emulate the old by incorporating all the geographically bounded people into national agendas of education, health care, a legal code, production, trade, taxation, a common language, national loyalty, and identity. Change, even in recently traditional societies, has become the norm.

The HIV/AIDS epidemic began in a remote area of Central Africa. Its virulence, however, was a product of modern life—particularly transportation of commercial goods that took the disease from its points of origin and spread it to cities via long-haul truck drivers. In time, the disease boarded jetliners and spread throughout the world, but nowhere is it more deadly than in Sub-Saharan Africa. This is not because of its place of origin as much as the failure to staunch it early, provide effective education and other preventive measures, diagnose those infected, and make available the drugs to slow or stop its deadly progress. Many nations badly afflicted by HIV/AIDS have been on a downward slope of material change for a generation, as their most productive young adults are decimated, leaving the nations' store of human capital badly depleted as the burdens of orphans and illness mount.

On a more positive note, the near eradication of smallpox, polio, tuberculosis, and rinderpest has had a dramatic effect on poor countries. Not only has immunization for DPT (diphtheria, pertussis, tetanus) and other diseases reduced human suffering; it has provided more certainty that infants would live to adulthood, reducing the insecurity of old age and thus provided an incentive to having fewer children.

The transition from a rate of five to that of two [births per woman], which took 130 years in Britain—from 1800 to 1930—took just 20 years—from 1965 to 1985—in South Korea. Mothers in developing countries today can expect to have three children. Their mothers had six. In some countries the speed of decline in the fertility rate has been astonishing. In Iran, it dropped from seven in 1984 to 1.9 in 2006 . . . That is about as fast as social change can happen. (Economist 2009a, 15)

Human beings are major change agents, though they are neither always rational nor all powerful. As Karl Marx's words remind us, "Men make their own history, but they do not make it as they please; they do not make it under self-selected circumstances, but under circumstances existing already, given and transmitted from the past" (Marx 1964, 1). There are significant limits to what human beings can do, but there are also opportunities to direct social change. Some fascinating questions involve human agency, i.e., what can and what should be done in order to influence the speed, scope, and direction of social change. In today's world, understanding social change in order to effect change is more important than ever.

INDIVIDUALS, GROUPS, SOCIAL STRUCTURE, AND AGENCY

Individuals. How much can we know about social change if we study individuals engaged in the experience of social change? Raymond Boudon (1986: 59) is unequivocal: "Social change must be seen as produced by the aggregation of individual actions, and what else could bring it about?" Certainly peoples' biographies, especially changes in their personal circumstances, behaviors, and outlooks, are what we see as social change. When the time and place of the individual are part of the story, the observer feels as if she understands the changes going on. The narrative of Iris Summers' life is just such an account. Philosophers of science call this "methodological individualism." The individual—Iris Summers and all those around her—constitutes the social life of a given time and place. To understand this social life as the accumulated experiences of individuals is to understand social change for a time and place.

Groups. Methodological individualism is usually compared to another approach, "holism" or social realism. It adds another mode to understanding why people do what they do. This is the study of groups. The early sociologist Emile Durkheim (1858–1917) developed the idea that social life is a thing *sui generis* (in and of itself), meaning that it has its own qualities and characteristics in addition to those of the individuals who compose a society. Durkheim ([1938] 1950; [1933] 1956) called the features of social groupings "social facts." Society is more than the sum of its parts (individuals), according to those who adopt the approach of social realism. It, too, must be studied as a thing, with its unique features constituting the data of social science.

Take, for example, a water droplet composed of hydrogen and oxygen. No amount of knowledge about these two elements can convey the sensation of a water droplet splashing on your face. Similarly, no matter how much you may know about thread, the aerodynamics of a woven sheet billowing as a sail could never be calculated. In this view, water and cloth are more than the sum of their parts. Going further, to know about water itself is not enough to understand the enjoyment of walking in the rain. A sail, too, is more than cloth. To appreciate the beauty of sails for centuries of painters, you have to grasp the aesthetics of sails and sailboats, not just the tensile strength or other features of cloth. The whole—including groups of people—*is* greater than, or at least is not the same thing as, its parts.

Social Structure. George Simmel (1858–1918), a major figure in nineteenth-century social science, similarly sought to understand social life by analyzing the characteristics of groups along with the characteristics of individuals. He is among the originators of "formal sociology" or the study of social forms of interaction (Coser 1971: 179–182; Simmel 1950). This is often called a structural approach to understanding why people do what they do. Structuralists do not dismiss the experience of individuals as unimportant, but emphasize the significance of the dimensions and characteristics of groups and the need to include them in any understanding of social life and social change.

An example is Karl Marx's (1818–1883) concept of alienation. When the young Marx wrote about alienation he was referring to the situation of people who work in factories organized for mass production and owned by those who purchase and profit from others' labor. Individuals leave their homes and go to work in a place not of their making or design, in which they use another's tools (the owner's) to construct items they will not use. Most importantly, the assembly line production relegates them to one aspect of the production process. The item's origin and completion are beyond their control. It will be sold by someone else to someone the workers never see for a price determined not by them but by someone else (Marx 1967: 287–301).

The people in Marx's account are working in a *structure* of meaninglessness and powerlessness. The work process and rules prevent

(Continued)

(Continued)

them from engaging in normal, healthy intercourse with fellow workers. Most importantly, the laws and norms of capitalism allow those for whom they work to pay the lowest wage the market will bear. This, in turn, enriches the buyers of their labor and increases the already asymmetrical power between owners and workers. In Marx's view, alienated labor is one of the structures of industrial capitalism; understanding alienation is fundamental to understanding capitalism's transformation.[1]

Agency. Social change is often described as something that happens to people, as if people were passive recipients of whatever life holds for them. That is rarely the case. People incessantly modify and invent new uses and meanings for things. The home computer industry could never have imagined the uses to which young people immediately put their product. They turned the World Wide Web into a vehicle for new forms of communication, self-expression, and relationships undreamed of by its creators. Take another example. As more and more health problems are traced to contemporary industrially produced food, people are stepping back and seeking new approaches, food networks, and healthier diets.[2] This is human agency creating social change.

People "are not simply passive recipients of culture" is how Dwayne Alwin and Ryan McCammon (2004: 44) put it. The McDonald's Corporation's business model takes any surprises away from a cross-country traveler who stops in to eat. Not so across the globe. McDonald's is very popular, but it has been forced to operate differently from what

---

[1]When Karl Marx's concept of alienation became popular among American sociologists there was a tendency to see it in psychological terms: alienation meant that workers *felt* powerless, their work *felt* meaningless, and they *felt* estranged from other workers (Seeman 1959). John Horton (1964) wrote convincingly that Marx's meaning of alienation was linked to the facts of exploitation in the capitalist labor process and a power inequity that is structured into the worker-owner relationship.

[2]*The Fight Over Food* (Wright and Middendorf 2008) is an excellent collection of essays on this topic, giving special emphasis to human agency in the face of global corporate power to shape everyday life. Michael Pollan (2010) provides an update on this in "The Food Movement, Rising."

is found in Detroit or Miami. The people of Hong Kong have a long tradition of "fast food" with its own rules and meanings. Despite McDonald's best efforts, the people of Hong Kong continue to believe a sandwich cannot be dinner. A McDonald's restaurant is likely to be occupied for long afternoon hours by socializing teenagers, and the food purchased by a doting grandparent for his grandchild is a snack, never a meal. The people of Hong Kong are active agents, transforming McDonald's food culture into their own (Watson 1997).

There are moments in peoples' lives when something is wrong. A personal problem is recognized as a larger issue. People become both aware and involved, looking for opportunities to make social change happen. As agents of social change, people seek to affect not just their personal lives but the social fabric in which they live.

Even a profound change in one individual's behavior is not enough to effect social change. With mobilization of many people in response to a grievance, the widespread adoption of a new technology, or an insight that shifts state or corporate resources in a new direction, however, social change becomes real.

The Enigma of Time

Is time a thing, or is time an idea? Is time an immutable feature of sensory reality that can be measured like width, depth, and height? Or, is it, like love, variable by place, mood, and movement? St. Augustine reasoned that "time was a property of the universe that God created" (Hawking 1996: 13). We often talk of time as a thing, an object to be manipulated, created, packaged, and disposed of. We save time, make time, divide our time, and divvy it out. We give it a value, as Benjamin Franklin counseled when he said that time is money. Time can be an almost unbearable burden when waiting for Christmas morning or the school day to end. Prisoners are sentenced to do time, and many lose their minds under the burden of time (Cohen and Taylor 1990: 647).

Images of Time

Conceptions of time have taken many forms, expressed in a variety of metaphors: an arrow, a stream, a force moving within an apparently stationary reality. If time is a river, as the ancient Greek philosopher Heraclites said, we can never step into the same moment twice. In this view, social change is the natural state of things; nothing can ever be repeated. This contrasts with Aristotle's view that reality is fundamentally stationary and moves only when a force is applied. Social change is understood as the consequence of a force happening outside the normally stable state of affairs. Time, in effect, stands still until moved. Newton's first and second laws of thermodynamics put this notion to rest; the real effect of a force is to *change* the speed of a body, rather than to set it moving (Hawking 1996: 23).

Questions of time raise other questions about beginnings and endings. If time is a fixed entity that can be calibrated, the past is a similarly designated entity with a beginning and end (the present). Religions "of the Book," i.e., Judaism, Christianity, and Islam, revere the story of Genesis, seeing God as having started it all. In this view history is linear, a progression with occasional setbacks. The present is a unique moment along a path leading, perhaps, from darkness to light. Many other religions have no sense of a beginning. "It always was" is the answer to "When did it begin?" Life is a continuation of eternity or possibly a cycle with a succession of new beginnings, always to return again and again. Time is a wheel, or time is meaningless in the larger understanding of life and death. The Neur of Sub-Saharan Africa are not alone in having no word for time (Roy 2001: 45).[3]

The question of time inevitably raises metaphysical questions that have challenged thinkers for millennia and are beyond the scope of inquiries into social change. These differing conceptions, and many others, however, remind us that how time is understood both reflects people's circumstances and provides a vocabulary of motives for the way they pursue their lives.

Measuring Time

The division of days, lives, tasks, activities, and accomplishments into measured units of time is so commonplace that it hardly bears mentioning.

---

[3]Pierre Bourdieu's conclusions from his anthropological research among the Kaybal of Algeria, subsistence herders and farmers, is relevant for thousands of groups of people over the past 10,000 years. The future, beyond the immediate horizon, is beyond knowing. It is as much a fantasy to talk about the future as are the wildest stories that entertain children. The Kaybal's view of time and the calculation of their life course reflect the elements they must deal with, being "bound up in immediate attachments to the directly perceived present." Theirs is "an attitude of nonchalant indifference to [time], the passage of which no one dreams of mastering, using up, saving . . . All the acts of life are free from the limitations of the timetable" (Bourdieu 1990: 221).

But it was not always this way. For most of human history the astronomical measure of time—the basis of today's clockwork—was much less precise and far less intrusive. People used the sun and moon to help gauge the seasons, but rainfall and vegetation, the ebb and flow of warm and cold weather, the movement of animals, and human endurance played larger roles in the decisions they made than did the clock.

Depending on the topic of change (e.g., social, geologic, celestial) one kind of time may be more useful than another. It is almost beyond the power of most of us to imagine light years, perhaps because the distance light travels in millions of years is so beyond anything we can experience. So, too, is the effect of speed on time, as revealed by Einstein's theory of relativity. At the speed of light, time as we know it stops. Of course, nothing with mass can travel that fast, but at a speed approaching the speed of light the clock being held by the traveler slows down proportionally, relative to the nontraveler. As Stephen Hawking (1996: 32) concludes, Einstein's "theory of relativity put an end to the idea of absolute time."

Geologic time is similarly difficult to fathom. We can appreciate the last ice age ten thousand years ago, but when handed a stone that originated three billion years ago and told that it came from a succession of uplifts forming mountains, intruded by volcanic magma, each stone weathered and washed into its current shape over tens of millions of years, we can hardly conjure in our imagination the time of these transformations and those of the Earth's crust. Who can fully grasp the geologic time of Paul Hawken's (1993: 21–22) observation: "Every day the worldwide economy [utilities, cars, houses, factories, and farms] burns an amount of energy the planet required ten thousand days to create." It staggers, if not completely eludes, the imagination.[4]

At the other end of the time spectrum are split seconds and the measures of quantum mechanics: nanoseconds and such. These times are more accessible to us because they involve parsing a larger unit of time we readily grasp. If athletic feats are measured in hundredths of a second and the brain's capacity to capture an image is measured in thousandths of a second, it is possible to think of and appreciate the more refined demarcations, even if we cannot feel them or they pass in less than the blink of an eye.

---

[4]Equally staggering is Wes Jackson's (2010: 6–7) observation that a person twenty-two years old "has lived through over half of all the oil ever burned; the ten-year-old, a quarter."

HISTORICAL ?

VAGUE

Social Time

?

American cultural expressions substitute time for geography. People living in an area with a great distance between towns may describe the next town as half an hour or two hours away. A "twenty-minute neighborhood" is one where most things can be obtained in a twenty-minute walk (or a five-minute bike ride). These expressions of social time focus on an experience, an occasion of social life, somewhat vague or ambiguous from a strictly measurement point of view. They have a personal reference: my isolation, my way of driving, my walk. When we talk about dinnertime and when we say "It's been a long day" or "It's been too long since I've seen you" the reference is to an experience (anticipating good food, being tired, or missing someone) more than the clock.

Even more vague, but socially significant, are common references to events by social time. To say that something happened after we were married, to look forward to a long weekend, or to measure your years as a young adult by semesters is comprehensible only by those who recognize the referenced events. This way of understanding time subordinates astronomical calculations to personal experiences. The Bible counsels "a time to be born a time to die/ a time to plant a time to sow/ a time to cast away stones/ a time to gather stones together" (Ecclesiastes 3: 1–8). The conception of time people hold is critical to understanding their "hopes, yearnings, and purposes" (Mannheim 1936). When protesters in Cairo's Tahrir Square chanted, "It's time to take back our dignity!" they were speaking the language of social time. Peoples' view of time—whether as something immutable, full of potential, inscrutable, or foreboding—has a great deal to do with their actions to effect and their responses to social change. Social time is, as George Gurivitch (1964) observed, the most important form of time for social change.

BIOGRAPHICAL

Making Sense of Large-Scale Social Change

In criminal court, the state's prosecutor explains why an accused individual is guilty. She arranges the facts, imputes a motivating reason for the crime, and links these together in a narrative that makes seemingly obvious connections. She tells a compelling story the jury will find believable. In response, the defense attorney explains the facts differently, adds additional, mitigating facts, challenges the quality of the prosecutor's evidence, and most importantly challenges the prosecutor's narrative and the purported links between pieces of evidence. Expert witnesses are often called into court to explain the evidence or answer questions of motivation. The hope is that the accused will be convicted if guilty and cleared if innocent. Court proceedings can be thought of as competing narratives seeking the best possible version of the truth. In the end, the narratives will be weighed by the jury, and one will prevail.

Theory as a Narrative

In its effort to get to the bottom of things, theory is not dramatically different from the court proceedings just described. Theory, too, provides competing narratives and ways of understanding. Theory raises questions. It focuses attention on important facts rather than all the possible facts. It provides concepts that give meaning to the facts, and it draws on the best possible research to establish connections between facts. In showing the processes that connect the facts, it offers an explanation.

Unlike a prosecutor or defense attorney, however, good theory explores the possibility that its conclusions could be wrong. It invites other understandings and answers to the questions posed. Using the same logic as good research and hypothesis testing, we don't prove a theory to be true. Rather, we fail to *disprove* a theory, and this increases our belief that we have it right. The understanding it provides prevails until a better theory or new information comes along to knock it off its pedestal.[5] In a nutshell, theory is understanding. It is a way to organize what is known and what can be assumed, to get an answer to a question or at least to gain clarity about something perplexing.

SOCIAL HISTORY AND SOCIAL CHANGE

History as a discipline does invite grubbing for detail…[Fortunately] most historians are concerned with…interpreting such facts, usually by means of narratives.

—C. Wright Mills (1959: 144).

Getting the facts straight is more than getting them correct and in chronological order. It involves connecting the facts into a coherent

(Continued)

---

[5]Thomas Kuhn's (1970) famous "paradigm of scientific revolutions" takes a somewhat different and quite intriguing approach to disproof. It emphasizes the point in time when an existing theory is no longer able to reconcile new information in a coherent narrative, setting up a crisis that impels those working on a problem to devise a new, often more parsimonious or elegant theory.

(Continued)

narrative, a story that begins to explain what happened, to answer the question of why? Arthur Stinchcombe, a social historian, explains that "the main tool of the historian [is] a narrative of a sequence of events." An individual interested in social change, according to Stinchcombe (1978: 13), wants to understand *why* things unfolded in the way they did. "The test of any theory of social change is its ability to analyze such narrative sequences."

The social historian's story, interesting and new to us, is not always in accord with our common-sense ideas about things. All of us have misconceptions and erroneous notions about the way the world works. A good social historian reminds us of this and helps us develop a better understanding, based on the facts, as well as a more sophisticated understanding of social processes, than we might otherwise have.

Stephanie Coontz's *The Way We Never Were* (1992) scrutinizes several assumptions and shows how these are largely convenient fabrications with little factual basis. A combination of selective examples, wishful thinking, and outright propaganda (often manufactured by advertisers) present idyllic images of family life that never existed for most people. When compared to the manifest problems routinely aired today—divorce and spousal abuse, irregular working hours and economic stress, the seeming abandonment of the family meal, and so forth—a longing for the good ol' days is inevitable. Coontz sets the record straight by enhancing our understanding of how things have changed, and not always for the worse.

When historians interject an explanation, it is rarely from an explicit theoretical or easily discernable perspective, though a critical reader may recognize a viewpoint in favor of certain facts and a hidden agenda guiding the reader toward the author's preferred conclusion. Most historians leave the search for theoretical explanations to other social scientists[6] and in fact many historians reject the label of social scientist in favor of associating their work with the humanities, more akin to writing nonfiction prose. Other historians, however, use concepts and theories to make sense of social change in ways very familiar to other social scientists.

---

[6]David Hackett Fischer points out in his very readable *Historians' Fallacies* (1970) that "unspoken explanations" are often erroneous; better to make explanations explicit and let the reader decide on their veracity. His chapter "Fallacies of Narration," on mistakes in imputing explanation in historical writing, is well done.

Montesquieu, Adam Smith, Karl Marx, Max Weber, Emile Durkheim, and other eighteenth- and nineteenth-century moral philosophers and scholars whose ideas and perspectives remain important for understanding social change were acutely aware of history. They studied it well, finding in history an understanding that made sense of their present circumstances. They lived, as we all should, in interesting times. The deep divide between rich and poor, recurrent wars, growing population and urbanization, and popular calls for radical social transformation were very much on their mind. Most pressing for them were the questions of the two "master narratives" discussed in Chapter 1: capitalism and the national state.

Given the tremendous possibilities to be found in new inventions and technologies, and more so in the organizational forms of the national state and corporate organizations, there was the inevitable question of how these new powers would be used. It was clear by 1880 that they were transforming social, economic, and political life. Long-standing and traditional culture was giving way to new forms of thought and social practices, identities, and attachments. The future might hold great promise, but present conditions seemed ominous, and the ability to control and direct the new powers was anything but certain.

This was particularly true for the conditions of everyday life experienced by the swelling urban population, the work life of the now proletarianized masses, and the political influence of elites and citizens in a democracy. Not only was there anxiety about the use of the new technical and organizational powers to amass great wealth and mobilize human resources, but there were also many questions about the very control of them. Who and what was directing them? Perhaps they had eclipsed their creators and would no longer answer to human efforts to use them for good rather than ill.

The men and women who tackled these questions were intellectual giants whose ideas are still read and studied today. Unlike most social theorists today, they were less hesitant about applying the bold stroke, the universal generalization, sweeping statements and vivid imagery, and were more prone to use grand metaphors that captured the reader's or listener's imagination. They were, as well, committed researchers, meticulous and seemingly untiring in their pursuit of historical and contemporary information. They knew that the way to understanding was through theory in order to make sense of the facts but that theories were useless and likely to be flawed guides if not rigorously and persistently tested against facts.

A Way of Understanding or Ways of Understanding?

The story goes that Herbert Marcuse, a German intellectual who fled to the United States ahead of the Holocaust, was given a faculty position, first at the New School for Social Research, then at Columbia University. Marcuse offered to teach social change but was informed that Neil Smelser, a prominent sociology professor at Columbia, was already teaching a course called theories of social change. "Well," responded Marcuse, "Professor Smelser can continue to teach theories of social change. I will teach *the* theory of social change."[7]

More sympathetic to Smelser than Marcuse, Robert Lauer (1991) in his survey of "perspectives on social change" describes several ways of seeing and understanding social change. Each is an approach to answering important questions about the human condition and the shifting sands of social life. All of them provide a framework for addressing specific questions of social change, such as the reasons one society embraces a particular technology and another ignores it.

In Lauer's view, each perspective has been useful in advancing social science understanding with regard to questions about social change. This more ecumenical view accepts the competitive framework of science and trusts its practitioners to sort through conflicting claims and approaches, make adjustments, innovate, improve, and move science along. None of the perspectives are fully developed, rigorous theories but are ways of thinking about and making sense of modern times. Lauer arranges them into four groups:

- Cyclic theories
- Developmental theories — *Evol. theories?*
- Structural-functional theories
- Social-psychological theories

The first two provide metaphors or models of human history. Cyclic theories apply the image of the wheel, the pendulum, and the oscillating wave: history repeats itself in new terrain, history moves back and forth, or it moves up and down. The second type, developmental theories, are usually more linear, with one dominant roadway or a series of alternate routes taking humankind on its way to a convergent or common destination (Eisenstadt 1964).

---

[7] I attribute this story to one of my undergraduate teachers, either Dr. Robert Parinbanayagam or Dr. Leonard Boonin, both formerly at the University of Missouri, but I may be wrong.

The third type, structural-functional theories, offers a set of principles to explain the formation and change of societies. This perspective emphasizes the way social organization and culture—including technology and belief systems—solve human problems, especially those associated with adaptation to the physical environment.

Social-psychological theories pay close attention to the thinking, attitudes, and self-images of people at particular junctures in human history, and especially the things that motivate them. By analyzing the attitudes and outlooks of successful people or the dominant attitudes and outlooks of many people in a given society, these theories purport to focus on the main reasons for social change. For example, the adoption of new value orientations and ambitions propels some people in traditional societies toward modernity, while the failure to do this consigns others to a low standard of living and limited involvement in their changing society.[8]

Organizing scholars' work in this way gives everyone a seat at the table. The chairs, however, are often shared by more than one perspective. That is, there is often a blending and fusing of perspectives, depending on the topic being studied, e.g., the degree of democracy likely to be embraced by religious elites or the enthusiasm with which some countries adopt private economic incentives and others do not. By recognizing the fusing and melding of perspectives, ways of understanding social change can be organized into two rather than four perspectives that highlight the way theory approaches questions of social change.

Society as an Evolving System

Nineteenth-century intellectuals' adoption of evolutionary theory in geology, biology, astronomy, and zoology was followed by widespread evolutionary thinking in economics and the other social sciences and remains popular today. Relying heavily on empirical observation, historical records, and comparative analysis, evolutionary systems theories of social change provide a set of conceptual tools and theoretical reasoning to make sense of well-researched processes. The scholarly analysis of societies as evolving systems is much more sophisticated than it was a hundred years ago, but its systems perspective—parts (organs) working together in a coordinated fashion, with communication (nerves) and the exertion of power (muscles), with its focus on growth and reproduction—continues to invoke the image of a living being or a mechanical device.

---

[8]David McClelland's *The Achieving Society* and Alex Inkeles and David Smith's *Becoming Modern* are among the most prominent in using a social-psychological approach to social change.

Just as a plant is composed of parts that perform various necessary functions for it to grow and reproduce, society's parts are recognized for their functionality. Not only do the parts work together; they are interdependent. Change in one part of a social system requires changes in other parts, or the whole will fail, weaken, and perish. Evolutionary theory focuses on change as a gradual process. Evolutionary theorists tend to be conservative about proposals to deliberately create social change, emphasizing the complexity introduced by change and voicing skepticism that significant or radical efforts to create social change can withstand the test of adaptability, the key to survival.

In this theoretical perspective, the social system (society) is made up of parts that must fit together in order to work effectively. Talcott Parsons (1951) conceptualized the parts of the social system in his fourfold LIGA configuration of functions and institutions: (L) latency/family; (I) integration/religion; (G) goal attainment/economy; and (A) adaptation/education.[9] Gerhard Lenski's five components—culture, population, material products, social organization, and social institutions—similarly paints a picture of a coordinated system (Nolan and Lenski 2009). Change in one component necessitates changes in the others. For example, population growth puts pressure on the environment, requiring a more efficient way of procuring material products through enhanced social organization. This puts pressure on traditional bonds that can be replaced by more contractual and rational relations among individuals and groups and is often accompanied by a more worldly rather than mystical religion.

Evolutionary social system theorists tend to describe eras, stages, steps, and social types that characterize the persistence of patterns of economic and social life. They seek to explain why one era or stage passes into the next (Eisenstadt 1964). Generations of anthropologists have categorized societies by level of complexity, seen in evolutionary perspective. A simpler form of political organization and economic activity (Stage 1) gave way to more complex forms (Stage 2) that favored some societies over those that retained less-complex forms. States, in particular, are distinguished from tribal societies governed by chiefdoms. Among states are many qualitative gradations of authority, e.g., violently seized and coercively held, hereditary,

---

[9]A synonym for latency, in Parsons' usage, could be persistence or continuation from one generation to another. Similarly, integration is the incorporation of individuals into the whole, most effectively by their sharing the same beliefs. Goal attainment is the organizing of efforts to do what the social system needs doing. And adaptation is the molding of individuals to the contours of the social system. For a clear explication of Parsons' system, see Jackson (1977).

religiously sanctioned, authoritarian and controlling, democratically elected. Most important are the variously complex ways the state channels resources to itself, organizes large-scale activities that become state functions, and administers its rule through an established legal code (Tainter 1988: 23–31), all giving them an advantage over nonstate societies.

EVOLUTIONARY CHANGE IN SPENCER, VEBLEN, AND SOROKIN

Charles Darwin's *Origin of the Species* in 1849 and his later books were published during the greatest expansion of human power the world has ever known. No land or sea was too remote or threatening to prevent human incursion. This age of global imperialism and discovery, along with the developments of science and industry, linked a popular and simplistic version of social change through an evolutionary process to an agenda for societal improvement. It is a curiosity that social improvement was so enthusiastically embraced by people who, using today's yardstick, would be considered ultraconservatives. Nineteenth-century scholars, Herbert Spencer in England and William Graham Sumner in the United States, unlike their more liberal colleagues, saw in evolution not only a pattern of social change but a morality or set of lessons for human behavior and political action.

With capabilities previously unavailable, human beings should heed the message that only the strong survive. The "absurd effort to make the world over," as Sumner (1954) put it, was doomed to fail. It ignored the elementary facts that the poor are poor because they are less fit to survive. The state is acting contrary to natural laws when it seeks to provide education for everyone, build urban sanitation systems, and protect peoples' safety on the job. Extreme individualists, their view of evolution was one of progress, but only if it was gained through individual effort. In an industrialized society, that meant wealth for the few who, by their success, had proved their fitness and hence their right to control resources and make decisions for the rest (Germino 1972).

Quite a different view of evolutionary change was advanced by one of the most interesting figures of the early twentieth century, Thorstein Veblen. Veblen, a Midwesterner and itinerant professor whose first language was Norwegian, died alone in a small cabin above Palo Alto, California. His evolutionary perspective reversed earlier social

(Continued)

(Continued)

evolutionists' belief that the strongest survive in a war "red in tooth and claw," as Herbert Spencer termed it. Rather, evolution was a process of technological development through human inventiveness that often provided opportunities for wasteful preening and systemic boasting (think of Donald Trump), contradictory trends that could not last but would surely be repeated again.

Veblen's writings may seem obscure, but terms that he originated, such as "conspicuous consumption," remain staples of the language a century later. He cast an original and often ironic eye on the world of economics and dissected the foibles as well as the major trends of his time. Why do people so value a swath of green lawn when the space could as well grow useful fruit and vegetables? How is small-town gossip a substitute for buying things you don't need? The answers Veblen offered said a great deal about social class and the less-than-rational ways organizations operate. Economic efficiency is largely a myth in Veblen's view. Rather, wealth bestows capabilities for claiming social status. The cultural manifestations of status are largely wasteful, as he shows page after page in his classic *Theory of the Leisure Class* (1899). Veblen was among the first to broach the idea that managers, not owners, were taking control of large corporations. They were running them with little of the innovativeness and vitality of entrepreneurs. Universities, too, were becoming victims of the same phenomenon.

A less biting but equally critical evolutionary view is found in the work of Russian-born Pitirim Sorokin who perceived evolution as oscillating cycles in the pattern of dominant culture. Through eons, culture shifts between two contrary poles: the hedonistic and the ascetic. Analyzing material culture and belief systems, embodied in architecture and other art forms, theologies, and laws, Sorokin proposed that a guiding set of principles—a cultural ethos—typifies a society. At its most hedonistic, all things are directed toward bodily pleasure and physical satisfaction, what he called sensate culture. Feelings are paramount, and ecstatic experiences are highly valued as sources of inspiration and understanding. At the other pole is ideational culture. Its ascetic ethos shuns whatever satisfies the senses as false pleasures and pursues a spiritual but reasoned and deliberate lifestyle.

Sorokin's *Social and Cultural Dynamics* (1937–41) views social change as pendulum swings. At the top of one sweep is a society whose religion, art, and social honor revolve around strong physical sensations, beauty, strength, and endurance. At the top of the opposite stroke is a radically different society, with religion, art, and social honor expressing the value of mental and spiritual depth, a calm and reasoned perspective, and a deep appreciation for what cannot be seen and felt. Sorokin drew, with enormous amounts of research, a picture of nonlinear evolution, with human societies moving back and forth between the two extremes, and usually somewhere between the two. His condemnation of the society to which he immigrated, the United States, as recklessly pursuing a shallow and ultimately disastrous path of materialism and consumerism, is a sentiment that keeps Sorokin's thought alive today.

The evolutionary perspective has become much more sophisticated and nuanced since Spencer, Veblen, and Sorokin, and the data that are best understood by evolutionary theory are far more reliable and voluminous. The basic principle of reproductive success as well as the genetics underlying the development of biological forms have become a standard part of the life sciences as well as the social sciences. While earlier social evolutionism uncritically adopted the idea of social progress and improvement, by the late twentieth century more skeptical appraisals of where human ingenuity was taking us, in light of the human-induced catastrophes of war and environmental damage, were in order.

Intellectual disputes over how quickly evolutionary changes occur and the pattern of evolution (e.g., Stephen J. Gould's [1980] idea of "punctuated equilibrium" posits a pattern of slow change and bursts of new forms) are much debated. Arguments about the capabilities for living species to adapt to unprecedented alterations in the physical environment—e.g., to urban life and genetically modified foods, global climate change, and a world of heretofore unknown chemicals—challenge visions of evolutionary change as inevitably positive or even capable of operating in the face of human pursuits in today's world.[10]

---

[10]The interested student can learn much more about these ideas in any number of scholarly texts. For starters, a text that locates Spencer and Veblen in historical context is Lewis Coser's *Masters of Sociological Thought* (1971). Sorokin is introduced well in Ray Cuzzort and Edith King's *Twentieth Century Social Thought* (2002).

Growth, Complexity, and Advantage. Those who hold the perspective of society as a changing system emphasize growth and complexity, both in the size of societies and in the network of interdependencies linking society's parts. Compare a small group to a large group when trying to get something done, for example, a school putting on a musical. A smaller group or small school needs everyone to be doing lots of different things. Everybody must be able to sing a little, dance a little, and help with the scenery, jumping in when something is needed. Even the star performer might help paint the sets, sell tickets, and work the lights if not on stage. In the large group or large school, everyone has a more specific role to play; they are much more specialized. This is sometimes described as a more developed division of labor.[11] Coordination is the key to this more complex production and is assigned to one person and his or her staff. Some evolutionists would predict better reviews (superiority) for the large school's musical than the small school's.

Social systems have evolved through time. The adaptive advantage has increasingly—though not always—gone to larger and more complex systems. Growth itself is not highly advantageous, but when social organizations can capitalize on the opportunities of more people doing specialized things, it gains an advantage. For instance, a large family in a very precarious natural environment means more mouths to feed. But perhaps ironically, a large family in that environment may have an advantage over a small family and be better able to sustain its members.

A larger family in agrarian societies as are found throughout Africa can be organized to pursue a variety of economic activities as a kind of insurance against drought or disease. A larger family among people who both grow crops and tend herds of several kinds of animals can assign the youngest children to scaring away birds from crops while they watch over docile herd animals like sheep. Older children help with crop planting, weeding, and harvesting as well as tending larger animals such as cattle. The adolescent children or young adults, especially males, may be sent into the bush with more difficult animals such as camels for months at a time, living off natural forage. If the family's crops fail because of drought or disease strikes one species of animal, other food sources are still available. As a last resort, camels can return home to nurture the family with milk and meat. A smaller

---

[11]Emile Durkheim (1858–1917), France's first great sociologist, wrote *The Division of Labor in Society* (1956) as an analysis of how social stability can be maintained in modern societies where weak interpersonal relations and social divisions, including class and status inequality, threaten disorder. The key is interdependence, i.e., people rely on others for the things they cannot do themselves.

family cannot diversify its household economy and is thus more vulnerable to the vicissitudes of weather and disease. Polygany, i.e., a man having two or more wives simultaneously, makes sense in the agrarian society as a means of establishing larger, flexible families (Massey 1986).[12]

Culture and Social Systems. Key to the evolution of the social system is culture, a human creation that makes adaptation possible and gives some human groups reproductive success while others decline and disappear. Nonhuman species (plants and animals) depend on their genetic makeup and geography for their destiny, evolving adaptive mutations both in physical characteristics and capabilities, as well as a social organization that enhances reproductive success. This is true for human beings as well, but humans have the additional tool of culture to overcome what would otherwise be obstacles to survival.[13]

While culture may be seen as a storehouse of received wisdom and unfriendly to new ideas and practices that would upset the systemic balance of a complex social system, one part of culture provides the driving force for change: technology. Technology is the practical outcome of borrowed knowledge, experimentation, and discovery; it is "that part of society's store of cultural information that enables its members to convert the resources available in their environment into the material products they need or desire" (Nolan and Lenski 2009: 57). In most evolutionary systems explanations, technology is an aid for adaptation by setting in motion a series of changes throughout the social system. The tension between culture as a conservative force and the possibilities of new technology form a dynamic of social change that challenges the tendency of social institutions toward stasis (Chirot 1994: 125–128). In this perspective, societies that adopt new, effective technological approaches tend to be the survivors.

---

[12]Similarly, polyandry (a woman having more than one husband), though this is seldom practiced today, helped meet many of the challenges of pastoralists in areas of Central Asia (Goldstein 1976).

[13]Jared Diamond's *Collapse* (2005) shows how a well-developed culture also can be a hindrance to adaptation, especially when the environment in which the culture developed changes. Diamond's discussion of deeply held values among the Norse Greenland colonies from 984 to 1335 CE explains how some values can be incompatible with survival. Chapter 14 of *Collapse*, "Why Do Some Societies Make Disastrous Decisions?" is especially enlightening and relevant today. See also Joseph Tainter's *The Collapse of Complex Societies* (1988). Tainter proposes a theory of "marginal productivity of socioeconomic change," i.e., that complexity has limits when adaptation becomes too expensive for some societies.

Human History as Systems of Evolutionary Change. As was discussed in Chapter 1, tens of thousands of years of human history saw very slow change in how people lived: usually in small bands and family groups. Adults lived only long enough to reproduce their kind, acquire the means of subsistence until their offspring could take over, and pass on the store of knowledge required to survive in the environment. When some groups began planting crops and herding animals, however, they increased their numbers and in time put pressure on hunters and gatherers who were themselves forced to become more sedentary, turning to crops for an increasing portion of their livelihood.

Technology, especially metal smelting, gave an advantage to some groups who developed superior weapons and digging tools that increased their agricultural productivity. This, in turn, required new forms of social organization, social rankings, and authority structures. Being better fed, the population increased along with the capacity to support experimentation and the creation of new knowledge and technologies. More complex social organization, communication networks, geographic mobility, and a unifying ideology (usually based on religion) allowed them to vanquish and absorb others. Though some became larger and capable of creating empires, these societies remained agricultural based. By means of a gradual process of plant and animal breeding and domestication, inventions that facilitated more widespread transportation of goods and people (especially warriors), and the accumulation of information (often through trade), agricultural societies were able to support even larger populations who increasingly lived in cities. In turn, cities became the incubators of science and technology, more sophisticated warfare, and far-flung financial relations.

In recent centuries, cities came to dominate the world. The modern world system, as Immanuel Wallerstein (1974) conceptualized it, is composed of centers of commerce, finance, science, and political decision making. On their peripheries are slower-changing regions supporting the centers with food, raw materials, and surplus population that in time becomes an industrial workforce. Andre Gunder Frank (1966) was among the first to recognize that the core-periphery relationship was not so much one of the more powerful and affluent areas stimulating the development and improvement of the periphery. Rather, the core transformed the periphery into dependent colonies, in effect creating "underdevelopment."

The evolution to industrialization and the kind of societies that dominate the world today are, in this perspective, the outcome of innovation and the adoption of superior technologies that could be reconciled with evolving cultural beliefs and practices. Societies that resisted change found themselves at a disadvantage, and thousands of them disappeared, if not literally, at least from all but the archeological record. The much more rapid rate of change in industrial society since 1500, chronicled by those who have an evolutionary perspective, is largely attributed to the explosion in scientific knowledge and its translation into technologies, supported by a largely secular ideology that values material abundance and embraces an expansive capitalist economy.

For most scholars who use the evolutionary systems perspective, ancient war and strife are signals that something was amiss in the otherwise smooth functioning of preindustrial societies. In modern times, war is increasingly broad in scope and environmentally disastrous in its conduct. Still, some systems evolutionary theorists view it more as a test "of a society's ability to survive" (Chirot 1994: 122–123). In light of modern war's incredible destructiveness, however, it stretches credulity to think it is testimony to an evolutionary advantage for humankind.

An evolutionary systems perspective seems to suggest that social change happens by the same fundamental process everywhere. Its practitioners emphasize social strains and breakdowns in coordination that necessitate adjustments and corrections in order to survive. Often sounding like after-the-fact explanations, when a society is vanquished, declines in size and power, or collapses, this is due to its failures to adjust or make the necessary changes to accommodate new circumstances. In the evolutionary systems perspective, the way things are has a kind of inevitability. The status quo is the natural and preferred state of affairs. Features of societies like political, class, and ethnic inequality are more likely to be seen as adaptations rather than being problematic. Others beg to differ.

Society as a Site of Conflict, Power, and the Resolution of Contradictions

In the evolutionary systems perspective, war may be seen as one means to achieve adaptive success. An alternative perspective, shared by many researchers, sees war and most processes of social change as ongoing competition and conflict over power and institutional control, consuming and seizing from others the products of human effort. This is usually called the conflict perspective.[14]

---

[14]Randall Collins' *Conflict Sociology* (1975) explores at length this approach to social inquiry. Two other very readable and resilient descriptions of conflict and change, the first comparing an evolutionary systems perspective to a conflict perspective and the second describing the synthesis of many approaches to understanding how conflict generates solutions and syntheses, are John Rex's *Key Problems of Sociological Theory* (1961) and Lewis Coser's "Social Conflict and the Theory of Social Change" (1957).

Rather than seeing society as a system of interdependent parts functioning to meet the needs of the population, the conflict perspective offers a more volatile image of social forces engaged in a continuing contest in pursuit of their interests. This is not to say that people are in pitched battle 24/7, but harmony and stability are always in tension. Social order in periods of calm is held in place by a balance of forces that are constantly shifting. "As the configuration of power changes, so moves history" (Roy 1997: 273).

Social Divides. Revolutionary movements and their leaders embrace an extreme version of the conflict perspective, as will be seen in Chapter 6. They seek to show how the revolution is embracing and forcing the inevitable; those opposing the revolution are seen as standing in the way of the future. The revolution is a consequence of the divisions in the society, asymmetrical power and privilege, which differentiate perpetrators and victims, exploiters and the exploited, the sinful and the faithful, the corrupt and the law abiding. Marx and Engels put it this way in their famous passage from *The Communist Manifesto*, written in 1848:

> The history of all hitherto existing society is the history of class struggles. Freeman and slave, patrician and plebian, lord and serf, guild-master and journeyman, in a word, oppressor and oppressed, stood in constant opposition to one another, carried on an uninterrupted, now hidden, now open fight . . . Our epoch . . . has simplified the class antagonisms: Society as a whole is more and more splitting up into two great hostile camps, into two great classes directly facing each other. (Marx and Engels 1978: 473–474)

In the conflict perspective, not every examination of social change is seen through such a radical lens whereby the transformation of society is a violent outcome of adversarial social divides. The perspective does emphasize unequal power, privilege, and material well-being as sources of change. Clashing interests and opposing forces, sometimes openly at odds but often quiescent, poised to challenge one another are embedded in the structure of societies. The outcome will determine something more personal and concrete than the adaptive ability of the society as a whole. It will determine who will rule, who will take an oversized portion of the spoils, and who will articulate what constitute the dominant attitudes, opinions, and ideology, at least for a time.

Conflict theorists are more prone to examine social movements as resistance that cuts against the grain of the status quo. Resistance to authority, through mobilization of resources and effective organization, poses a challenge to powerful elites and their interests. How social movements are confronted by those with power helps to define the tactics the social movement will develop to advance its cause. The conflict perspective sees in everyday life an ongoing competition, rarely open conflict, that colors many if not most social arrangements, protected by the more privileged and challenged by the less powerful.

CONFLICT PERSPECTIVES OF KARL MARX, C. WRIGHT MILLS, AND GEORG SIMMEL

It might seem strange, even wrong, to be reading about theories of social change and find out a person like Karl Marx has influenced how social scientists try to understand social change. Isn't Marx the man behind communism, celebrated in the former Soviet Union and the People's Republic of China, Vietnam, and Cuba? They are or were countries toward which the United States carried on either long and costly wars or a cold war of protracted efforts at diplomatic and economic isolation and defeat. This has little to do, however, with the quality of Marx's scholarship. It has remained a source of insight for 150 years of research, and the scholarly thinking his work spawned continues to inform the way we can understand how social change happens.

Marx, in fact, wrote very little about communism, and did so reluctantly.[15] His scholarly topic was capitalism and the way it transformed societies in Europe. This was the research subject of many other economists and social philosophers of the late eighteenth and nineteenth centuries. Like those who formulated an evolutionary systems perspective, Marx saw capitalism as a historical phase that would be replaced by something more progressive, i.e., less warlike, providing more freedom of choice, guided by rational thinking, and requiring less human drudgery to live a life of abundance in which all basic needs for everyone would be satisfied. The forces of industrial production were celebrated. What Marx and others objected to was the power asymmetry in a capitalist mode of production. Capitalism vests power and authority in, and a disproportionate share of the fruits of labor to, an increasingly small group of people. They, in turn, use their

[15]Most of what Marx wrote about communism as a form of future society has been assembled in a few pages (Marx 1956, Part 5). His most-often-cited remarks about communist society, "Critique of the Gotha Programme," were written at the request of others who wanted his support for the 1871 popular uprising in Paris.

(Continued)

(Continued)

wealth and power to promote ideas and cultural practices (e.g., materialistic values and social honor conferred on wealth and consumption) and are able to dominate the state in order to have laws, judicial decisions, and enforcement favor their interests.

The way Marx reached his conclusions, through careful study of the records of industrial investment and the expansion of trade, was equal or superior to other research at the time. The conceptual framework Marx used, growing out of his training as a doctoral student in philosophy, emphasized the dynamic, often contradictory, forces at work and how their conflict leads to a resolution—often a compromise or synthesis, but also through violence and destruction—which is something new in human history and superior to what came before. Marx was optimistic about human progress. The values behind his efforts continue to resonate: human potentiality, freely developed associations of people who work for the betterment of their lives and others, an end to exploitation and misery, and a culture that reflects authentic needs and aspirations. This, of course, meant a rejection of much of what capitalism was doing during his lifetime; hence the radical agenda to change capitalism made Marx's ideas the guiding light of later revolutionary thinkers like Nikolai Lenin, Rosa Luxenburg, Ho Chi Minh, Emma Goldman, Fidel Castro, and Mao Zedong.[16]

Early twentieth-century European scholars in the United States were familiar with Marx's scholarship, but few mainstream American social scientists made a serious effort to study his ideas. When translations began to appear by midcentury, some young social scientists took note, most prominently in the United States a Columbia University sociologist, C. Wright Mills. Mills, who described himself as a "plain Marxist," was an iconoclastic but strong scholar raising questions about power and social class in the 1950s. He was, along with those writing for *Dissent*,

---

[16]One more caveat is in order. Twentieth-century political leaders well recognized for their dictatorial authority and brutality, like Joseph Stalin in the USSR, Pol Pot in Cambodia, and Nicolas Ceausescu in Romania, considered themselves Marxists. This association made it difficult for many scholars to work comfortably with Marx's and other Marxist scholars' ideas.

the *Nation*, and a few other magazines, a voice of skepticism and criticism about growing corporate and military power. His *White Collar* and *The Power Elite* paved the way for the much more robust analysis of American society and its global economic and military reach by young sociologists, political scientists, geographers, historians, and economists. At a time of concern about the proliferation and testing of nuclear weapons, a growing civil rights movement, the personal and social costs of patriarchy, and growing awareness of environmental pollution, the conflict perspective's focus on vested power and social change provided encouragement for social activism. In the university, the scholarship fueled the emergence of women's studies, ethnic studies, peace studies, and antiwar activism.

A different strand of the conflict perspective, that of Georg Simmel (1858–1918), influenced social science scholarship many years after Marx's death. A product of the German university system, Simmel wrote about a wide range of topics throughout his lifetime pursuit of an understanding of the dynamics of social life. His philosophies of money and human vitality are less well known today than is his analysis of the structure of social forms, discussed earlier in this chapter. Guiding his dynamic analysis of relationships was the idea of contradiction, that what is immediately apparent obscures or hides its opposite. His studies included character types that embody recurrent, regularized but contradictory ways in which individuals and groups relate to one another across many different contexts. "The stranger," for instance, is both apart and near, remote and familiar. Simmel treated conflict as a natural, even positive, force that reveals and resolves contradictions in social relationships. Resolutions are always transitory, however, as new contradictions emerge and, in turn, evoke conflict that initiates a new process of resolution (Levine 1971).

Ideology and Power. Unlike the evolutionary systems perspective, the conflict perspective is less likely to treat culture as a societal-wide adaptive device. Rather, it sees elites justifying their dominant position in society by creating and promoting culture that describes their privilege and power as natural, fair, beneficial, and morally correct. They do this by controlling the religious system, education, political ideologies, and popular narratives about social inequality, status hierarchies, and opportunity structures.

While technological innovation, adaptation, and reproductive success are key concepts for evolutionary theories, the key concept for conflict theorists is power. The sources of power are many, ranging from cultural values that promote acceptance or extol resistance, to social norms prescribing behavior that facilitates predictable, orderly interaction or legitimizes oppositional subcultural behavior. Power also resides in the control of resources, whether these are the means of physical health and comfort, education and training, wealth, the legislative and judicial systems, jobs, or an ideology projecting a worldview upon which people draw for understanding, inspiration, or defiance (Mannheim 1936).

Contradictions as Activators of Social Change. The dynamic of social change, and the central feature distinguishing the conflict perspective from the evolutionary systems perspective, lies in what conflict theorists see as the friction between various social forces. Often referred to as societal contradictions, conflict theorists analyze paradoxes and the emergence of dilemmas that pose challenges and pit groups against one another.

Change through the resolution of contradictions provides a way to understand the changes in women's labor force participation in the past half century. Looking rather uncritically at women's work in the early 1950s, social scientists were content to accept the statistic that most women in the United States were "not working." Two-thirds of adult women ages 18–64 were not in the paid labor force. What it hid from view was the tremendous amount of work women were doing, much of it outside the market economy but critical to the well being of their families and the economy. Their household labor alone was worth billions of dollars. Looked at another way, corporations were able to "externalize" many of their costs by having wives and partners not in the workforce pay for them with unpaid labor.

Women's time and effort supported men and firms by absorbing much of the expense of doing business, at no cost to employers. Among other things, women worked hard and without pay to ensure the health and success of men in their families so they could be productive employees. They provided free child care and took care of sick children at no cost to the company. When feminist writers began pointing out these hidden contributions, this provided a strong argument for public policy and legislation regarding divorce settlements that divided family resources.

As personal income growth slowed in the 1970s, U.S. families found themselves unable to achieve improvements in the lifestyles they had anticipated. To get ahead required more than one household earner, and wives began flocking into the labor force. Women of color and mothers of single-headed households had always worked outside the home in large numbers, as had unmarried women, but their situation was largely ignored until millions more women became new wage earners.

This evoked another contradiction: what Arlie Hochschild (1989) called the "second shift." Wives in the paid workforce continued to do nearly the same amount of housework as they had when not employed. They were doing two jobs, one at work and a second at home. The gendered division of labor became a subject of scrutiny that led to demands for change not only across the society but at the most intimate level, i.e., between household partners.

As more and more women took jobs outside the home, the sexism that remained a dominant theme in the culture generated other contradictions, including the disparate treatment of men and women in the workplace and the way the paid work women did was defined. Women, including those in professional positions, were expected to take care of the routine tasks in the workplace that looked like "women's work," such as making coffee, cleaning up the break room, and listening patiently to the complaints of fellow male workers without expressing their own opinions or describing the particulars of their circumstances. Women's wages were considered supplemental to their husband's, expressed as a wage gap for comparable work. Women in professions, retail, and service work were passed over and tracked by companies in ways that precluded their being considered for positions of authority (and more pay).

Polling data in the 1950s found that a majority of young women rated meeting and marrying a young man as the most important reason for going to college. This attitude was not necessarily inappropriate, but it suggested that women's college education was less important than the opportunities for pursuing a satisfactory marriage and family life. It, too, contradicted the emerging reality of the second half of the twentieth century. Young women needed college education as much as young men, and for the same reasons.

In these and many other ways, the circumstances of families and of women confronted the sexism of the dominant culture and the male privileges enshrined in organizational practices. These contradictions provided the dynamic for social change. The women's movement captured much of this dynamism, becoming a visible force for change. It led to changes in less obvious ways as well. For example, educators began taking seriously the declining math and science scores of young girls as they passed through K–12 education. They initiated successful programs to reverse this, in anticipation of young women pursuing careers in medicine, engineering, and research. Today, women's numbers are growing rapidly in most previously male-dominated fields, including law, the biological and physical sciences, and business.

A conflict perspective emphasizes how groups hold on to their privileges and power and the conflict this generates with those having less or no power. In the resolution of the contradictions between the status quo and emerging circumstances, there is no sense that things may finally be put right once and for all. Social change is a constant condition of the working out of competing interests, shifting power, and problem solving, sometimes requiring innovative, unsettling, and even radical solutions.

Understanding Social Change: Two Explanations of a War

Understanding vis-à-vis the narrative of theory, when followed with the logic and rigor developed through nearly two centuries of social science research, can make vexing events and changing social life intelligible. It conceptually organizes empirical data in a way that satisfies our most critical eye or, at the least, is superior to any other way of making sense of them. The following extended discussion of the effort to understand the wars in the former country of Yugoslavia illustrates this process.

All of the facts needed to explain the wars were not available when the wars broke out, but information gathered before, during, and after the wars contradicted what had been a compelling earlier explanation. Ethnic hatred was said to be the cause of the wars. It was based on, as described early in this chapter, popular assumptions about the way things are and the way things happen. If these were right, the highest levels of ethnic hatred should have corresponded to the worst of the wars. The accumulation of data from Yugoslavia showed this not to be the case. Research on ethnic prejudice done in midcentury, the "contact hypothesis" offered a rebuttal of the ethnic hatred explanation, requiring an alternative explanation for the wars. This explanation focuses on political processes that accelerated the breakdown of the state, dragging the people of Yugoslavia into war and generating ethnic hatred as a consequence, not a causal factor.

Civil wars twenty years ago led to terrible destruction and suffering in the former Yugoslavia. The country was formed after World War II by consolidating the south Slavic people of the Balkan Peninsula. It was an effort to unify a geographic region composed of people who spoke several different languages, had regionally specific dominant religions (Catholicism, Islam, and Orthodox Christianity), and had different historical experiences under both the Ottoman Empire and the world wars of the twentieth

Figure 3.1   Map of the former Yugoslavia

century.[17] Some memories of World War II cast the various nationalities in the roles of oppressors and victims, rivals, antagonists, and enemies.

The Initial Explanation: Ethnic Hatred

It is a familiar observation that ethnic, national, and racial conflicts are laced with intergroup distrust, animosity, and even hatred. The language of ethnic conflict is filled with stereotypes and degrading caricatures. The Hutu propaganda preceding the 1994 genocide in Rwanda, where nearly 800,000 people were killed, called their rival Tutsis "cockroaches." It urged Hutus to mercilessly stamp out the Tutsis. When people take up arms against neighbors with whom they have lived for generations, stand by idly or even lend support while gangs and paramilitary forces destroy their

---

[17]A similar effort was made after World War I and failed by 1936.

neighbors' homes, drive them away, murder them, and imprison them in concentration camps, it is only reasonable to assume that there is great hatred between the groups. At the least, something must have happened to ignite this hatred and push people to commit barbarous acts.

When the wars that dismembered Yugoslavia took place, first in the republic of Slovenia, then in Croatia and Montenegro, most tragically in Bosnia (1991–1995), and finally in Kosovo (1998–1999), the conventional wisdom explained this in terms of ethnic hatred. Robert Kaplan's *Balkan Ghosts*, read by then president Bill Clinton, explained how ethnic hatreds were suppressed by the communist government, "thereby creating a kind of multiplier effect for violence" (Kaplan 1993: 30). This was also the opinion of British politicians, including then prime minister John Major and his foreign secretary Douglas Hurd (Ramet 2004: 740–741). It was widely endorsed by influential journalists like Elizabeth Drew: "One of the things one learns from actually being here is that the fears and hatreds which have been unleashed are absolutely formidable" (Drew 1992: 70). This "seething caldron" metaphor, that simmering emotions finally boiled over and blew the lid, spilling and spewing noxious violence across the land, was a popular explanation (Sekulić 2011). It was, however, wrong.

Indeed, the facts were wrong. It mistook a symptom of ethnic conflict— the expression of hatred—for a cause. Most tragically, it offered misdirected policies for dealing with the conflict and similar conflicts elsewhere. These ranged between doing nothing by accepting the inevitability of ethnic hatred and policies that furthered the practice of "ethnic cleansing" by physically— if less violently—relocating families to ethnically distinct enclaves.

A Theory of Interpersonal Contact. In 1954 Gordon Allport, a psychology professor at Harvard University, published *The Nature of Prejudice*. It synthesized years of studies of attitudes, perceptions, and behaviors of dominant groups toward racial, ethnic, religious, and national minorities. Allport put forth the contact theory of ethnic relations. Could it be applicable to understanding the events in war-torn Yugoslavia? Allport knew from a substantial body of previous research that prejudice is an outcome of the amount and quality of interaction between groups. Importantly for improving intergroup relations, he reasoned that the amount and type of contact between groups can significantly influence whether or not a person becomes or remains prejudiced.

From his own and others' research, Allport stipulated the conditions under which contact *improves* intergroup relations:

- Sustained acquaintanceship is better than superficial and "tourist mode" contact.
- Pursuing a common effort is better than being in competition.
- Familiarity as neighbors is better than a situation defined by a neighborhood as "under threat of encroaching minorities."
- A shared, egalitarian status is better than working in a superior or inferior relationship.
- Having supportive leadership is better than having leaders who express skepticism or hostility.

Allport put forth the idea—actually a "middle-range theory" of small-group behavior: interaction between members of different ethnic groups, under favorable conditions, tends to result in positive feelings, can reverse previous feelings of prejudice, and is a prophylaxis against demagoguery and racist sentiments that otherwise would drive a wedge between ethnic groups.

To understand Allport's contact hypothesis, imagine two segregated high schools that are crosstown rivals—one predominately white and the other predominately black. The schools compete on athletic fields, courts, and tracks, and for all the other prizes in high school competition. A vocabulary of motives that attributes both wins and losses to supposed racial or ethnic differences could only be expected. Would it be any wonder that the rivalry sometimes creates bad feelings and spills over into name calling and violence between the schools, one white and one black? Bad feelings easily translate into racial stereotypes and prejudice.

Now, imagine integrated high schools. Black and white players, thespians, debaters, singers, and musicians compete with their crosstown rival teams that are also black and white. According to Allport's theory, when on the same team and engaged in a mutually beneficial activity, the result is feelings of trust and common purpose. People of another race or ethnicity now know one another as individuals, dispelling myths and stereotypes. Principals, teachers, and staff give voice to amicable relations and are intolerant of racial slurs and jokes. Interracial dialogue is encouraged. Insipient problems are dealt with quickly and fairly.

Allport's contact theory has been upheld in hundreds of studies. Ethnic tolerance is best fostered when common goals are being pursued, groups are of relatively equal status, and those in authority support intergroup harmony. It remains today a powerful explanation of intergroup behavior (Hewstone and Brown 1986) and, like any good theory, is useful in making sense of situations with similar features in other times and places.

Ethnic Contact and the Wars in Yugoslavia. A short-lived civil war broke out in 1990 in the republic of Slovenia. Ninety-five percent of Slovenia's population was ethnic Slovenes. Contact between Slovenes and nonethnic Slovenes was minimal. Next, a longer and more deadly war broke out in Croatia where 81 percent of the population was ethnic Croats. Finally, the most destructive war broke out in Bosnia, which was the most ethnically heterogeneous republic in Yugoslavia: 40 percent Bosniak/Muslim, 32 percent Serb, and 21 percent Croat. There was a direct relationship between ethnic diversity (i.e., contact) and the level of conflict, the opposite of what Allport's theory predicts.

The contact hypothesis would predict that, if the conflict was based on ethnic prejudice and feelings of hatred, the war in Bosnia should have been least destructive. Bosnia was the most ethnically diverse and integrated republic. Contact among ethnic group members was greatest, and intergroup familiarity was highest. On the contrary, the war in Bosnia, in terms of duration, level of interpersonal violence, and loss of life and property, was terrible. It was the most destructive war in Europe since World War II, with more than 100,000 people dying, more than half of whom were innocent civilians. Media told the world of concentration and rape camps. More than a million people were forced to flee their homes. As the old saying goes, "Familiarity breeds contempt." Perhaps in Bosnia, Serbs, Croats, and Muslims knew each other better and developed a dislike for one another more virulent than elsewhere in Yugoslavia.

Because it would predict greater interethnic tolerance and less prejudice in Bosnia than elsewhere—yet war there was the worst of contemporary Yugoslavia's many wars—it may be that Allport's theory is wrong. One would have to reject Allport's theory as an explanation of the wars in Yugoslavia *if* their cause was ethnic hatred. Was this actually the case? Was ethnic hatred strongest in Bosnia, the opposite of what Allport's contact theory would predict? Empirical data on ethnic tolerance or prejudice for the warring republics were needed in order to infer a causal relationship between feelings and actions (hatred and fighting). Facts were needed, or Allport's theory would be thrown into serious doubt, lending credence to Kaplan and others' view of ethnic hostility as the explanation for violence and war.

The facts were available from a random-sample survey of 13,422 people in the former Yugoslavia done a year before conflict began. Of all the Yugoslav republics, the data show that ethnic tolerance was highest in Bosnia. The data show across Yugoslavia that as ethnic diversity declines, ethnic tolerance also declines. The greater one republic is numerically dominated by a single ethnic group (e.g., in Slovenia), the higher the level of ethnic *in*tolerance, just as Allport's theory predicts (Hodson, Sekulić, and Massey 1994). In Bosnia, interethnic contact was highest and ethnic tolerance was greatest. Because Allport's prediction regarding contact and prejudice was confirmed by the facts, the assumed connection between ethnic hatred and violent conflict was thrown into question. What *did* cause the wars?

A Better Explanation: Elite Manipulation

The facts about the wars in the former Yugoslavia required a rejection of ethnic hatred as the reason for the wars. A new line of thinking was required to explain what happened to violently tear the country apart. It could not ignore the actions of individuals and their feelings of insecurity, resentment, and in-group loyalty. Rather, to understand the conflict required a structural analysis, in this case the political and economic processes unfolding around the people of Yugoslavia, filling them with fear and uncertainty and in many cases engulfing them in a conflict they did not want or welcome.

A theory of ethnic conflict is not the same as a theory of ethnic hatred. Conflict has many sources, and there are other theories that can offer a cogent explanation for the wars. One, a theory of ethnic competition, looks at economic factors, in particular rivalry for resources that may lead to violent clashes when other avenues to settle the dispute are unavailable. By way of illustration, the genocide in Rwanda has been explained as a combination of scarce available land for farming and one of the highest levels of population density in the world. In the case of Yugoslavia's wars, however, there is little evidence to support the theory of ethnic competition as an explanation.

A better explanation for ethnic violence in the Yugoslavia conflict focuses on "political entrepreneurship" or elite manipulation. When the Soviet Union was beginning its change of system (*perestroika*) and Mikhail Gorbachev as prime minister was encouraging a more open government (*glasnost*), countries like Yugoslavia, with state-run economic systems, were undergoing a similar self-examination. The command economy of state socialism was proving increasingly inefficient, and the legitimacy for the rule of the communist party was eroding. The impending collapse of the Soviet Union emboldened critics of the government in other countries to propose radical changes in their economy, polity, and society. Change vis-à-vis the collapse of weakened states was in the air in Poland, Hungary, and Yugoslavia, soon to be followed by Czechoslovakia, Bulgaria, and Romania.

In the former Yugoslavia, a growing economic crisis of the 1980s raised popular anxieties and a widespread sense of foreboding. The federal government was unable to implement policies to alleviate economic problems and assuage public anxiety. Serb, Croat, and Slovenian nationalist politicians, sidelined for decades, seized the opportunity to direct popular sentiments toward an ideology of ethnic resentment and defensive ethnic nationalism. Popular anxiety became increasingly visible and well organized, with meetings, marches, strikes, protests, pamphlets, and newspapers. They expressed the need to trust only those of one's ethnicity and blamed other ethnic groups for the country's problems. The social movements culminated in the formation of political parties calling for secession from the Yugoslav nation state, initiating the civil wars that disassembled the nation.

In the absence of established political parties and the coalition of civic groups aligned in support of one party or another, Yugoslavs faced a political vacuum. A weakened state, due to serious economic problems and conflicting political agendas, opened up space for some communist leaders like Slobodan Milosović and others, e.g., Croatia's Franjo Tudjman who previously had espoused an ethnic nationalist agenda, to pursue power for themselves. Like peddlers rushing with their goods into a commodity-starved town, political entrepreneurs quickly filled the political vacuum with messages accusing other ethnic groups or republics of having a dangerous agenda. Reminders of past conflict and the construction of myths fanned the flames of resentment and distrust.

Every ethnic group was encouraged by nationalist political entrepreneurs to feel it had been victimized and would only be safe if it protected itself from the others. Who was to blame? Who could you trust? Who should you fear? In this atmosphere of fear, the only persons to be trusted are those of one's own ethnicity, including those would-be leaders of a particular ethnicity. The appeal to ethnic solidarity resonated with claims that ethnicity is a "natural bond" and the proper form for a national state. Those of another ethnicity should be designated as "minority groups," with restricted rights. It is not hard to see how a backlash and violent resistance to such measures would be forthcoming.

Once war began and the killing and atrocities mounted, ethnic hatred did increase dramatically. This fact is apparent from a longitudinal series of four cross-sectional random surveys in Croatia that measured ethnic tolerance. The first was done in 1985, six years before the war. The second, part of the survey discussed earlier, was carried out one year before the war. The third was done soon after the war subsided, and the fourth was done several years later. In 1985, the level of ethnic tolerance was fairly high in Croatia. Tolerance was slightly lower in 1989, just prior to the war, but dramatically lower in 1996, three years after the war ended. By 2003, ethnic tolerance was rising, but was not as high as its 1985 level. These data and other analyses show that the conflict itself generates ethnic intolerance, distrust, and hatred as a consequence of experiencing violence, personal loss, and wartime propaganda (Ringdal and Ringdal 2011). Ethnic nationalists encouraged and played on popular anxiety, and they pursued power by stoking fear and the promise of protection through ethnic solidarity. Ethnic hatred was not the cause but an outcome of the wars (Sekulić, Massey, and Hodson 2006; see also Smith 1981).

Making Sense of Modern Times

The subjects of the next five chapters—technology, social movements, war, corporations, and the state—were discussed at the end of Chapter 1. They were described as mechanisms, to use Merton's term, that "move the social order along the path" of change (Lauer 1991: 130). It is a mistake to imagine that social life is unchanging until one of these mechanisms creates change. Social order is a dynamic system in continual change. The question is, What are the greatest influences on the speed, scope, and direction of change? Technology, social movements, war, and the actions of corporations and states are the major drivers of social change today. Understanding how they do this helps unscramble situations of social change that might otherwise seem totally inexplicable.

Social change at a particular time and place comes about through a combination of factors. The mechanisms are often linked together or have their greatest impact sequentially. In order to think about and clarify their influence on social change, they can be discussed as five drivers, but none is solely responsible for the way social change occurs. In the following chapters it becomes clear, for example, that the state's behavior cannot be understood independent of the economic context, today dominated by corporations. The sources of new knowledge, invention, and technology are in the public (e.g., state) and private (e.g., corporate) spheres where their priorities often guide what is studied, what problems are solved, and who benefits or suffers from the consequences. How war effects social change cannot be understood independent of the state, corporations, technology, and social movements.

Establishing causality is an inevitably complex endeavor. To say, for example, that technology *causes* social change glosses over the many processes—structural and social-psychological—that link technology to change. It may also suggest that understanding can be independent of the historical context and the myriad unique facts about circumstances and conditions that must be considered in any explanation. The next five chapters are a modest attempt to help you begin thinking about social change by clarifying the mechanisms or complex processes of change in today's world. It is essentially an effort not at explanation but rather to make sense of modern times.[18]

Topics for Discussion and Activities for Further Study

Topics for Discussion

1. Take sides, role-play, or work in groups to discuss agency and social structure. Someone or some group can take a very extreme position that people have unlimited free choice. In expressing this, with relevant examples, others should write down objections. What constrains choices? Who or what constructs the lists of options from which we choose? What choices are not even considered but might have been in another time and place? Discuss these.

2. We offer explanations all the time. They may be only loosely based on facts, but they almost always reveal something about the way we look at the world. Pick a topic that involves change, e.g., why the school's team has gotten worse or better than in the past, why there are more vegetarians and vegans today than ten years ago, or why young people are more comfortable in interracial situations than were their grandparents. Listen carefully to what people are saying. Keep a list that distinguishes between the biological, psychological, and social explanations. How often does stating "the facts" substitute for an explanation? What thinking, assumption, or common-sense reasoning lie behind a statement of facts that purports to explain?

3. This chapter offered an explanation of women's increasing participation in the paid workforce. Offer a systems-evolution explanation that considers interdependent parts, adjustments and adaptation, growth and complexity, and reproductive success as part of the explanation.

4. Think about an organization you are familiar with, say, your high school or college. Describe it as a social system. Then, examine your school as a site of competitive interests, power, contradictions, and conflict. Can you apply one of these to an actual situation or event in your school in order to explain what happened?

Activities for Further Study

1. Science fiction uses the theme of time in very creative ways. For those of you who read sci-fi, collect several ways in which time figures into plots. Why does time seem to be such a popular trope in this kind of literature?

---

[18]Max Weber, a giant figure in the founding of the social sciences, concluded that clarification may be all we can hope for in our efforts to understand (Weber [1904] 1949).

2. Is globalization social change or an explanation of social change? Globalization is understood in many different ways, and there is disagreement about what it is. There is even more disagreement about how it is changing the way people work, recreate, and think about themselves. And there is disagreement about what lies behind it, i.e., the mechanism impelling it, speeding it up, and making it more pervasive worldwide. Do some reading about globalization, beyond what you can find on Wikipedia. Decide for yourself: is globalization social change or does it explain social change?

3. Find an explanation for something that is now thoroughly discredited. For example, phrenologists used to explain criminal behavior (and many other things) by the contours of the cranium. (Stephen J. Gould's *The Mismeasure of Man* [1981] tells this story well.) Medicine was previously based on humors, i.e., four fluids of the body. Corporal punishment in schools was thought to improve learning. Find out why an idea (an explanation for something that was of broad interest) was discarded. What does this tell you about the way we go about understanding something?

4. The professional sociology association journal *Contexts* includes articles by very good social scientists that any student can read and understand, unlike a lot of articles in scholarly journals. If your school library carries *Contexts*, find an article in it on a topic that interests you. Pick an article that explains something. How does it do this? What kind of an explanation does this seem to be?

4

Technology, Science, and Innovation

The Social Consequences of New Knowledge and New Ways to Do Things

Imagine that you cannot read. You look at shapes of thin black lines on a white page and cannot comprehend their writer's intention. You cannot decipher the figures' meaning and translate the lines on the page into images or ideas. You cannot conjure up sounds to read aloud to a listener. Looking at the script that follows, imagine—if you don't read Chinese, Hebrew, Russian (Cyrillic), or Farsi (Persian)—how incomprehensible the English script you are reading can be.

你需要打電話給你的母親，告訴她你愛她

אתה צריך להתקשר לאמא שלך ולהגיד לה שאתה אוהב אותה

Вам необходимо позвонить матери и сказать ей, что ты ее любишь

شما باید با مادرت و به او بگو او شما را دوست دارم[1]

You cannot read simple instructions and must depend on the diagrams. You cannot read signs and must depend on the simple picture. You cannot read a magazine but only scan the photos and remain perplexed by the cartoons. You cannot write notes to yourself. You cannot compose text messages or e-mails. The Internet is a purely hit-and-miss jumble of links. Envelopes come in the mail: are they for you? Someone puts a piece of paper in front of you and says, "Sign this." Not only can you not sign your name; you don't even know what you're supposed to be signing!

What does the world look like with no understanding of writing? It might help to remind ourselves that for most of human history people did not read or write. Today, 500 million adults are illiterate. *Your* social world, fortunately, is literate. Writing is everywhere, and a high level of literacy is assumed. To be illiterate in a literate society is to be cut off from learning, published information, poetry and literature, the world of books, most of the Internet, and the protocols and regimens of bureaucracy that govern public life. You cannot apply for a job. What work can you do if you can't read and write?

The Technology of Literacy

A World Without Writing

Imagine for a moment an entirely nonliterate world. What kind of world is it where the written word is not the dominant form of communication? What can and cannot be done? Large bureaucracies are out. So is written law. Libraries? Archives? Gone. So are the Internet, texting, and many of the other ways we communicate with each other. Schools as we know them don't exist. Instructions are memorized and conveyed verbally. Knowledge can be recorded only with images. Is the smartphone possible? Perhaps it can be operated, but it can't be well utilized or maintained without writing.

---

[1]Each line translates into English as "You need to call your mother and tell her you love her."

It's almost certain that everyone's social world would be smaller. Face-to-face contact and communication would be required for both practical and emotional interaction. Writing is so integral to the operation of large-scale organizations that mass-transit systems, the production and distribution of automobiles and petroleum products, the maintenance of highways and traffic safety, to say nothing of retail stores, hospitals, electric and other utility companies, and the postal service (Oh well, who's sending or receiving letters?) would be impossible in their present form. Virtual communities and social networking using the web would cease. The "cloud" would evaporate.

Literacy and Power

Imagine a different world, one in which only a small portion of the people are literate. Everyone else depends on them for writing. The elite communicate with each other through the written word; their world is built and supported by writing. For everyone else, participation in this world depends on the consent and help of this elite. The literate elite hold the power. Now, consider the opposite: the literate are a tiny minority of ordinary people with a special ability to read and write, a set of skills needed by the elite. How can the elite make the literate few do what they cannot do, for their benefit? The literates could be enslaved, made a special caste with limited rights, with privileges forthcoming only when they do the elite's bidding. Even using brute force, this would be a difficult relationship to maintain.

For many centuries in ancient China and other empires and in Medieval Europe, only a small, privileged minority were literate. Religious coteries benefited by having this skill and occupied an influential place. Those who were not literate worked for the benefit of the literate elite. This began to change when Johannes Gutenberg first printed the Bible in 1454. For the next hundred years the Bible's increasing availability became a stimulus to widespread literacy.[2] From the fifteenth century on, thanks to the printing press, varying religious interpretations proliferated, along with sects and religious faiths. They pursued their own interpretation of a Bible they could now read. Social change speeded up.

Literacy can be a technology of power. Recognizing this, in 1825 Chief Sequoyah created a syllable-based alphabet, as well as a literacy program. It was adopted by his Cherokee Nation to provide protection against the

---

[2]Andrew Pettegree's (2010) history of the first years of printing shows how less edifying uses of the new technology included dubious certificates, family documents, posters, bits of news, almanacs, poems, and hoaxes. Only about 5 percent of the books printed were by well-regarded authors.

ravages of government policies and the drive to take over the Indians' territory, with limited success. Illiteracy, the denial of this technology, is a means of social control. Is it any wonder that in the states of America where slavery was widely practiced, it was illegal to teach slaves to write? The slave revolts, like the one led by Nat Turner in 1831, and the anxiety provoked by a growing movement to abolish slavery led to the enactment of even more laws to suppress literacy among slaves. In the fifteen years before the Civil War, all education for persons of African descent, free and slave, was prohibited in the southern states (Genovese 1976).

Literacy and Social Change

The fight against European colonialism throughout Africa and Asia was led by those few indigenous people who could read and write, individuals who read about and absorbed the ideals of freedom, democracy, and national independence. Highly literate leaders like Julius Nyerere in Tanganika, Mohandas Ghandi in Bengal, Kwame Nkrumah in Ghana, and Ho Chi Minh in Indochina (Vietnam) could not only articulate but could write about and publish their grievances and ideas for their nation's future, developing and communicating revolutionary agendas at home and abroad.

Today, among young women ages 15–24, almost one in three in sub-Saharan Africa and one in four in South Asia are not able to read and write. Illiteracy has contributed to their subjugation and the denial of their rights. For the hundreds of millions of illiterate and poor people worldwide, especially women, in low-income countries today, the situation is dire, but it is promising. In Lesotho, 92 women out of 100 were literate in 2000, and 98.5 percent of women ages 15 to 24 had basic literacy. The trend is positive for other African countries as well. The improved status and prospects for women are changing, driven by their increasing literacy. In India, Kenya, and elsewhere, rising literacy among women is a major contributor to their adoption of modern birth control methods and a precipitous decline in birth rates so important to the improvement in their life chances. When people become literate their lives change.

Iris Summers and those around her knew their future depended on getting an education, i.e., enhancing their literacy. Participation in civic life, economic improvement, social respect, and the ability to decide for oneself depends on possessing the technology of literacy. Every person, group of people, and nation recognizes this. Supporters of education, voiced in 1850 by the movement for universal primary education in the United States, emphasize literacy and numeracy as critical to democratic participation and a healthy economy. The high level of literacy in Taiwan, South Korea,

Hong Kong, Japan, and Singapore in the 1950s and 1960s was critical to their economies' adoption of manufacturing technologies and participation in global markets that stimulated rapid economic growth and a dramatic rise in their standards of living. Every nation, from Mexico to Mozambique, Cuba to Kazakhstan, and Tunisia to Thailand, sees literacy as a first and vital step in improved well-being, economic development, and participation in a rapidly changing society.

The strongest economies today are not only those that process raw materials and make things. They also provide services largely based on the use of information. Sometimes called a postindustrial economy, as many as 70 percent of the working people in affluent countries are engaged in some form of *information* manipulation or communication rather than manufacturing, extraction of raw materials, agriculture, or construction. The information is written, if not on paper, then digitally. Digital communication has changed the way we communicate to the point that our lives are truly different from before. It builds on the legacy of social history in which literacy has been a vital technology.[3]

Understanding Technology as an Agent of Social Change

Technology looms across disciplines as a source of social, economic, and political change. It is often the master variable that explains everything (Herrera 2006: 3).

China invented and developed a wide array of devices and other technologies, later adopted in Europe, that were critical for technology-driven social change: the compass, gunpowder, chain-driven transmissions, spinning and paddle wheels, iron suspension bridges, printing, paper, and many more (Mokyr 1990: 209; Volti 2001: 69). This contrasts to Europe where "during the thousand year period from the fourth to the fourteenth century, technique progressed little at all" (Herrera 2006: 119). That was to change dramatically during the last centuries of the millennium, with Europe, the United States, Japan and other industrializing nations becoming the initiators of technologically driven social

---

[3]Marshall McLuhan (1962), one of the most creative, insightful, and provocative observers of information and social life, was equally aware of the negative consequences of our reliance on the written word, arguing that it fragments knowledge and promotes an abstract relationship between people and toward the sensual world. McLuhan, a writer by trade, coined the term "alphabetic man" to express his skepticism.

change worldwide. Much of this was a consequence of the growth of science and its transformative application into technology.

From Stirrups to Cities

The history of the stirrup, the loop or pedal hung from the sides of a saddle to support the rider's foot, is the quintessential favorite case study of technology and social change. As the historical record shows, saddles without stirrups are a very old device, dating several centuries before the birth of Christ. Saddles were introduced to the Roman Empire from the East and replaced the horse blanket and riding cushions during the first century CE, but armies used only small numbers of horses in battle for several more centuries. Large numbers of cavalry did not become the dominant feature of war until the eighth century. The turning point, according to historian Lynn White (1962), came with the development of the stirrup for military purposes. The social changes this wrought transformed the lives of everyone in Europe and reconfigured Europe in a prelude to the formation of modern nations.

The impact this device had on social change is a dramatic story, though not a sudden event. The stirrup's development covered a great deal of territory. The foot stirrup is found in archeological records from South Asia dating 2,500 years ago, but it is to China (in the fifth century of the Middle Kingdom) that the invention of a stirrup useful for military combat is credited. Its use spread through what is now Afghanistan and Turkistan, from China to Korea and Japan, on to Turkey, and by the seventh century to Persia (modern-day Iran). White (1962: 24) concludes that the stirrup's use spread across present-day Europe to Iberia (present-day Spain and Portugal) as well as Byzantium in the early eighth century. The stirrup's military use is important because it facilitated the seizure of vast tracts of land then owned by the Church of Rome. This seizure of land marks the onset of the form of social, economic, and political organization that dominated Europe for the next centuries: feudalism.

It is no coincidence. The stirrup made possible a significantly more effective means of coercion, but not without new costs. Using horses specifically trained for warfare made necessary large landholdings for hay and pasture. Land that had belonged to the church was confiscated by secular warlords and given to warriors who, in turn, pledged to provide mounted warriors (knights) when called upon by their lord. The feudal lord owned the knights' land and commanded authority by letting his knights use the land. From this emerged the codes of chivalry (honor and prowess in combat) and fealty (loyalty) that guided professional soldiering into the twentieth century. "The duty of knight's service is the key to feudal institutions. It is 'the touchstone of feudalism . . . Its acceptance as the determining principle of land-tenure involved a social revolution'" (White 1962: 31; quoting Cronne 1939: 253).

The story of the stirrup illustrates only one reason for a technology's adoption: the coercive advantage it provided. It points to the central role of violence and war in the formation and perpetuation of political units, e.g., the Mongol Empire across Asia and the monarchical states in Europe. The stirrup gave power to an elite group, a social class who then held sway over the mass of peasants and through the centuries instilled in them an ideology of legitimate rule. In time, the stirrup was matched and bettered by new technologies of war, and it was refined for less bellicose endeavors. Swords into plowshares? Ironically, the sword's development helped to stimulate the working of iron itself, which was tapped for agriculture through the development of the heavy plow.

In feudal society, land was wealth. Working the land occupied nearly everyone. Those in authority were the few who controlled and lived off the agricultural labor of others. Obligations tied everyone together. Often termed the manorial system, peasants (serfs) were tied to the land they farmed and obligated (taxed) by working on roads and other basic infrastructures. Landholders were obligated to those in greater authority for taxes and providing troops—including mounted knights—for battle. It was the ability to coerce others and the obligation to join in the coercive endeavors that held feudal society together.

Feudalism, originating in the eighth to tenth centuries and lasting in some areas into the eighteenth century, began with the stirrup and was solidified with the iron plow. By harnessing several oxen to a deep-digging plow, heavy, fertile soil could be farmed. This yielded a higher productivity that allowed horses rather than oxen to be used (breeds similar to the heavy war horse used in "mounted shock combat"), in turn increasing agricultural yields. Along with the development of the three-field rotation system of winter planting, spring planting, and fallowing, fields were enlarged and planted in long strips rather than small plots near peasants' homes.

Horses gave both greater productivity and greater mobility. With the adoption of the shoulder harness (originating in China) that replaced the wooden ox yoke, horses could be hooked to wagons. Goods could be moved more speedily over longer distances, and people were able to move from hamlets of five and six homes to small towns from which they traveled to their fields, often by horsepower. Some scholars, including Lynn White (1962: 66–67), speculate that this is the origin of European urbanization.

The Twentieth Century, an Age of Technological Change

Iris Summers' life was filled with a seemingly endless stream of technological innovations. When she was born, radio was a distant luxury, television was something for science fiction, and proto-computers were mechanical devices with no practical use. The airplane and automobile were in their infancy. Roads were primarily for horse-drawn wagons, carriages, and people on foot. Movies were silent, short, and black and white. Most babies were born at home, childhood immunizations were rare, and the cause of death for most elderly persons was "old age." Albert Einstein had just quit his job in a patent office to take a university professorship. No one had split the atom; nuclear power and nuclear weapons were unimagined. In football there was no forward pass, and there was no jump shot in basketball.

Though industrial production was overtaking agriculture in the national economy of Iris Summers' childhood, Clair Brown describes the "bleakness of everyday material life" for most people in the early part of the twentieth century. The decades of Iris Summers' life involved dramatic changes in living standards, much of it driven by "mass production and continuous automation; technological innovation; credit and mass marketing; worldwide communication and information systems; [and] integration of the global economy" (Brown 1994: 6).

Automobiles and their demand for roads, parking space, servicing, and fuel came to dominate the man-made landscape. In homes, wood-burning stoves gave way to gas stoves, electrification brought lights, washing machines, refrigerators, and a seemingly endless parade of appliances to sweep the floor, make coffee, play music, and announce the news.[4] Bathrooms gradually moved indoors, and hot water became available at the turn of a tap. Air conditioning invited millions of people to move to Florida and the Southwest. Telephones provided instant communication over thousands of miles, and air travel became commonplace. In her lifetime, Iris Summers saw a man walk on the moon. Hubble's photos revealed billions of galaxies and unimaginably huge star nebulae in an expanding universe at least fourteen billion years old. Medical technology kept Iris Summers alive when her

---

[4]In 1939, the first year of electricity in rural Henry County, Indiana, only 11 percent of farms had indoor toilets, but two-thirds acquired washing machines. Nationally, half of all newly electrified farm families almost immediately bought washing machines and a quarter bought refrigerators. While urban women "worked just as many hours after electrification . . . farm women permanently escaped from much heavy labor" (Nye 1990: 319, 326).

mother and grandmother would have perished. Her life was saved, as well, on film by friends, children, and grandchildren wielding tiny cameras that replayed images—moving and still—instantly.

Technology is an integral part of human history and provides the markers for eras and periods dear to the evolutionary-systems perspective: e.g., the stone (tools) age, irrigated agriculture, the (smelting of) bronze age, the iron (plow) age, steam (industrial age) power, the internal combustion engine, and digital (information age) technology. These technologies were so powerful in their ability to provide advantages over their predecessors and had such a ripple effect on social and economic activities that they seem to mark unmistakable levels of progress in the development of civilization. Indeed, technologies seemed for centuries to be moving in a linear path, from simple to complex, and to be vastly superior to what came before. Only with a more circumspect view, and one that questions the idea of modernity itself, could the image of technological progress be questioned or challenged.

The twentieth century can be thought of as the age in which technological democracy flourished. The widespread and popular acquisition of new consumer products became a central focus of economic life and leisure. It was not without a price, however, and many of the challenges of the twenty-first century will involve a reorientation in the use of energy, a reversal of patterns of environmental depletion, damage, and pollution, the disposal and recycling of manufactured objects, and a rethinking of the relationship of technology to quality of life as people across the globe acquire and increase their capability to participate in new technologies.

A Changing Social Reality

One of the most fascinating features of people and central to their social world is the many different ways individuals see and think about the world around them. The way we speak, act, and think is "bound up with the adoption of new instruments" according to philosopher J. Z. Young (in McLuhan 1962: 6). When technology alters their world, perceptions change. What was once unimaginable becomes commonplace and taken for granted. We see with new eyes.

In some respects technology has a way of discarding differences and homogenizing culture and social life. Grandparents in Germany watch their grandchildren frolic before the Christmas tree in a living room in the United States. When the children look at the large flat screen that is the computer monitor, they see their grandparents watching them. The children shout out their excitement with anticipation of opening their presents. They talk back and forth; the grandparents speak in German, the children understand but answer in English. Everything in the children's room, everything in the kitchen, and everything in the living room is familiar to the grandparents. The children could be opening their presents in Munich, and the picture would look much the same.

In recent decades the pace of life and expectations, especially for communication, have quickened dramatically. Thirty years ago an American professor on sabbatical in East Africa could count on six weeks between sending a letter to colleagues and getting a reply. Now she Skypes in real-time video nearly every morning with someone at her department in the States. She and her colleagues use tiny, hand-held computers for work and play. They communicate endlessly by e-mail, text messages, and social networking sites. Has the world gotten smaller? Physically? Socially? Culturally? Personally? It seems so, but their access to the world makes their world larger, too.

Technology as Device, Activity, and Social Organization

What Is Technology?

Technology is the means to act upon the social and physical environment in order to solve problems or meet a felt need. In its simplest form, technology is "the application of knowledge to the achievement of particular goals or to the solution of particular problems" (Moore 1972: 5). Three things constitute technology as understood in social science. First, some technologies are physical objects (e.g., a smartphone) that embody a great deal of knowledge. We usually think of technology in this sense, as a physical thing, a tool, instrument, machine, or device. It is, but it is more than this.

Technology includes the knowledge to operate devices as well as skills, routines, and methods. The second form of technology is technique, a way of doing something, such as communicating by written language, using a method for growing crops, or building resonant violins based on well-kept secrets. "Like science, culture, and art, technology is something we *know*, and technological change should be regarded properly as a set of changes in our knowledge" (Mokyr 1990: 276). Finally, some technologies are an organization of practices. Housing loans and mortgages are useful devices for allowing people to simultaneously occupy a house, condominium, or apartment and pay for it at the same time. The mortgage financing system, too, is a technology.

Technology as a System. Technologies are best understood as agents of social change by thinking of them as systems, not as objects, bits of knowledge, or discrete practices. As well, no technology operates outside a system that supports it—in its development, operation, and maintenance.

Rudi Volti, in *Society and Technological Change*, explains that technology is "a system based on the application of knowledge, manifested in physical objects and organizational forms, for the attainment of specific goals." As a *system* it is "not just material artifacts [but includes] human skills, organizational patterns, and attitudes," specifically a rational outlook that is "essentially optimistic" (Volti 2001: 6, 11).[5] A laser scanner, as much as an iron plow, is imbedded in a network of social and economic relations that make its use possible and effective. This network, as well as the skills to construct, maintain, and use it are all part of laser-scanner technology as a system. Langdon Winner similarly sees technology as interlocking material objects, activities, and social organizations, networks, and social arrangements. "Apparatus, technique, and organization are interdependent, that is, reciprocally necessary for each other's successful operation" (Winner 1977: 200).

Geoffrey Herrera's description of the railroad as a technology that drove social change is worth quoting at length. The railroad is

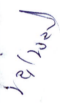

> a complex sociological-technological system . . . [T]he railroad as a technology includes not just the wooden ties, steel track, locomotive, and rolling stock, but also the legislation that creates the funding, the timetables and schedules, the market and logistics for the procurement of supplies, coordination between state and civilian authorities, and so on. (Herrera 2006: 35)

Technology as a Social Phenomenon. In this chapter, the role of technology as a force for social change is examined and analyzed. Any technology includes a social network of creators, suppliers, adapters, users, and the much larger number of people whose lives it affects. These people and their organizations operate in a network of resources, skills, applications, economic and political support, and cultural understandings. The *social* nature of technology is emphasized throughout the chapter, including the social context from which new ideas, devices, and skills emerge. Computers and the Internet remind us that a technology acquires new meaning and is adapted to the active interests and imagination of its users, i.e., through their agency. People have an uncanny knack for finding

---

[5]Langdon Winner summarizes the "technological habit of mind" as (1) seeing systems and parts, (2) using reason, (3) and attitudes that one can make things work right and one can do it better (Winner 1977: 200).

a use for something that its originators did not intend. As mentioned earlier, young people had ideas about using the first personal computers far different from the electrical engineers, mathematicians, and military strategists who developed the computer. Computer games and social networking were the furthest thing from these forbearers' minds.

Technological Change and Social Change

How do technologies change social life? Each technology tells a story somewhat different from the next, but there are significant commonalities in the ways social life can be driven by technology. One way is the changes in technology that alter the use of time and effort. Think of the washing machine that replaced scrubbing boards and wash tubs. The time (and effort) saved, especially for rural women, was dramatic. Computer word processing eliminated the retyping of entire legal documents that had to be error free. The power saw, photocopier, front-end loader...; the list of time- and labor-saving devices is endless.

Technology expands and creates new demands and opportunities for new commodities. The player piano, first run by a foot treadle, spawned a huge market for sheet music. Soon every middle-class home could have live music. Interestingly, as the popularity of the piano increased, it was replaced by the violin as the preferred instrument for young middle-class boys, and girls became the more likely recipients of piano lessons.[6] Rudi Volti gives the example of the radio, whose "development was 'pulled' along by the demand for better methods of communication" (Volti 2001: 195). Radio's potential to entertain and inform greatly stimulated its development into an affordable commodity, to be enjoyed at home and in cars, and within a few years—with the development of transistors—everywhere radio waves could reach.

New economic activities and occupations emerge as technology creates value. Something useful (and marketable) is done that wasn't done before. The U.S. *Directory of Occupations* offers a fascinating walk through technological change. A lighting assistant on movie sets is a "best boy." (You have probably seen this in movie titles.) The name comes from a master craftsmen's chief apprentice; it migrated into the age of electricity and then movies. The names of those who make scaffolding, handle booms, and

---

[6]Coincidentally, at about the same time the preferred colors for boys and girls, pink and blue, switched. Now, 150 years later, eyebrows would be raised if a boy baby in the United States was regularly dressed in pink.

construct rigging in moviemaking, "grip" and "key grip," come from the circus and earlier Vaudeville stage performances. In contrast, "computer data librarian" is a new occupation and even newer is "videographer." The *Directory of Occupations* fifth edition for the first time divides recycling workers into specialists and laborers, a sign of changing times.

Technologies create new sources of power, and the control of science and technology enhances the power of those who possess them (Moore 2008). The stirrup initially shook up the distribution of power by investing its users with a better way to deliver violence. The technology of nuclear weapons had much the same effect in international relations. The prospect of deterring aggression by possessing nuclear weapons makes them a desirable addition to the arsenal of some countries, including those that consider themselves vulnerable to attack. In turn, nuclear weapons nonproliferation treaties, the U.N.'s International Atomic Energy Agency, and the International Campaign to Abolish Nuclear Weapons challenge such efforts, fearing the prospects of nuclear war.

Mushroom cloud of an above-ground nuclear weapons test.
Source: Thinkstock/Photodisc/Digital Vision.

It is important not to exaggerate technology to the point where it alone, outside the cultural, political, and economic forces impressing themselves on us, appears to make people change both their behavior and views of one another. David Brooks comes close to doing just that in his discussion of cell phones and courtship. Reading about sex diaries published by an online magazine, he discovered how people, especially young adults, use cell phones to text "multiple possible partners in search of the best arrangements." The cell phone is not just a device for doing what people want to do. It induces a lamentable shift in social norms.

> Technology . . . dissolves obstacles . . . The opportunity to contact many people at once seems to encourage compartmentalization, as people try to establish different kinds of romantic attachments with different people at the same time . . . It seems to encourage an attitude of contingency. If you have several options perpetually before you, and if technology makes it easier to jump from one option to another, you will naturally adopt the mentality of a comparison shopper. (Brooks 2009)

Sometimes called technological determinism, Brooks underestimates the role human agency plays. Rather than talking about technology "naturally" causing the adoption of new behaviors, it is more accurate to consider how technologies provide new things to do. People recognize new possibilities in a technology and pursue them, sometimes to their benefit, sometimes to the benefit of others, and sometimes to no one's benefit—an outcome obvious only in hindsight.

Though we often speak as if a technological innovation comes into being and thereby changes the way we live, this is too simple a way to think about technology and social change. "We see reciprocal, reinforcing relationships between technological change on the one hand and social change on the other" (Volti 2001: 188). Social change is an ongoing process, whether the change is rapid or slow. Technologies emerge in changing circumstances and influence the speed, scope, and course of social change, based not only on technological possibilities but also human agency and the constellation of interests—often commercial, sometimes military—that are in a position to guide a technology's development and adoption.

David Noble's *America by Design* examines the rapid development of "science-based industry" including the creation of "scientific technical education and professional engineering" in the latter part of the nineteenth and early twentieth centuries. Edison's, Bell's, and others' laboratories provided much of the driving force behind the economic growth built around new technologies.

The electrical and chemical industries formed the vanguard of modern technology in America . . . they fostered the gradual electrification and chemicalization of the older craft-based industries, [e.g.] . . . petroleum refining, wood distillation, extractive and metallurgical, sugar refining, rubber, canning, paper and pulp, photography, cement, lime and plasters, fertilizers. Later these were followed by steel, ceramics and glass, paints and varnishes, soap, leather, textiles, and vegetable oils. (Noble 1977: 18–19)

Corporations, able to afford the research and powerful enough to shift university education and political practices in their favor, cultivated these technologies and in so doing contributed much to the social changes of the twentieth century.

INSTRUMENTAL AND TECHNICAL RATIONALITY

Abundance in affluent societies, made possible by technologies, is increasingly contradicted by the depletion of nonrenewable resources, environmental pollution, and personal questions about satisfaction and identity in a material-driven life. In the last century, Max Weber and more recently Herbert Marcuse (1964) and Jurgen Habermas (1970) have written about the disparity between instrumental and substantive rationality that helps to sort through this conundrum and understand why people who have fewer things often seem to live much better than those who have more.

Technology is an excellent way to get from point A to point B. There is water several feet underground; a pump can bring it to the surface for a garden and household needs. This is *instrumental rationality*. In military affairs and international relations, it is equivalent to tactical reasoning. Making a safer, more affordable, more easily ridden bicycle involved thousands of decisions made on the basis of instrumental reasoning.

Reasoning about the "bigger picture" involves *substantive rationality* or, in military and diplomatic terms, strategic reasoning. There is more ambiguity and uncertainty in making these kinds of decisions. In particular, questions arise about the wisdom or value of getting from point A to point B. Trying to reason about this requires going beyond what *can* be done—with materials and know-how—and enters the area of what *should* be done. Much of the decision making about technology assumes that whatever can be done should be done. Yes, it is possible to get more water out of the ground, but is this always a good idea?

Technology and Science

Pure and Applied Research

Social change does not spring directly from scientific understanding. Gradually throughout the nineteenth and twentieth centuries, scientific research provided the knowledge that became part of new technologies. Technologies themselves, often worked out in small workshops of clever tinkerers, mechanics, and inventors, were built into the machinery that ran early factories and the goods factories produced. The products became the items of everyday life—embedded in appliances, buildings, luxury goods, apparel, and, of course, trains, automobiles, and airplanes. Because the products of scientific applications have had such a strong influence on the changing world, and because we now live so intimately with science, it is important to understand something about scientific inquiry in a changing world.

Pure Science. A great deal of scientific research has been carried out over the past three hundred years with no intention of having a practical application, and no research is guaranteed to generate something useful. Scientific inquiry is guided by curiosity and creativity to explore the human condition and beyond, from subatomic physics to intergalactic astronomy. While there are occasional lucky guesses, most research is carried out with dogged efforts that have gone down many wrong paths and generated far more failures than successes. It is not a very efficient process, but the sum total of scientific accomplishment has opened up a vastly more intelligent and informed view of who we are and what is going on around us.

New research gathers information, reconfigures what is already known, and fits discoveries into the webs of conceptual schemes, providing understandings of what makes things work. The findings may have a very small audience of highly specialized scientists. The complexity of the methodology and interpretation of the results are buried in scientific jargon and mathematical models, and they go largely unnoticed by the public. Its questions are those posed by the scientific enterprise itself, not questions raised by the proverbial "man on the street," the practitioner who needs immediate answers, or the would-be entrepreneur looking for that next big idea. At some point, however, there may be a breakthrough that gets the public's attention.[7]

---

[7]In order to procure funding, even pure research may need to publicize its findings, trying to convince others of the promise it holds for revealing something of general interest.

Typical is this case of biochemical research. For decades the atomic structure of glutamates, chemical messenger molecules that open channels of communication between brain cells—critical to all brain activity, including memory—eluded scientists. In 2009, three researchers finally mapped the entire structure or "tangle of protein strands coiled and folded together to form a molecule machine." The research was publicly funded and published in the journal *Nature*. The writers explained that after six years of work they began to make headway: "Then we worked like crazy men for about five years." They have since "deposited all of the structural data in a freely accessible public database." As reported, "Knowing the structure could help research find new treatments" for Alzheimer's disease, epilepsy, and schizophrenia (Rojas-Burke 2009: C2), but after all these years of work, even that is uncertain.

Science Applied. Scientific research is a luxury made possible by the affluence of the United States, France, and other wealthy countries. Poorer countries depend on this, sending their brightest and most ambitious citizens abroad for quality graduate education and professional training where cutting-edge science is done. The expectation is that this experience will be of use in the development of technologies that can solve problems and benefit the economy, health, and infrastructure of not just the affluent but those who need it the most.

When scientific research is directed toward solving a problem, it is usually referred to as applied science. The Office of Scientific Research and Development (OSRD), begun during the Second World War, undertook the Manhattan Project. The problem was to win a war, and the science explicitly focused on developing a nuclear weapon. As part of the U.S. and its allies' war effort, atomic bombs were created and detonated. They destroyed the Japanese cities of Hiroshima and Nagasaki in 1945, killing 400,000 Japanese civilians and effectively forcing the Japanese leadership to surrender.

In 1958, the U.S. Department of Defense established the Defense Advanced Research Project Agency (DARPA) in response to the Soviet Union's launching of *Sputnik*, the first orbiting spacecraft. The space race was on, and the United States was determined not to cede space exploration to its Cold War rival. Soon the National Aerospace Science Agency (NASA) was established to conduct most of this applied-science work. Hundreds of billions of dollars have gone into research and the various technologies of space exploration, many of which have generated devices and applications for other uses, with some having commercial value. Probably the best known of these is Teflon, the nonstick material applied to tiles of a spacecraft's outer surface, now found on the surface of pans and skillets.

State Funding for Science

From 1975 to 1988, Senator William Proxmire of Wisconsin held a monthly news conference to announce his selection of the Golden Fleece Award. The iconoclastic Proxmire made a reputation as a political maverick by highlighting what he saw as frivolous, bizarre, incomprehensible, or just plain silly research paid for with taxpayers' money. "Trying to find out if drunken fish are more aggressive than sober fish" was how he characterized one funded research project. Another study tried to find out "why inmates want to escape from prison" (Rudin and Chadwick 2005). Research studies of emotions, and especially love, were frequently objects of his ridicule, as was one that sought to "teach college students how to watch TV." The awards were always good for a laugh by striking the anti-intellectual funny bone. The laugh was on science's seemingly clueless practitioners.

Geoff Herrera (2006: 190) observed that "the development of new technologies under the impact of the science-industry-state collaboration has become routinized and state directed." The size and scope of this collaboration makes the late Senator Proxmire's concerns seem trivial or small-minded.

NASA and the National Science Foundation (NSF), National Institute of Health (NIH), the Environmental Protection Agency (EPA), the Department of Energy, to say nothing of research funded by DARPA, award tens of billions of dollars for research each year. Joyce Appleby (2009) estimates that between 1941 and 1960, U.S. government funding for research and development increased thirteenfold. By some estimates, the federal government funds 64 percent of all research done in the United States.

One example is medical research. Of the $94 billion spent on biomedical studies in 2003, the drug and medical industries spent $54 billion on drug, biotechnology, and medical-device research; the remainder was funded by NIH ($26.4 billion) and other public entities. NIH was the largest global contributor ($1.3 billion) in 2009 of funding for serious diseases—malaria, HIV/AIDS, tuberculosis, meningitis, leprosy, and dengue fever—in poor countries. The pharmaceutical industry ranked third in giving, after national funding (Sweden, the EU, Brazil, India, the United States, etc.) and the Gates Foundation (McNeil 2009).

Innovations like the Internet, created as a tool of national defense, were produced with government sponsorship and funding. Most funding with public money is in direct response to congressionally mandated issues, such as cardiovascular disease, clean energy, and the sequestration of carbon dioxide generated by burning coal for electricity. Funding for military-related research is a significant portion of the federal budget and has generated some important innovations and technologies for nonmilitary use. In car navigation systems are indirectly an outcome of military research on satellite communication. The design of commercial jets owes much to research and development of military aircraft.

The Science-Practice-Technology Nexus

The relationship among science, practice, and technology is a fascinating one. While curiosity in science and the complexity of building scientific understanding often seem far removed from everyday practices and needs, it is vital that much of science be done without too heavy an obligation to show immediate or obviously useful results. On the other hand, technology development has benefited from idle curiosity and very practically oriented mechanics who make things that work.

Science may play no immediate role in explaining how a technology is accomplished. In time, however, scientific work may inadvertently provide an answer to such questions. Volti's examples of iron smelting and aircraft flight are revealing.

> For decades iron was produced in blast furnaces even though existing levels of metallurgical knowledge provided few clues about what was actually occurring inside the furnace . . . Even today, successful aircraft are built in the absence of completely adequate theories of air turbulence or compressibility that would allow the design to optimal configurations. (Volti 2001: 57)

Many home repair novices have found themselves admiring their work and saying, "It works, but I'm not sure why." We can be fairly sure, however, that *someone* knows, could know, or will know in time.

Bacteriology. The operation of a particular technology, though perhaps quite satisfactory, may at some point give rise to questions that motivate inquiry under the controlled conditions of scientific research. Medical practice and public health are good cases in point. Predating the European plagues, contagion was believed to be the reason for the spread of disease. Ancient Greek and Arab scholars proposed the spread of disease by means of minute living creatures, too small to see, from an infected person to a healthy person. Ibn Khatima proposed the same process during the Black Death bubonic plague. Isolation and quarantine, and even inoculation, were practiced centuries before a germ theory of infection was established by Louis Pasteur and others in the 1890s. Without understanding the actual reasons for contagion, however, many bogus explanations were promoted that recommended useless or even counterproductive methods of prevention and treatment.

During the London cholera epidemics of the nineteenth century, the cause of illness was thought to be impure air, miasma. Miasma theory's proponents encouraged the city to drain cesspools of human waste, then collecting in yards and alleys, into the Thames River. This reduced the urban stink (cleaned the air) but thoroughly polluted the drinking water of hundreds of thousands of Londoners. Water-born cholera bacterium (germs) had a far easier time spreading from diseased victims to new victims, killing fifty thousand people in one epidemic.[8]

The clinical practices of Pasteur, Lister, and others stimulated scientific research that led to the field of bacteriology and the growth of chemistry and biology. Joseph Ben-David has explored this connection and found a consistent pattern of medical practitioners such as Koch and Semmelweis who had a background in scientific research performing experiments on their patients or analyzing their patients' cases in a rigorous scientific manner. He refers to their "hybrid roles" as a source of rapid discovery in medical science, possible only in places such as England where the line between university-supported research and medical practice could be breached in a kind of cross-fertilization (Ben-David 1960).[9]

Bicycles and Cars. Science is only one source of knowledge and understanding embodied in technology. Much discovery is neither accomplished through scientific research nor by people who consider themselves

---

[8]A fascinating account of the discovery that cholera is a water-born bacterium is Stephen Johnson's *Ghost Maps* (2006). Edward Tufte explains the brilliance of John Snow's identifying the disease by means of mapping in the second chapter of his *Visual Explanations* (1997).

[9]David Noble's (1977: 20–32) history of engineering education in the United States in the late eighteenth century recounts the university's blending of practical experience and rapidly developing scientific fields like chemistry and electricity, especially as a consequence of the 1862 Morrill Act that established land-grant colleges to advance agriculture and mechanical arts. Joel Mokyer (1990: 232) similarly describes how British science was "predominantly experimental and mechanical, whereas French science was largely mathematical and deductive . . . In the early stages of the Industrial Revolution, there was an advantage in having a science that was applied and down-to-earth and a scientific community that maintained close ties with engineers and manufacturers."

scientists. Typical is the bicycle's fascinating and surprisingly long history of development. Very little of this involved scientists directly. Instead, mechanics, engineers, and crafters in workshops in France, Germany, England, and elsewhere tried different designs, materials, devices, and mechanisms to create some very funny, dangerous, fashionable, fast, and infuriating two-wheeled machines that took a hundred years to gain wide popularity (Bijker 1995; Pinch and Bijker 1987).

In the process of finding a design that worked best for the general population, science played a part. Rubber tires were developed, as was a pedal-driven chain drive, front-fork-bearing steering, and dependable brakes. Accommodating women's clothing was a major challenge overcome by both bicycle design and new fashions. Trial and error, serendipity, common-sense reasoning, and the hope of financial success, not the scientific method alone, guided the bicycle's early history.

The new woman and her bicycle. A comic look from the magazine *Puck*, 1895. Library of Congress Prints and Photographs Division, Washington, D.C.

The demand for safe, maneuverable, and relatively comfortable bicycles became great enough to economically justify mass production. Their manufacture moved from small workshops to new factories that, in turn, developed techniques such as metal stamping rather than forging to produce bicycles more economically. Factory mass production lowered the cost and further added to the bicycle's popularity. (Ironically, within a decade the system of bicycle manufacture became the model for automobile production.) What had been developed as a sport vehicle emphasizing speed and skill became an item of everyday life by the turn of the nineteenth century. It not only provided a new leisure activity but also became a staple of the economy by moving people more quickly to and from work, facilitating jobs that had been done on foot or with a hand cart, and giving rise to businesses that supported bicycle production as well as the leisure of bicycling.

Considerable scientific research went into the metal alloys, the pneumatic tire, and lubricants that were the technology of the machine. Most of this was unseen by users, and much of the know-how that went into the bicycle had long been practiced in other applications. The same could not be said as confidently about the automobile.

The complexity of automobile technology, not only the machine that is driven but the entire technology of production, distribution, and use, depended on thousands of scientific experiments and discoveries. The operation of the internal combustion engine and the means of attaining optimal power in the combustion of fuel, the transmission and drivetrain, suspension, and electrical and other systems drew on decades of research on materials and mechanics. Efforts to improve the performance, durability, and production costs have continued to benefit from scientific inquiry and knowledge, much of which was a direct result of questions raised by the operation of automobile technology itself, as well as pure research that initially seemed to have no bearing on automobiles.

Innovation and Social Change

Technologies may emerge from a clearly understood need that motivates people to seek a better way to do what they are already doing or to find a solution to a recognized problem. The desire, by means of the blacksmith's forge, to produce blades that could better hold an edge, balance weight and strength, and take a blow without breaking led to centuries of experimentation with heat and cooling, ways of folding heated metal, mixing alloys, and adding ores that became the backbone of metallurgical technologies. The dominance of armored horsemen spurred the desire to find an innovative equalizer, in time satisfied by the longbow and the pike. An arrow from a longbow could pierce the horse's and horseman's armor; the pike could outreach his swooping sword and pull him to the ground where his heavy armor made him an enfeebled victim for slaughter.

Diffusion of Innovations

> *The diffusion of innovations explains social change, one of the most fundamental of human processes.*
>
> —Everett Rogers, *Diffusion of Innovations* (2003)

Several decades ago agricultural scientists began to systematically do what peasants and farmers have always done: select seeds with the best qualities for planting the next year's crop. The difference in the twentieth century was the commercial creation of hybrid varieties, based on Mendel's nineteenth-century plant experiments. The scientific development of new seed varieties held out possibilities for vastly greater yields per acre of land. In time this technology linked farmers to corporations that would supply the seeds each year, a harbinger of contract farming (a form of sharecropping) that dominates much of family farming today.

In 1928, Iowa growers were offered for the first time hybridized corn, a seed that was deliberately created on research farms. Farmers were encouraged to try the new seeds. Only a few of them did, on a small scale, but this provided encouragement for others. In time, neighbors planted the new seeds, and within a decade the hybrid varieties were widely used.

Bryce Ryan and Neal Gross (1943) studied two Iowa communities in 1941, by which time nearly all of the farmers were planting only hybrid seeds.[10] Ryan and Gross used retrospective accounts (recall) and public and private records to find out how the technology diffusion process had occurred. They postulated the famous "S curve" or sigmoid model of the

---

[10]Hybrid varieties can have several advantages. Plants are bred to be stronger (more plant vigor), more resistant to drought or disease, and faster growing, with greater uniformity that makes mechanical harvesting and processing easier. They can provide larger and more abundant harvests. Their main drawback is that they cannot reliably yield seeds that will guarantee a repeat performance for next year's crop. Rather than allowing "open pollination," seed production must occur for hybrids under controlled conditions, at research and development sites of universities and corporations. New seeds, not those of last year's harvest, must be planted to get the benefits of hybridization.

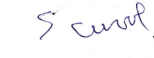

diffusion of innovation (Rogers 2003: 272–273). At first a few people adopted the hybrid seeds and most others watched from the sidelines. Then there was a rapid increase in adoptions, followed by a slower rate of adoption of the last holdouts.[11]

A study of the new tetracycline antibiotic in four Illinois towns in the 1950s (Coleman, Katz, and Menzel 1966) was considered for years a strong replication of the S curve process of the diffusion of innovation. In a replication of the tetracycline study, with more powerful statistical methods, van den Bulte and Lilien (2001) showed something else. In fact, antibiotics were not new to doctors in 1953. The maker of tetracycline convinced doctors to switch the antibiotic they were prescribing to patients. Aggressive advertising and inducements to doctors offered by drug company sales personnel, rather than the previously thought influences of social networks of physicians, were the most significant reasons for its adoption. Understanding technological change today cannot fail to take account of the efforts of large corporations that stand to gain from a technology's adoption. We return to this topic in Chapter 7.   Comp. Causes

Technology Adoption and Systems of Social Change. The adoption of new technology begins the process of social change, but other factors contribute heavily to the direction, speed, and scope of change. The growing of new varieties of hybrid seeds illustrates this.

Farmers cannot control the pollination process and must buy new hybrid seeds each year.[12] This, plus the pursuit of higher yields that require controlled inputs of water, fertilizer, insecticides, and herbicides for successful cultivation of the new seeds, significantly changed U.S. farming. Production costs increased rapidly, and huge yields depressed market prices. The plant technology became part of a complex technology system, resulting in significant social change far from the first Iowa cornfields.

---

[11]This S curve process, replicated in Deutschmann and Fals Borda's (1962) study of six innovations among Columbian peasants and farmers, stimulated hundreds of studies that supported the S curve model as a general picture of the process of technological diffusion (Rogers 2003: 65–68, 202–204, 268–271, 326–330).

[12]Genetic modification of food has produced varieties that resemble hybrid seeds in their desired qualities but can effectively produce seeds that will grow into plants identical to their parent. This quality has caused seed producers to restrict the saving of seeds, by both contracting with farmers to stipulate that purchasers would not save seeds and having laws passed that outlaw seed saving, both in the United States and in poor countries where international development projects promote genetically engineered varieties.

For centuries, farmers all over the world saved some of their annual harvest for next year's planting. They shared and traded varieties of seed grain with others in order to best accommodate environmental conditions and family needs. In the 1950s and 1960s hybrid varieties became the staple worldwide and the centerpiece of the "Green Revolution" that is credited with saving millions of people from starvation. Abandoning the long-held practice of saving seeds by planting hybrid varieties set off a worldwide transformation in farming, as it had done in the United States.

Hybrid seeds favored ever larger production and monocropping with larger and more expensive machinery. They helped push the price of land higher and began the onset of industrial agriculture. The higher expense of farming and land consolidation exacerbated rural landlessness and spurred rural-to-urban migration throughout the world. With few demands for rural skills in the city, poverty and urban squalor replaced rural self-sufficiency for millions of people (Glaeser 1987; Griffin 1974).

In the United States, the relationship between farming and the state changed dramatically, including the creation of a government price support system, with subsidies running into the tens of billions of dollars. Low prices to farmers that resulted from the surpluses of food also led to new legislation. It authorized surplus grain to be sent overseas as part of a revamped U.S. foreign policy following the Second World War. Public Law 480, often referred to as Food for Peace, became important in the mid-1950s for American efforts to win favor abroad—especially in poor countries—as part of the Cold War.

Food shipments had always been part of famine relief. PL480 was a new step. By sending surplus grain to governments overseas, it could then be sold in local markets to fund development projects, i.e., to build or improve schools, clinics, roads, businesses, and governance. Unfortunately, using food as foreign aid was sometimes devastating to local agricultural producers. Peasant farmers saw the market for their own commodities plummet. In many areas the food preferences of millions of people were permanently altered.

Technology and Western "Exceptionalism"

Why the West?

What were the "the major forces driving [and] . . . propelling Europe into commercial expansion and industrial capitalism?" This question, posed by Eric Wolf (1982: ix) in his *Europe and the People Without History*, has occupied social science inquiry—especially economics, sociology, and political science—for nearly two hundred years. Why did the speed of social change in the West (Europe and then North America), followed in the twentieth century by the Soviet Union, Japan, and other parts of Asia, accelerate at a speed unknown in earlier human history? Chinese and Muslim empires had sufficient protean technologies to launch economic expansion and industry. Why didn't they use them to develop the engine of economic growth, capitalism? Why did it happen in the less-technologically sophisticated West?

Sometimes called the riddle of Western "exceptionalism," several answers have been offered. In the West the power of governing and religious elites was relatively weak, providing opportunities for new social classes to pursue wealth through trade and commerce. Other regions are more vulnerable to external threats and challenges, but Europe's geography enabled the most ambitious men and women to take advantage of its favorable location and natural defenses. In Europe, new ideas were being formulated about the individual, society, and the state, ideas that encouraged the growth of liberty, representative democracy, and an economy that rewards capital accumulation, innovation, and risk taking.

Joel Mokyr, in *The Lever of Riches*, emphasizes the changing ethos, orientation, and outlook that had for millennia guided people in their relationship with the natural elements.

> After 1400 Europe's attitudes to the material world grew increasingly exploitative. . . . Neither the submission of humans to overwhelming forces of nature, nor their unquestioned dominance over nature in the anthropocentric view held by Western civilizations was prevalent in China. (Mokyr 1990: 227–228)

In light of this powerful underlying motivation, convenient geographic circumstances, weak conservative elites, and a level of structural instability that favored social change, technology was harnessed in pursuit of economic gain. Like the sorcerer's apprentice, it proliferated and became the means for creating and pursuing economic gain at a pace the world had never seen.

MAX WEBER ON THE MORALITY OF WORK

Why did Europe, at a particular time, actively pursue invention, science, and new technologies? What habits of mind, moral principles, spiritual values, personal identities, and ways of seeing the place of human

(Continued)

(Continued)

beings in the natural environment characterize such people and groups? Joel Mokyr attributes Europe's economic expansion to an "exploitative" attitude. Was there more to it than this? Max Weber's (1864–1920) famous thesis on the Protestant ethic and the spirit of capitalism has been a very influential, if often criticized, analysis that goes some way in answering this question. Weber attributes the West's exceptionalism to a unique religious doctrine, but not to that alone or for long.

The rudiments of modern banking, the legal framework for corporations and the stock exchange, and the institutional capacity to borrow massive amounts of capital required monetary, financial, and organizational innovations that greatly expanded the reach of national states and individual fortunes. These technologies, exercised with an eye toward amassing and making use of wealth, were to a significant degree contrary to Christian ways of thinking prior to the Protestant Reformation. For centuries the church didn't recognize the concept of private property, and it actively forbade usury.[13]

In *The Protestant Ethic and the Spirit of Capitalism*, written as a series of essays in 1904–05, Weber ([1920] 1958) describes how the Protestant reformer John Calvin (1509–1564) put forward a view of humankind's relationship with God that emphasized each person's spiritual accountability to God. The earth itself was God's, full of his wonders, riches, and great potential. Most significantly, Calvin taught that everyone born on the earth was either predestined to eternal life or not. In Calvin's notion of the "elect," not everyone could find everlasting life; only a select few were summoned to heaven.

After Calvin's death, in the teachings of Richard Baxter and others, it was widely believed that the righteous "elect" could be identified, most tellingly, by their pious lifestyle and the zeal with which they carried out God's intention to improve upon the world. God expected people to apply their efforts to his earthly gifts and in so doing to glorify him. Increasing the abundance of nature became a mission of devotion involving tireless work to build up and advance the production

---

[13]The Roman Catholic Church accepted the legitimacy of private property in 1136, and in 1563 the recently established Church of England did the same.

of worldly goods. Caution, however, had to be taken not to succumb to human temptations made possible by success in accumulating wealth. Not only idleness, the proverbial "devil's playground," but personal enjoyment of one's wealth was to be minimized. God had intended humankind to be the stewards of God's bounty. What was built was his, not for personal pleasure or to be used in hedonistic pursuits and an ostentatious lifestyle. Hard work became synonymous with "the Protestant ethic."

Though a great deal of twentieth-century historical scholarship has found flaws in Weber's thesis (e.g., Samuelsson 1961; Tawney [1926] 1954), its emphasis on an emerging habit of mind that emphasized hard work far beyond what is necessary for a secure life, and most importantly the accumulation and investment of capital in ventures that would generate wealth, became the ideal of successful business practice. To a great degree, by the turn of the nineteenth century hard work and thrift had become moral virtues for many Europeans. The seemingly ceaseless efforts in invention, science, and intellectual inquiry were very much part of the biographies of Darwin, Fulton, Bell, Huxley, and Max Weber himself. Some worked tirelessly for fame, some for power, and others like Darwin worked to solve the mysteries of nature. Work and success, not wealth, became the secular symbols of devotion and righteousness.

They, and many others in whom the spirit of enterprise was strong, were not Calvinists or even particularly religious, but they inherited a powerful sense of the value of work. In Max Weber's terms, nothing better exhibits the Protestant ethic than the legacy of intrepid discoveries, endless tinkering to make improvements, and long hours in the laboratory, store, office, and shop. The moral value associated with work has guided public policy since the industrial revolution for the great majority of people. The poor are commonly seen as people who don't work hard and therefore deserve their travails. "What does he *do*?" This is a question that probes the worth of an individual. Accomplishments are the measure of a person's character and social standing. No longer dominated by Calvinist asceticism, enjoying being rich, a possibility much enhanced by industrialization and the corporate form, took on new significance.

The pursuit of economic growth, which at the personal level is the pursuit of a higher standard of living, became more likely after the fifteenth century in Europe and more widely and strongly desired. It is not the desire alone, nor the means alone, but the combination and strength of the two that has driven much of modern social change. Daniel Headrick (1981: 9–10), in writing about technology, concludes that capitalism resulted "from both appropriate motives and adequate means . . . [T]he appearance of a new technology can trigger or reinforce a motive by making the desired end possible or acceptably inexpensive." Equally important, "a motive can occasion a search for appropriate means," i.e., technology.

As discussed earlier, Europe's empires prior to mercantile capitalism regularly engaged in plunder, land taxation, and tariffs on trade. The political system was organized for conquest and coercive domination. This is made apparent in the state's interest in developing military technologies. Conversely, the church in Europe, a wealthy landowning institution, had an interest in glorious monuments to its power. It used its wealth to foster architectural engineering and materials technologies that built the grand churches.

Productive technologies that would benefit peasant producers who were the vast majority of the population were slower to develop. As discussed earlier, these included the use of draft animals—especially horses—the means to efficiently use animal power with well-designed plows and harnesses, and a more productive technique for farming using the three-field rotation system. Once adopted, these technologies enabled higher agricultural yields and greater ease of transportation.

Towns and cities grew and fostered a more autonomous class of artisans and craftspeople, bankers and real estate proprietors, and merchants and manufacturers. They formed an emerging political class whose elite supported new forms of state authority. The new classes challenged the absolutist state, dissolving it or weakening it by setting off democratic revolutions (Moore 1966). Means and motive coalesced in a spiral of invention, discovery, technology, and science that has become the hallmark of modern times.

Technology and Economic Growth

"The majority of technological innovations are adopted by private business pursuing their economic goals" (Volti 2001: 79). Economic interests, especially those of profit-seeking corporations, play a major role in selecting, generating, shaping, and applying technologies of production and the technologies made available for purchase. As Noble (1977) shows, that has been the case in the United States for at least the past 130 years.

Human labor, skills, and expertise are the essence of economic activity. Labor's value depends on how well it is used, how much its products are needed and wanted, and how effectively it can be preserved in the products it creates. In the pursuit of economic gain, technology is favored that makes labor more effective and efficient. The process of economizing—getting the most output for a given amount of inputs—is enhanced by using technology. This is what makes the investment in technology so important in the expansion and profitability of corporations.

Robert Solow, Nobel laureate in economics, explained more than fifty years ago how technological change that increases the productivity of labor is the basis for most economic growth. Technology economizes on labor and should result in greater output and a higher standard of living. Quite simply, if a worker can produce more by using a new or better technology, and she continues to work as much as before, there will be more to enjoy and sustain her life and that of her family (Solow 1957). In William Easterly's interpretation of Solow's law, "It was the rate of technological progress that determined long-range growth of income per person . . . The only way that workers' incomes can keep increasing in the long-run: labor saving technological progress" (Easterly 2001: 52, 54).

Technology and Social Change in the Periphery

Imperialism and the Quest for Colonies

How can a nation grow larger? It can move its geographic borders outward, of course, though this will probably entail war with one's neighbors. Another way is to extend the political and economic reach of a nation by coercion in order to trade and acquire markets, without incorporating new territory into the nation. The British could boast that "the sun never sets on the British Empire," though the sun sets on the British Isles every evening. Eric Wolf (1982) dates Europe's expansion through imperial conquest from 1415 with the Portuguese seizure of Cueta on the African side of the Straits of Gibraltar. Columbus' voyages, followed by Spanish and Portuguese expeditions to the Caribbean, and subsequent voyages and conquests to what became Central and South America and the Philippines over the next 150 years returned with precious metals and gems that provided personal and state wealth. Imperialism spread as countries claimed the sole right to exploit regions far from their shores.

Not to be outdone, the Dutch chartered the East India and West India Companies and seized much of the Portuguese trade. The British formed their own East India Company and challenged other European powers' trade monopolies in South Asia, the Caribbean, Africa, North America, and later in China. Ports and sea routes were contested among Europe's powers, with vast armadas constructed to ensure safe passage of a nation's merchant fleet. Only very limited excursions were made into the interior of these regions; local groups were enlisted to extract wealth upriver. As brutal as these military conquests and engagements were, the greatest damage was in the millions of people abducted and sold into slavery as well as the scourge of disease spread by Europeans that killed millions of indigenous people.

What had begun as European trade in silk, tea, spices, and beautiful objects produced by the Mughal and Qing empires shifted to organized cultivation in South Asia of cotton, sugar, rice, indigo, and opium, products Britain needed to feed its growing urban workforce and products (especially opium) to be sold, often at gunpoint, in China. The British East India Company, initially granted a monopolistic trade charter, by 1765 took control of the civil administration of the British India Empire or Raj and evolved into an arm of the British government. It was disbanded in 1800 as Britain began to reorganize its global holdings (Wolf 1982: 244 ff.).

The date usually attributed to the onset of the Industrial Revolution, 1775, marks what Herrera calls "the second great wave of European imperialism . . . European powers conducted their last and most sweeping makeover of the world." By 1800, Europe's claims to other regions included 35 percent of Africa, mostly along the coasts, and Asia, mostly Indonesia, Bengal, and Southeast Asia. "By 1878, that figure had climbed to sixty-seven percent [of the world], and in 1914 it stood at eighty-four percent" (Herrera 2006: 106).

Technologies of Colonization. The pursuit of wealth motivated the initial plunder, conducted in the form of trade. The maturing of capitalism linked to imperialism shifted the strategy from plunder to colonization. Technology played significantly in this, allowing greater geographic penetration of Africa, Asia, and North and South America. Technologies facilitated the effective defeat and subjugation of the kingdoms and peoples who resisted, as well as effective coordination of military and commercial activities through new means of communication and travel. Beyond trade, imperial powers reorganized much of the societies they conquered in order to provide the raw materials to fuel the rising industrial system back home. Colonies became a source of new markets. The colonial enterprise itself became a major economic activity with demands to be satisfied by industry, including technical innovations that created entirely new industries.

Daniel Headrick's *The Tools of Empire* (1981) explains the importance of technology in three phases of European imperialism. Steamboats, gunboats, and quinine for the control of malaria were the most important in Europe's initial exploration and penetration of Africa and Asia. Next, improvements in guns facilitated the imperial conquest and rule of large parts of these regions. Finally, the telegraph and greater speed of ocean travel, along with railroads, were enlisted in the long-term exploitation of these areas for the benefit of rapidly growing industrial economies. Two of these illustrate the role technology played in this quest for wealth: guns and railroads.

Guns had changed surprisingly little between the first muzzle-loaded flint-action muskets in 1550 and muskets in 1800, despite the development of smokeless gunpowder and improved firing mechanisms. All of this changed by 1840 with the introduction of breech loaders, along with the development of percussion caps, rifling of barrels, and oblong bullets.

In attacking and occupying the Sudan, British troops confronted forty thousand Dervish fighters, fierce and skilled but without the quality of rifles used by the British. As chronicled by the young Winston Churchill, "the battle of Obdurman—the most signal triumph ever gained by the arms of science over barbarians" left eleven thousand Dervish dead. Churchill compared this slaughter favorably to the loss of only twenty British soldiers and twenty of their Egyptian allies (quoted in Headrick 1981: 118).

The railroad, too, was an enormous boon to colonialism. Geoffrey Herrera describes how

> . . . the railroad allowed European imperialists to create and secure markets and pacify large expanses of territory. Without it, the extension of European political and economic influence into the interiors of Africa, Asia, and Latin America . . . would have been all but impossible. (Herrera 2006: 113)

The railroads' purpose was to move raw materials from the interior to ports or navigable rivers from where they were shipped to Europe and to move manufactured goods into the interior for sale or barter in exchange for raw materials. Conveniently, in Africa, China, and India, troops could be quickly brought to areas where local people resisted the terms of trade that were highly favorable to the imperial powers' commerce.

The creation of a vast network of railroads at home and abroad required major support from the state. In Britain, legislation was used to create corporations, limited-liability companies that could raise private capital without risk to the individual fortunes of stockholders. Where this was insufficient to gather needed investment funds, governments built and became direct owners of railroads. This was done especially when financial returns from commerce were low, but an external power (e.g., Russia in northern Afghanistan) was seen as potentially threatening to the empire. More than economics, the railroad was a technology of state power.

Resistance to Technology or Resistance to Change

Resistance to technology is often personal. Gregory Han, a writer and editor who lives in Los Angeles, doesn't use a cell phone. He can certainly afford one, and his work could be done more conveniently if he had nearly instantaneous communication with those with whom he works. He hasn't rejected digital communication; he "makes a living blogging about interior design and tech gadgets," but he has chosen to be among the 15 percent of American adults who don't have a cell phone and the possibly 5 percent who do so deliberately. "It's a luxury not to be reached when I'm out and about . . . I feel I benefit by living in the moment and not having a ring or a buzz or an inclination to always look at the screen" (Miller 2009: B5). Han is bucking the tide and probably won't see his personal decision have a major impact on society.

In many regions of the United States and around the world, religious communities live with a minimum of modern technologies. The Amish in Lancaster County, Pennsylvania, the Amana Colonies in Iowa, Mennonites in Kansas, and Hutterites in Canada made an explicit decision hundreds of years ago, reaffirmed with each generation, to reject many of the devices and conveniences people elsewhere take for granted. Some have adopted electricity and the machinery that makes their livestock healthier, though they continue to drive buggies rather than automobiles and shun radios, television, computers, and cell phones.

Langdon Winner's description of Christian righteousness a thousand years ago might speak for the view of these contemporary groups. "That something might be useful or profitable to men did not make it right and just. It has to fit a precise conception of justice before God. When an element of technique appeared to be righteous from every point of view, it was adopted" (Winner 1977: 120). Otherwise, it can be done without.

UTOPIA, DYSTOPIA, AND THE LESSONS OF DR. FRANKENSTEIN

Utopia is nowhere. The word comes from the two Greek words for "not" and "a place." For more than five hundred years, philosophers and social critics have used the literary device of a traveler returning from a far-off place. Thomas More (1478–1535) did this in the first book by the name, *Utopia*, and so did the French utopians, including Denis Diderot (1713–1784) in his *Voyage to Bougainville*. Another device, a dream that seems like reality, also has been a popular way to critique and explore alternatives to the current reality, as William Morris did in his 1890 novel *News from Nowhere*. Edward Bellamy's *Looking Backward*, published in 1888, was enormously popular as a design for superior living through reason, practicality, and humane technology. A plan to actually create a utopia is an ambition in perfection.[14] Jeremy Bentham (1748–1832) gave it a try with his New Harmony community in southern Indiana. Religious groups like the Shakers established preferred ways of living communally; most were short lived (Nordhoff [1875] 1966). A standard feature of attempted utopian communities has been the limitation placed on technology, restricting its scope and role in favor of work, contemplation, and community.

The twentieth century has also been prone to utopia's opposite, dystopia (a bad place). Rather than limiting or humanizing technology, dystopias were imagined to be technological nightmares. They are political black holes with crumbling technology, like George Orwell's *1984*, but often they describe well-intended efforts in social change gone horribly awry, sometimes because the technology doesn't work as planned and more often because a rush to technology can only dehumanize, enslave, and find nefarious uses by evildoers. H. G. Wells' *Island of Dr. Moreau*, Yevgeny Zamyatin's *We*, and Ray Bradbury's *Fahrenheit 451* are of this mind. Aldous Huxley's *Brave New World* and *Brave New World Revisited* tapped a popular sentiment that technology itself is the barrier to a better life.

Why does utopia restrict technology, and why is it so much a part of dystopia? The dystopian sentiment, shared by many reasonable people, is highly skeptical about the ability of science and technology to produce genuine abundance with greater ease, open new vistas and modes of expression, and end the scourges that rob people of health and a long life.[15]

---

[14]In *Technological Utopianism in American Culture* Howard Segal (1985) examines twenty-five utopias, many of them inspired by Bellamy's novel, that use the device of utopia to critique the way their current society was thought to be falling short of its potential. Written between 1883 and 1933, they all emphasize technology as the key to progress.

[15]Two significant exceptions are Henry Ford's experiment in Brazil, Fordlandia, and Thomas Edison's plans to build a model community, complete with electricity of course, in Muscle Shoals, Alabama. These inventive technologists were anything but dystopian about what technology could do.

(Continued)

(Continued)

Perhaps the first, and certainly the first famous, technological dystopia, Mary Shelley's *Frankenstein; or, the Modern Prometheus,* offers an answer. Dr. Frankenstein's monster was more than a threat to life and limb. His existence embodied the hubris of science. To a romantic like Mary Shelley, science's advocates would do whatever they could, heedless of moral boundaries and the responsibility for technology gone haywire. A version of this sentiment continues to animate debates today over genetic engineering, cloning, and other biotechnologies.

Three things are captured in Mary Shelley's novel and repeated in science fiction today. First, technology is power. Like any power, it is subject to abuse. Second, technology has unintended consequences. It's a good idea to "look before you leap," as Iris Summers liked to say, but rarely can anyone see very far ahead where science and technology are concerned. And finally, technology can diminish the human character. There is something healthy, even glorious, in working hard, using mind and body, at one with the elements. Technology alienates individuals from one another and what Marx called their "species being," i.e., the positive things that make us human. It distracts and distorts and in time enfeebles us to the point that we become the subjects of technology, its captives.[16]

George Kateb (1972: 79–80) observes that it is technology's unknown and perhaps uncontrollable capacity for change that is its most frightening quality. Not knowing the future that science and technology seem eager to beckon is what makes Dr. Frankenstein's experiments so scary.

---

[16]A rather incredible, articulate rant on digital technology and its capacity to dehumanize is Hal Crowther's (2010) "One Hundred Fears of Solitude."

Japan's Return to the Sword

Technological change is often a subversive process that results in the modification or destruction of established social roles, relationships and values. . . . Technological changes, both major and minor, often lead to a restructuring of power relations, the redistribution of wealth and income, and an alteration of human relationships (Volti 2001: 18).

Just as religious leaders may successfully challenge a technology and block its spread to their flock, political elites may see a machine or device as a threat to their rule. The reversion to the sword, spear, and bow and arrow in Japan after the country was unified under the Tokugawa Shogunate is a classic case in point and one of the most-often-cited examples of social changes made possible by technology but rejected in favor of maintaining stability and order, for a time.

Firearms—guns and cannons, as well as land mines—were introduced to the Japanese islands in 1543 during the long power struggle among Japan's warlords (1490 to 1600). The Japanese's manufacturing works, skill in metallurgy, and level of technical expertise quickly focused on firearm production to develop high-quality armaments. By 1560 firearms were a central part of battles. The warlord Hideyoshi's invasion of Korea in 1590 included as many as 300,000 men, of whom a quarter had matchlock rifles (Perrin 1979: 27).

When the Tokugawa clan gained dominance over the other warlords, the unflattering notion that "efficient weapons tend to overshadow men who use them" (Perrin 1979: 23–24) gave impetus to policies ridding the country of firearms. The first to lose them were civilians who were compelled to contribute their guns to the effort to construct a massive statue of Buddha. Soon the new government gained enough control to regulate arms production and ceased issuing manufacturing permits. Its parallel decision to stop ordering firearms for its army essentially eliminated the market. As compensation, the manufacturers were offered social security and the honorific title of samurai.

After the Shimabara Rebellion (1637) in which upwards of twenty thousand Christian converts using firearms were defeated, the Tokugawa Shogunate ruled over two hundred years of peace. It closed Japan to most foreign trade and foreign visitors (especially targeting Christian missionaries) and remained without guns until the early nineteenth century. After attacks by Commodore Perry's fleet in 1854 that forced Japan to accept foreign trade, the country again began manufacturing guns and cannons, and by 1900 Japan was a heavily militarized nation capable of battling European nations for its interests.

It is unusual for a dominant elite to so effectively reject and undermine a once-used and effective technology. Key to the decision was that a small but powerful group of Samurai families strongly supported the rejection of firearms and a return to the sword. It was possible, in large part, because Japan is a series of islands that, at the time, could be defended without firearms. By emphasizing the cultural significance of the sword and through effective government steps to eliminate the firearms market—as well as cultivating the rejection of Western influences symbolized in firearms—the Tokugawa removed a major threat to their dynastic rule. For two centuries they slowed, but could not stop, social change.

Conservative Peasants

Since the 1950s, international-development specialists from wealthy countries have sought to convince peasants in poor countries to farm differently. Though barely scraping by, peasant farmers are often very reluctant to change their practices, adopt new seeds, designate a large portion of their harvest for sale, and with money from their sales purchase the fertilizer, herbicides, and insecticides the new seeds need to survive. The "conservative peasant" thesis was advanced to explain this seemingly irrational behavior: an unwillingness to accept the technologies of change, even if they promised more food, more money to spend, and a higher standard of living.

As James Scott's studies of Southeast Asian societies make clear, there are good reasons why poor peasants are skeptical of technological innovation. Their work is a "moral economy" that extends beyond the rational, contractual, and impersonal boundaries of the market (Scott 1976; 1985). It extends to issues of fairness, family obligation, and community well-being. They live, as R. H. Tawney described, like "a man standing permanently up to the neck in water, so that even a ripple is sufficient to drown him" (Tawney [1932] 1966: 77). They are skeptical or even hostile to those from outside their orbit who try to change their life but would suffer none of the consequences of failure. The authority of the past is strong, and it grips the present so as to preclude potentially threatening experiments and alterations in the way things have always been done. For people who have so little control over their lives, trying to plan for and control the future is only an invitation to disappointment. Worse, it could be fatal.

The beautiful terraced hills of Bali, an island in Indonesia that produces enough paddy rice to support millions of people, could be much more productive if new varieties of rice were planted. This was tried in the early 1970s in some areas, with disastrous results (Lansing 1987). Farmers quickly returned to their earlier varieties along with the previous mode of organizing, planting, watering, and harvesting coordinated by the local Hindu priests in the "water-temple system." The farmers' resistance was well reasoned, based on the sharing of water among many peasants farming a hillside, as well as the need to mutually control insects and rodent pests. The complexity of the agricultural system, though based on seemingly rudimentary technologies, required the close coordination of traditional experts—Hindu priests—and the willingness of everyone to do what was asked.

For thousands of years, traditional agriculture found ways to work with the vicissitudes of weather and disease in order to provide a satisfactory level of subsistence. Usually supported by a belief system embedded in local religion, communities vested authority in elders. There were extremely complex systems for allocating land and labor, a diversity of foods that satisfied dietary needs and culinary tastes, and obligations of reciprocity among groups in times of stress.

Terraced paddy rice in Bali

Source: Thinkstock/Goodshoot/Goodshoot

This gave people a rough but sustainable life. Resistance to change, as it appeared to outsiders who failed to understand and appreciate the calculations that go into life and death decisions, seemed irrational (Nair 1979). More than a few leaders of new nations forced Western or Soviet farming methods on the peasants in their countries, often leading to ecological disasters, famine, and national bankruptcy.

THE TECHNOLOGICAL FIX AS RESISTANCE TO CHANGE

America will invent its way back to prosperity.

—Senators Patrick Leahy
and Jeff Sessions

The idea that technology can fix what ails us reverses the normal way we think about technology and social change. Technology is a force for change, with adaptation following adoption and a rippling of social, economic, and cultural consequences. The technological-fix approach turns this around. When things are changing, especially for the worse, the hope is that a newly devised technology will come to the rescue, allowing us to continue doing what we have been doing, leaving our lives little changed.

A prime example is contemporary climate change. There may be political dispute, but among climate and other scientists it is well recognized that the Earth's average temperatures are increasing at an unprecedented rate.[17] This is almost certainly due to the increasing concentration of carbon dioxide and other heat-trapping gasses in the atmosphere and increasing amounts of carbon dioxide in the oceans. The primary culprits of unsustainable levels of carbon dioxide and other greenhouse gas emissions are the burning of fossil fuels and the cutting and clearing of trees and leafy plants that for millennia have drawn carbon dioxide from the atmosphere. Climate change is in part or entirely anthropogenic, man-made. What is needed, some say, is a technological solution to fix this problem.

A technological fix is sought in reducing the pollution caused by burning fossil fuels, particularly coal and petroleum products. The energy from burning less fossil fuel would be replaced by sustainable energy sources. The challenge is to improve the efficiency of renewable energy generated by wind, the sun, water and waves, and possibly bio-fuels. It's important that these alternatives not be inconvenient, unsightly, or too costly. They might require modest changes in lifestyle, as would minimal efforts at energy conservation and a reduction in

---

[17]According to the National Climatic Data Center (NCDC 2011), "The decadal global and ocean average temperature anomaly for 2001–2010 was the warmest decade on record for the globe…This surpassed the previous decadal record (1991–2000)." Also see fn. 16, Chapter 2.

material consumption of things requiring high energy inputs to manufacture and transport.

A technological fix is sought with even more fervor in finding ways to continue with current lifestyles. Carbon dioxide sequestration, i.e., compressing CO_2 gas and storing it underground in formations that previously held natural gas and oil, is a popular idea. That way, coal can be burned, power plants can continue to operate, hydrogen gas can be generated to fuel cleaner-running cars and trucks, and life will change very little or not at all.

As another example, the major health problems in affluent countries have been met with amazing discoveries in the life sciences that have been applied to innovations in health care. Federal and private research has produced drugs to improve both the quality and longevity of life. Of course, good health is more than taking a pill. Lifestyle factors—sensible eating, not smoking, regular exercise, less stress at work, getting enough sleep—are important keys to a healthy life. Wouldn't it be nice, though, to keep eating and doing what we know is not good for us? Poor health could be circumvented if the right medication can be discovered. Are you interested in a pharmaceutical alternative to working out in the gym?

The Great Recession to which senators Leahy and Sessions earlier referred was the consequence of a financial crisis. Institutions entrusted with huge amounts of the public's money invented exotic financial "instruments"—e.g., credit default swaps, collateralized debt obligations, subprime mortgages—that made fund managers, brokers, and banking executives fabulously wealthy. To get everything back to where it was before the Great Recession, the senators are sure that new technologies will spring forth. We can only hope these will be better than the innovations that led to millions of lost jobs, billions of dollars of unpaid bills, millions of home foreclosures, billions of dollars in lost pensions, and trillions of dollars of public debt.

The Global Spread of Technology

The acceleration of social change, driven in part by science and technological innovation in the West, is now a worldwide phenomenon. *Globalization* is a shorthand term to describe the linkages, networks of influence, and flows of capital, goods, ideas, and images around the world that expedite rapid change (Brown 2004, Part 5). Technology is a major influence on the direction and forms this is taking and will take in the future. Today, power asymmetry favors the West. The rules of commerce, the preferred political forms, and the flow of capital all favor those who originally developed the engine of growth, capitalism. The ground is shifting, but technology will continue to be a vital source of change, for good and for ill.

Technology Transfer

In most major universities in the United States, an office of technology transfer looks for ways to apply research done in university laboratories. Most desirable are applications of science that have commercial value. That way, some or all of the proceeds from the sale of the research findings, whether a new technique, device, or discovery, come back to the university and help support its budget. Technology transfer offices exist in governments as well as corporations for the same purpose. In an increasingly interconnected world, the transfers are often around the globe.

Technology transfer is only a special case of the diffusion of innovation. In fact, the stirrup qualifies as a technology transfer. Brought by nomadic invaders, it caught on wherever it went, giving the same advantage to those who adopted it. Archeologists are fond of mapping the diffusion of ancient innovations. So are art historians who study the materials used in works of art, paleobotanists interested in the spread of plant husbandry techniques, and architectural historians who trace the changing practices of building bridges, homes, churches, and fortifications.

In the world today, much of the workforce is involved in transferring information technology and providing information services. Giving expert advice and technical training, diagnosing problems and devising solutions, sharing ideas, and linking up those who know with those who need to know constitute a great many jobs, both in affluent countries and worldwide.

These jobs use, sell, and transfer information that is legally the property of whoever owns the rights to the information. The same goes for music, literature, and other products of creative and scientific endeavors. In the modern era, the transfer and use of technology is circumscribed by these commercially valuable rights. This is contested terrain around the world.

> The scope of the controversy [over ownership and use] is vast. It might encompass debates about ownership of the formula for an AIDS vaccine, a Miles Davis riff, a software algorithm, or a new way of uncorking a wine bottle. Each of these is an idea embodied in physical forms: formulas, notes, code, or drawings are turned into capsules, records, CD-ROMS, or corkscrews. The economic consequences of the dispute are immense. (Evans 2002: 160–161)

The Debate Over Technology Transfer

Because technology is a major driver of social change in poorer countries today, its transfer across international borders is a vital issue at the center of the ongoing debate about globalization. For convenience sake, two opposing positions about the direction in which social change is going have been simplified and designated as proponents and advocates. Any other term—e.g., *opponents, skeptics, critics*—would prejudice and misrepresent the differing positions, since both see their position as realistic, progressive, and welcoming of responsible social change.

Proponents. On one side are proponents of open markets and the minimization of obstacles to global trade and the flow of capital in search of raw materials, low-cost labor, assembly sites, and markets. Proponents tend to be multinational corporations and governments of affluent nations where most patents and copyrights originate, where research laboratories and commercial development spend billions of dollars and seek a profitable return on their investment. Proponents endorse what is sometimes called the Washington Consensus, guided by neoliberal economic practices that emerged with the implosion of the Soviet Union and the opening of countries to practices more amenable to global capitalism.[18]

Advocates. On the other side are advocates for poorer nations and rapidly growing non-Western countries, as well as international nongovernmental organizations and foundations committed to solving problems in poor countries. The advocates question or oppose the private ownership of many resources (e.g., water) and the commercialization of products of science needed to ensure basic human comfort and dignity. They endorse a state's regulation of the flow of capital and goods that might undermine or destabilize local or national economies and cultures. Some advocates are economic nationalists, looking out for their own economies by restricting what foreign corporations can do in their territory and using state planning to direct economic activity. Other advocates make a high priority of improving

---

[18]Among the ten principles proposed by John Williamson (1993), originator of the term *Washington Consensus*, are a "belt-tightening" of weak economies emphasizing tax reform, privatization of state-owned firms, ending regulations that restrict market entry or reduce competition, and legal protection for property rights. Often called neoliberal economics, it emphasizes the original philosophical liberalism that opposed interference by the sovereign (monarchs) in economic and private affairs. This can be compared to what has been dubbed the Beijing Consensus, discussed in Chapter 8.

the lives of the poorest of the world's people ahead of enhancing the affluence of others.

Proponents argue that international corporations will not support research and development in countries that do not ensure technology transfer will be backed by patent and copyright enforcement. Their investment will not provide a financial return needed to justify it. Poor countries will lose the benefit of the corporations' expertise and capital, jobs will not be created, and new goods and services will have to be imported rather than produced at home. If poor countries cannot afford the technologies, they can borrow needed funds, paying back the loan with the economic growth made possible by the technologies.

Advocates like Nobel Laureate Joseph Stiglitz (2002) argue that, because the playing field is not level, poor countries do not benefit by protecting others' copyrights and patents. Their own businesses are not able to compete with the more sophisticated and well-funded foreign corporations in their midst. Goods are more expensive when patents and copyrights give a monopoly to the (foreign) holder. People in poor countries cannot afford to buy more expensive goods, including pharmaceuticals, even when they need them. Finally, many things being patented should not be. The "global commons" should be available to all. This is especially the case with plants, seeds, and genetic material.

Much of the original material that has been turned into patented commodities of international agriculture originated in the southern hemisphere. The argument Bolivia, Brazil, and others make is that their vegetation was "stolen" and is now being sold back to them as genetically modified seeds and pharmaceuticals. Proponents and multinational corporations like Monsanto, Archer Daniels Midland, Roche Laboratories, and others argue that the cost of these commodities is in the value added by their research and development. Advocates in less-affluent countries have argued at every round of trade talks since 1994, and especially since the tumultuous 1999 WTO meeting in Seattle, that their own development should not be held hostage to legal decisions made by and for wealthy countries. To date these talks have stalled, in part over the issue of technology transfer.[19]

---

[19]The United Nations hosted two World Seed Conferences to find a way that patents can be honored in order to promote plant research and provide corporate access to genetic resources. Critics, such as those at the International Institute for Environment and Development in England, are advocating an International Seed Treaty that will protect local seed saving and plant diversity rather than promoting genetically modified crops.

To reconcile these often conflicting positions, international organizations, including the United Nation's World Intellectual Property Organization (WIPO) and TRIPS (Trade Aspects of Intellectual Property Rights), an agreement of the World Trade Organization (WTO), have been established. TRIPS stipulates that all internationally traded intellectual property must be in conformity with its guidelines for patents and copyrights and that trading nations must honor the property's legal status. The United States, European Union, and Japan are the loudest proponents for enforcement of the agreement. No wonder. Corporations based in their countries have the most to lose. U.S. companies earned $38 billion in royalties in 1996, and much more today, even while losing billions of dollars in patent and copyright infringements (Economist 2001).

The HIV/AIDS epidemic has been virulent in sub-Saharan Africa. For many years the globalization debate swirled around the provision of drugs to treat people infected with HIV and to prevent the transmission of the AIDS virus from pregnant mothers to their unborn children. Pharmaceutical companies hold the patents for the drugs, and the prices are often out of the reach of those who need them. Even with government subsidies and international support, the drugs are too expensive to afford. In addition, their expense drains away money needed to develop and maintain a sufficient medical distribution and support system for other medical needs.

In this case, some progress has been made. Drug companies have lowered their prices for drugs. They have worked out agreements that allow the sale of generic drugs, manufactured with their formulas but no longer under patent protection, alongside their own, and they have made large financial contributions to help fund the drugs' purchase and distribution. Like the opposition to the marketing of baby formula discussed in the next chapter, much of this controversy was led by the global social movement for AIDS prevention and treatment.

Trade and commerce, with the threat or use of the state's military force if impeded, retains many of the earlier features of mercantile capitalism and Western imperialism, hence the outcry often heard in poor countries: neoimperialism and neocolonialism. One important aspect of this, and an issue very much on the United Nations agenda, is the West's control of technology and the effect this has on economic growth in less-affluent countries. In the view of political economics, the tactics may have changed since the end of European, Russian, U.S., and Japanese imperialism, but the strategic advantages and interests of the West have not. Technology transfer is at the center of the debate over social change in poorer countries, and its resolution may require significant social and political changes in affluent countries.

International Development and Appropriate Technology

Several years ago an international aid organization decided to work on problems of food production in poor countries. It selected a region in West Africa populated by subsistence farmers whose gardens and fields produced only enough food for their family's needs. They used farming methods centuries old, without the benefit of draft animals or motor-powered implements. The foreigners saw this and decided that small garden plows with gasoline engines, rototillers, would help immensely. They brought rototillers to the poor farmers and taught the men how to use them. Almost immediately the size of people's gardens and fields expanded, promising more food for the family and extra food that could be marketed to raise cash for new purchases. It looked like a slam dunk for a new technology.

Some years later, representatives from the aid organization returned to the region to see how things were going. The rototillers were nowhere to be found. Had they been sold? Lost? Abandoned? The foreigners asked everyone and got no satisfactory answers. The fields looked well tended. They were small and hand cultivated, much as they were before the rototillers. What happened? After several days the answer came, with a surprise and a lesson about appropriate technology. The people from the well-meaning aid organization had made several mistakes.

First, they assumed that men do the farming. In fact, men help break the ground and might help in the weeding and harvesting, but farming is mostly women's work. In the household's division of labor, she raises the crops and feeds the family. He tends the livestock and is often gone with animals into the bush. Men had been taught by the aid organization how to use a machine for work they rarely did. The second error was to know little about what the people themselves thought they needed most. The experts saw small fields that could be much larger and people breaking the soil by hand when a machine could do it much faster. Did people need larger fields and labor-saving cultivation? No one asked them. Third, they were oblivious about the need for repairs, new parts, and the cost of operating gasoline-powered rototillers. People have little cash for fuel and even less to replace broken parts. There are no parts in the shops. The machines' complexity was overlooked by the aid organization; they are used to having repair services and plenty of money.

In time the donated machines were found. The engines had been removed and tossed into the undergrowth. The women had gotten local carvers to hollow out chunks of tree trunks to make wooden basins. They turned the rototiller's frame over, reset the wheels, attached the basins to the frame, and in that way had devised the technology they needed most: wheelbarrows. The carvers made wooden wheels when the rubber tires wore out, and a leather harness was designed that could be affixed to the handles. That way, a woman could push the cart and keep a hand free to hold and care for a child suspended on her waist in a sling of cloth. Women's loads became lighter, their work was more efficient, and as before, they could tote in their wheelbarrows what food the family could spare to markets miles away.

Introducing Technology Inappropriately. Whether or not this story is completely true, some version of it has occurred thousands of times. In an effort to improve peoples' lives, government agencies, corporations, and nonprofit organizations in affluent countries have encouraged poor people to use new technology. Unfortunately, too often this was done without understanding what local people needed, what they could do themselves, and the impact this would have on their current economic and social arrangements.

For example, when water is so valued that it is literally the most important currency between competing groups, its regulation and the political arrangements built around it cannot be overlooked. Thinking that people can have more water if only wells are drilled and basins are filled overlooks the complexity and fragility of water allocation practices that are the backbone of social, political, and economic relationships in many semiarid and drought-prone areas. Water development projects have drilled wells, set diesel pumps, built troughs and distribution canals, and gone away believing poor peoples' greatest problem had been solved. Often, conflict has broken out, people have died, pumps have been destroyed, and wells filled with sand. Less violently, pumps have broken down, and wells—drilled too deep and narrow for water to be drawn by hand—are abandoned. Without spare parts and a sustainable source of power, the wells and pumps are worthless.

Technologies that increase water quantity can only be appropriate when the people affected are themselves involved in the process. Water serves as a tie between people, a source of power, the obligation of decision-making bodies, and the measure of other things' value. Among the Rahanweyn in south-central Somalia, nonkin groups are allied by water-sharing agreements. The local people or particular families exercise power because the water is theirs. The selection of the people who regulate water's distribution is recognition of their status and others' respect (Massey 1986). Altering the calculus of value, power, status, and alliances can only be successful if those affected are very much involved. This, too, is a critical part of making technology appropriate.

Borrowed and introduced technology is inevitable, as it has been for thousands of years. The stories of missionaries bringing steel axes to indigenous people in Australia (Sharp 1952) and the introduction of snowmobiles to Lapps in northern Finland (Pelto 1973) are well-known and painful reminders, however, of the influence inappropriate technology can have on peoples' lives.[20] In today's global economy, where every corner of the world is open to digital communication and commercial culture, where the search for new markets sniffs out every possible household and community as a potential customer, and where industrial labor forces are being assembled in ever-poorer areas, the foisting of new technology is relentless (Chua 2003).

The most basic problems associated with shelter, food, water, disease, education, and work are addressed with the intention of helping poor people to be less poor and more able to make choices about their lives. A noncommercial motive is expected to guide development efforts. The work tends to focus on cleaner water and healthier, better-fed, and better-educated people with safer and more productive work and more access to the wider world. Technology plays a major positive role in making this happen. Unless the introduced technology is a good fit, its consequences are well understood, and people most affected have the resources to make adjustments to or mitigate its negative impacts, however, the outcome will be mixed at best, and possibly much worse.

Appropriate Technology: Microcredit. Since the end of the Second World War, banks and international governmental lending institutions like the World Bank and regional development banks have made loans to improve people's lives in poor countries and those ravaged by war and natural disasters. International banking organizations have usually provided funding for massive infrastructure projects such as hydroelectric dams, highways, refineries, and ports, as well as complex, expensive schemes to transform agriculture from subsistence farming to market-directed cash crops. Much of the technology required to do

---

[20]Jared Diamond (1997) provides an extremely interesting examination of technology and the environmental limits to its adoption among native Australians in Chapter 15 of his *Guns, Germs, and Steel*.

these things lies in rust and ruin today, but the money borrowed has to be repaid.[21]

Quite a different financial technology—a way to make and repay loans by organizations that hardly resemble banks—has proved to be much more appropriate. It begins with the idea that people know what their problems are and have good ideas for solving them. Where there is very little opportunity to get loans and money is controlled by a few individuals who operate more like the mafia than a legitimate business, most people—and especially the poor—are shut out. They cannot afford the initial start-up investment to begin improving their lives.

People know that with only a few dollars they can buy chicks and building materials to start a small egg and broiler business. A man can buy a sewing machine and set up a tailor shop, even if it's on the sidewalk or in an empty lot. A woman knows that she can sell minutes of calls to those in her village if only she can buy a cell phone and a solar panel to keep it charged. With just a few dollars someone can buy a saw, an adz, and a cart and can, with a great deal of work, craft lumber to be sold to others.

Muhammad Yunis received the 2006 Nobel Peace Prize and in 2009 the Presidential Medal of Freedom for his pioneering work in microcredit for the poor in rural Bangladesh. He established the Grameen Bank with his own very modest savings and built it into a model for microfinance worldwide. After finishing a degree in the United States, Yunis returned home and worked with groups mostly of women who, through mutual support, helped one another repay their loans. In addition to the changes made possible by microcredit, the self-empowering groups took up issues of health, childrearing, spousal abuse, and caste and gender discrimination, initiating dramatic positive changes in their lives.

Today the World Bank and hundreds of other nonprofit organizations provide microcredit. Because loan repayment rates are very positive,

---

[21]"Debt forgiveness" is one of the Millennium Development Goals (MDG) established by the United Nations in 2000. Since the mid-1980s, more capital has flowed out of developing countries as a whole to repay loans than has gone in to fund aid projects. The MDG of debt forgiveness was intended to reverse this trend. Many countries are too poor, and their economies are too weak, to keep up with debt repayment requirements. Dozens had essentially defaulted on their loans. Others were in a revolving door, with new development aid coming in while repayment funds were going out. In many cases countries had paid back an amount of money greater than the original loans but still owed a great deal of interest. In pursuit of the MDG of debt forgiveness, in 2005 the debt of the world's sixteen poorest nations was forgiven by the World Bank and the African Development Bank.

commercial banks have joined in to a limited extent. There is a palpable improvement in peoples' lives, in large part because the technology of micro-credit is appropriate. The technologies purchased with loans are also appro-priate, because they are well understood by borrowers to address their most pressing problems (Yunis 2009).

Sustainable Technology. Appropriate technology is sustainable. It is often constructed using knowledge and materials readily at hand, with skills and tools available to the persons using the technology, and it solves what they see as an important problem. People like to say, "Give a man fish today, and he'll be hungry tomorrow. Give him a fishing pole, and he'll eat better tomor-row." This may be true, but only (1) if the fishing pole is the right one for his conditions and (2) not so complex that he needs an outdoor sports shop nearby, (3) if fish are actually abundant enough for him to catch, (4) if he doesn't irreparably deplete the stock of fish, (5) if heavy metals haven't pol-luted the water and made eating the fish a health hazard, and (6) if his beliefs do not hold fish too sacred to eat. It would also be a good idea to bury the fish remains to enrich his fields or garden, the source of most of his nutrition.

In 1973, E. F. Schumacher wrote *Small Is Beautiful*, an economist's plea to think about economics "as if people mattered." As the title suggests, Schumacher wanted things to be done on a smaller "human scale" that allowed people to have control over their lives and the ability to guide social change. Borrowing from Gandhi and some principles of Buddhism, one of his central ideas was appropriate technology, or what he called inter-mediate technology.

Today there is an active appropriate-technology social movement, espe-cially around food production. University research, nongovernmental organizations, and businesses have responded with innovative designs, espe-cially as the environmental movement has gained strength. Things as simple as a windup flashlight, solar-powered batteries to run laptop computers, and harnesses for draft animals can change peoples' lives for the better, in ways they desire and with means they can control.

Topics for Discussion and Activities for Further Study

Topics for Discussion

1. What does it mean to say "technology is a system"? Explore this idea and map it on the board by taking a technology you are familiar with and looking at it as a system. Who and what is part of this technology system, and how are they connected? Some ties are pecuniary, others are bureaucratic, and some are informational.

2. It is very understandable to think that a technology is a device. It is, but it is many other things as well. Make a list of technologies that are not machines, devices, or implements. After you do this, think of the ways these nonmachine technologies may require machines and devices to make them work.

3. Max Weber concluded *The Protestant Ethic and the Spirit of Capitalism* with this statement: "The Puritan wanted to work in a calling; we are forced to do so . . . [T]he tremendous cosmos of the modern economic order . . . is now bound to technical and economic conditions . . . which determine the lives of all the individuals who are born into this mechanism . . . In Baxter's view the care for external goods should only lie on the shoulders of the 'saint like a light cloak' . . . But fate decreed that the cloak should become an iron cage."

What is Weber saying, and do you agree? Are we too bound to technology and its promises?

4. How are smartphones and other digital technologies changing lives, possibly in ways completely unintended by their creators? Are the behavioral and attitudinal changes among social networking technology users creating *social* change? How are nonusers of these devices affected by the changes? (You may want to read Crowther [2010], referenced in Footnote 16, before having this discussion.)

5. Many students transfer technologies for their own use without following the laws of patents and copyrights. Is this a problem? If so, does it have a technological solution? If technology offers a limited solution, what else could be done to address the phenomenon?

Activities for Further Study

1. Genetic modification, described in Footnote 12, is having a tremendous impact on food production worldwide. Look at one food item—a grain, a fruit, or a vegetable—and explore the social changes going on around this item, including the resistance to GM.

2. The idea of designing stoves for the world's poorest people, as appropriate technology, is often in the news. Do some research that explores this effort. What are the problems with many of the current stoves or ways of preparing food? How do new designs, often created by individuals with a great deal of experience in poor countries, address these problems? What makes one design more appropriate than another?

3. Technology is central to food: how it is produced, processed, sold, and eaten. Strawberries, milk, corn, peanuts, and other items are often transformed as they become the food we eat. Select one item. Look at its history and confrontation with technologies (including but not limited to devices or machinery). What choices were made, and why, that were most influential in determining its current life cycle?

4. What scientific research is finding applications today? Who is doing the research? Were these applications part of the agenda when the research began? Who will benefit from the applications? What social impact could the applications have, if any?

5. The diffusion of innovation is hardly studied by social scientists today, but it is a very hot topic in business schools. It is a big part of marketing research that targets specific groups of consumers. What are some recently devised tactics being used to "diffuse" commercial "innovations" to young children? Elderly people? Teenagers? Ethnic minorities? In what ways is this not actually the "diffusion of innovation"?

5

Social Movements

Human Agency and Mobilization for Change

T he headlines read,

Pro-Democracy Activist Sentenced to Prison

Demonstrators Demand Restoration of Electricity in Poorest Neighborhoods

Anti-Abortion Group Blocks Doors of Health Clinic

Boycott of Shell Petroleum Hurts Sales in Britain

Congress Deluged with Calls and Letters Opposing Health Legislation

Mass Arrests at Global Warming Summit in Copenhagen

Whaling Opponents' Boat Struck by Japanese Ship

Parents Demand Alternative to Teaching Evolution

Recall Petition Seeks Congressman's Ouster

It is impossible to read the day's news without coming across accounts of efforts to direct social change. Whether these attempts are conventional and nondisruptive or disruptive and even violent, seeking minor changes or major transformations of political, social, or economic affairs, they are part of everyday life in the twenty-first century. This is especially the case as political systems become more open to democratic participation and citizens' voices clash over public affairs. Veterans' groups; economic elites; union workers; the elderly; religious assemblies; human rights advocates; environmentalists and conservationists; lesbian, gay, bisexual, and transgendered (LGBT) activists; gun owners; the urban homeless; healthy food and public health proponents; antitax enthusiasts; nudists; anglers and hunters—they seek to make their voices heard. They work for action that will advance their interests and make social change happen.

Making Social Change Happen

There is hardly an issue of importance today that does not have at least one group, and in some cases dozens or even hundreds of groups, advocating for social change or trying to channel change in a direction they see as desirable. Democratic governments vary considerably in their openness to public opinion and the efforts of groups to affect legislation and its enforcement, but even authoritarian states are not immune from strongly felt demands that can mobilize large numbers of people and possibly influential elites, enlist international groups and public opinion, and sway their governments to act. Public outcry and mass demonstrations against authoritarian states and systemic corruption—as recent events in Tunisia, Egypt, Syria, Bahrain, Libya, and Yemen reveal—have the potential to topple repressive regimes.

Social movements exist in what is sometimes called civil society, the area of public life that involves groups of people in activities outside the formal arena of politics. It is there that social movements engage in contentious politics. Charles Tilly (1999) described social movement actions as "repertoires of contention" requiring people to publicly identify themselves in ways that expose them to criticism, rebuke, or worse. Making social change is not for the faint of heart.

There are many theories on the cause of social movements. David Snow and his colleagues coined the term "quotidian disruption" to describe the interruption and unsettling of everyday, taken-for-granted life that recognizes grievances and is a driving force of social movements (Snow et al. 1998). Ted Gurr's *Why Men Rebel* (1970) accounts for social movements and revolutionary upheaval in terms of the disparity between what people come to expect and what is actually the situation. The greater the disparity, the more likely people are to raise objections and possibly rebel. Social contradictions between expressed values like freedom and equality of opportunity and the realities of racial discrimination, social class inequality, patriarchy and gender prejudice, homophobia, and anti-Semitism and Islamaphobia generate not only appeals, but also actions to change the political, cultural, or economic status quo.

How Social Movements Matter

Much of what sociologists, political scientists, and journalists have written about social movements concentrates on the reasons groups mobilize in the first place. How they act in opposition to authority, make their demands heard, or seek access to the halls of power are also well-researched, fascinating topics. Social activists and people interested in social organization have focused their attention on the things that make one social movement successful and another a failure, offering something like a how-to primer. As this chapter discusses, what it means to be successful has often been defined by the character of the social movement organization itself: how well it frames and articulates issues and grievances; how well its leaders serve the organization; how well resources are gathered and used on behalf of the movement; and how well it gains acceptance and supportive public opinion.

Much less attention has been given to the attribution of desired outcomes of social movements. That is, how do we know that a social movement was the reason for a change? It is possible, if difficult, to trace this change to the actions of the movement. Arguably, in a complex society with many institutions and organizations vying to exert power, attributing causality to the actions of a social movement can be challenging. In some cases change occurs, but not the change that was intended by the social movement.

In most cases, social movements are not successful. They may be episodic bursts of energy or very short-lived endeavors. On the other hand, they may last a long time because they are very good from an organizational point of view. But they never reach their target, either because they focus too much on their internal needs as an organization and take their eye off the prize or because the opposition outflanks and effectively neutralizes them. In some cases events outpace a social movement, and the social change that occurs leaves the movement on the sidelines or with a hollow sounding message.

Social movements do matter, especially when they are a significant force for social change. A social movement's perspective on, or articulation of, social reality can gain the public's attention, helping to focus the discussion by framing issues, problems, and dilemmas. Its effect may be as intended or may be unintended, but the course of social change is different from what it would have been had people not pursued their cause. It may generate countermovements, be opposed and suppressed by authorities, be discredited and undermined, be forced to shift tactics, or accept an outcome less than what its participants had hoped for. This, too, matters.

Successful or not, participating in a social movement can have a profound and lifelong impact on how participants see the world around them, their own sense of personal efficacy, and their commitment to a lifetime of involvement, solving problems, and making change. Participating in a social movement can be frustrating and deeply disappointing, to say nothing of the hard work it requires to build and sustain a social movement organization. Involvement can also be exhilarating and life changing. In this way, accumulated personal change as a result of social movement participation, too, is a social change process.

The Significance of the Civil Rights Movement. Think for a moment about the U.S. civil rights movement. Iris Summers' older children could spend the morning playing baseball in a public park with any other child, but after lunch, when swimming lessons began in the pool across the street, several friends were left behind. They were barred from using the public swimming pool. As it was explained to her children, "They can swim in the river." Decades of effort, some costing people their lives, were required to change race relations across the nation, not just in the parts of the country with a history of slavery. Legally sanctioned discrimination and official segregation, everyday norms and public opinion, and deep-seated cultural prejudices were the reality that had grown pervasive in the United States in the hundred years following the abolition of slavery and the passage of the Civil War amendments to the U.S. Constitution.

Often referred to as Jim Crow, these laws, practices, and norms were the status quo. They would not be dismantled by the evolution of a postindustrial capitalist economy, the sometimes glacially slow actions and compromises of state legislatures and the federal government, or even the articulate writings and speeches of many admired individuals. Only a sustained social movement that mobilized and found support across a broad collection of social movement organizations, churches and synagogues, high schools and colleges, and committed individuals could successfully force the end of Jim Crow.

Over time, the movement shifted strategies and combined tactics as needed, enlisted allies, and changed public opinion to create the social change that began to rectify the sad history of human slavery and Jim Crow.[1] The thousands of people who braved insults and attacks, who wrote, spoke,

---

[1]Taeku Lee's *Mobilizing Public Opinion* (2002) traces and analyzes shifting public opinion toward the civil rights movement and documents the turning points in white attitudes against Jim Crow.

Young man drinking at a "colored" water cooler, Oklahoma City, Oklahoma. 1939. Russell Lee, Photographer. Farm Security Administration Photo Archives.

Source: Farm Security Administration. Photographic Archives. LC-USF33- 012327-M5 [P&P] | LC-DIG-fsa-8a26761. http://www.loc.gov/pictures/search/?q=8a26761&op=PHRASE&sg=true

marched, argued, and explained, who shared a common vision and sometimes fought among themselves, joined together in the civil rights movement. They forced its agenda into the public's consciousness and onto the political system. It very much mattered.

Understanding Social Movements as Change Agents

Social movements have fascinated writers and scholars for centuries but more often than not as anomalies and peculiar, threatening, unpredictable, and disorderly events. Perhaps in part because social movements have become such a pervasive aspect of modern societies, it has only been in the past few decades that social scientists have made significant headway in making sense of social movements' origin, form, and dynamics.[2]

Social movements are "politics by other means"—to paraphrase Clausewitz's description of war as diplomacy by other means. They exist outside the normal functioning of the state. Movement participants are outsiders who seek to influence those in authority to address their grievances. "Transgressive contention" is one way to think of social movements, in contrast to the "contained contention" that is the normal activity of courts and litigation, legislative debates and bill writing, interest groups, lobbying, the filing of complaints about public services, voting, political advertising, and the dozens of other activities that contribute to how laws are written, passed, and enforced (McAdam, Tarrow, and Tilly 2001: 7). Social movements are a means apart from this. In that fact lies their effectiveness in making change.

What Is a Social Movement?

Understanding social movements is a first step in evaluating their role in social change. A good place to begin is with the concept itself. What is a social movement, and what is not? David Snow and Sarah Soule (2010: 6) provide a useful conceptualization: "Social movements are collectivities acting with some degree of organization and continuity, partly outside institutional or organizational channels, for the purpose of challenging extant systems of authority, or resisting change in such systems."

They are "collectivities" or groups with a distinguishable identity. As movement participants become more articulate in expressing grievances and coalescing around an agenda of action, the movement becomes more organized. Individuals may come and go, but the group persists. The more "transgressive" the movement, the more it is outside institutional or organizational channels, but in time it may seek to work more closely within these channels in the same way an advocacy group, political party, or governmental commission or bureau does. In fact, access to and a bargaining position within established decision-making bodies is a goal of many social movements, but becoming an "insider" marks its end as a social movement engaged in politics by other means.

---

[2]Another factor could be that sociologists and political scientists in recent decades are more likely than were their professors to have participated in social movements themselves, motivating them to better understand the events and processes they were a part of.

Challenge to Authority. A modern complex society has many "systems of authorities" that exercise power over peoples' lives. Social movements often target the state "because they consider it either directly responsible for their grievances, or as the institution best suited to address them" (Kolb 2007: 5). Governments are systems of authority, but large corporations are, in many ways, even more powerful (Hertz 2010). For example, large corporations decide to locate, move, invest heavily, or lay off workers. With little or no public discussion, they can radically alter the economic health of communities by changing job opportunities, tax revenue, and need for public services. Private corporations develop and market products that become the environment of consumption. This includes everything from a bank's services to Internet access, prescription drugs, the safety features of cars, deodorants, and the sugar content in breakfast cereal. Increasingly, as will be seen, social movements are challenging corporations as systems of authority (Soule 2009).

Social movements often challenge the state and corporations and contest the decisions they make. Only rarely do social movements challenge the core idea that authority should be vested in corporations and the economic system they are a part of. Equally rare are social movements that challenge the foundations of the national state: to tax, make war, regulate commerce, issue currency, ensure a reasonable level of social security, and provide for public safety. These are agendas for revolutions that begin as social movements and, occasionally and very importantly, create social change, as will be seen in Chapter 6. The vast majority of social movements make more modest but still quite significant challenges to authority.

Collective Goods. Usually, but not always, social movements pursue collective goods. These are things that must be shared by a group or category of people, regardless of their own participation in the social movement. Because the state is often the dispenser of collective goods, many social movements focus their attention on government policies, including the guarantee of rights and the provision of benefits (Amenta 2006: 16). The U.S. pension movement that worked to establish a guaranteed cash payment for elderly citizens had success in 1935 with the creation of the Social Security Administration (Quadango 1984). By creating a trust fund of contributions from lifetime earnings, social security has been available to qualifying seniors ever since. Securing an old-age pension was a collective good that extended far beyond those who worked for its creation.

Social movements consider rights to be collective goods. If rights are secured by some, they are shared by all. If privacy rights activists can prevent the state, in lieu of a court order, from trawling through personal e-mail messages, everyone's e-mail communications are legally protected. If a group of workers successfully contest an employer's scrutiny of members' political contributions or a prospective employer's effort to find out a woman's intention to have children, this right can be recognized as legally applying to all workers.

Environmental movements, scientists, and supportive governments secured collective benefits when the Montreal Protocol, enacted in 1989, banned chlorofluorocarbons' discharge into the atmosphere. These gasses, most commonly found in refrigerator coolant and aerosols, were rapidly creating a "hole" by depleting the ozone layer of the Earth's atmosphere that is vital in limiting infrared light that reaches the Earth's surface and contributes significantly to skin cancer. It was a major victory for the environmental movement, but its benefits were not limited to the activists themselves. Everyone on earth benefited from their actions and this agreement.

COMMON GOODS AND FREE RIDERS

A useful way to think about common goods and collective benefits is to imagine a coastal fishing community that for years has seen its sons and daughters killed in shipwrecks on the shoals. Businesses have lost shipments, and a hoped-for tourist trade never materialized. The village council decided to build a lighthouse to make safer the passage into the harbor. Everyone had to give a cash contribution for materials, and all able-bodied adults had to contribute twenty hours of labor. The lighthouse was built. Most people think it was a good idea. Everyone now shares in the improved safety, sense of well-being, and the prosperity it brought. But there was opposition, both active and passive.

A few people in the town, particularly the morticians and the owner of the cabinet shop that makes coffins, thought the lighthouse was a bad idea. The morticians and coffin makers expressed their opposition to the city council. More than a few others simply wanted to opt out.

This was the rational thing to do. They reasoned, correctly, that if everyone else helps pay for it and does their share of the work, it will get built. Nobody will miss their meager contributions. The benefits of the lighthouse can't be denied to anyone in the village, even those who opted out. They will benefit, and they will have saved themselves some time and money.

That's why the village council levied a tax and a work requirement and created a penalty for noncompliance. A tax, a levy, a law, or a policy is required to overcome what Mancur Olson (1965) calls the "free rider problem," i.e., collective benefits that cannot be denied to people who would otherwise make no contribution. No one is fined or jailed for not joining a social movement, however much they will benefit from the collective goods that accrue. Are participants, then, acting irrationally?

Why do social movement activists pursue collective benefits? Why not let others do it? Especially curious are movement participants who do not directly benefit. Whites were very involved in the civil rights movement, seeking to secure equal opportunities and treatment for those who were discriminated against under Jim Crow. They already had these collective benefits. Why did they participate? Social movement scholars have taken the concept of common goods a step further to address this question.

One answer is that the activists *are* gaining a benefit simply through their participation. It might be the camaraderie that comes with involvement in a common effort. It might be the change in identity or improvement in self-worth that comes from trying to make a situation better. It could be that making an effort in pursuit of one's beliefs, values, or ideals is rewarding by giving greater meaning to a person's beliefs and values. This topic is taken up later in the chapter. Finally, it might be that social movement participants have a clearer idea than others about the kind of community, society, and world in which they want to live and are willing to work for it.

Grievances. Everyone can think of a personal grievance, a pet peeve, a gripe that they'd like to see something done about. That's life. Grievances that give rise to social movements are more likely to strike a deeper cord and be more unsettling and problematic. They are felt to be violations of the way things should and could be, and they seem to escape personal solutions. Sometimes expressed in terms of injustice or unfairness, grievances may pose a threat to one's livelihood, personal safety, and sense of well-being. A grievance can stem from the values and beliefs that reach most deeply into who we think we are, our humanity, and our sense of belonging to something greater than ourselves.

One of the immediate intentions of social movements is to be recognized as a group with a grievance. "I'm mad as hell, and I'm not gonna take it anymore!" Howard Beal's famous public outburst in the movie *Network* captures the histrionics of many early movements: a lone person standing up to authority. When aggrieved persons constitute a group, rather than a lone, alienated individual like Howard Beal, their displeasure expresses both a potential for action and points to the structural basis for the grievance. It is not merely the result of an individual's unique circumstance, bad luck, or poorly made decision. A grievance is lodged against a social circumstance, buttressed by some organization's practices that affect many people simultaneously.

Albert Hirschman's *Exit, Voice, and Loyalty* (1970) describes three possible responses to a sense that something is wrong or things are going in the wrong direction. To exit is to leave, drop out, or somehow isolate and insulate oneself from the threat or the source of grievance. Loyalty means joining up, embracing what seemed problematic, going with the flow, finding new opportunities or silver linings. Voice is contention or the response of resistance.[3] Critical to emerging social movements is the recognition that the grievance is shared by others, and its source is subject to change. It is the sense of sharing an injustice, a deep objection, and a sense of possibility that distinguishes social movement grievances from the more pedestrian, personal problems and fears we all face.

Who Are Social Movement Participants?

Though elected on a platform of social and political change, Barack Obama's first years in office were met with vociferous protests by people who objected to the direction of change his policies recommended. A participant at an oppositional rally organized by Tea Party supporters described herself as "'just a stay-at-home mom' who became agitated about the federal stimulus package" (Zernike 2010: A15).[4] Though perhaps more

---

[3]Robert Merton's *Social Structure and Anomie* (1968: 185–214) analyzes crime as deviance or a reaction to "anomie," i.e., a misalignment of desired outcomes and the means to achieve them. In Merton's view, some crimes are resistance—to deprivation, to unequal justice, or to an absence of legitimate avenues for success. Most people don't go the route of crime. Some find their "voice"—to use Hirschman's (1970) term, in another form of resistance: social movements.

[4]Nina Eliasoph's *Avoiding Politics* explains how participants often justify their involvement by making public references to their personal identity—using phrases like "just a mom"—and expressing personal concerns, rather than offering reasons based on the public interest or larger political context (Eliasoph 1998: 4 ff.).

"Astroturf" than grassroots,[5] her felt grievance motivated her participation. Active participants, however, are only a small fraction of those sharing a grievance. Where are the others? Why do some people, but not just anyone, join a social movement? There are at least three strongly predictive conditions, all of them relevant to explaining why Iris Summers was never an active social movement participant.

The first reason is social networks. People who interact with others who can put them in contact with a social movement, who may themselves be active or who support another person's involvement, are more likely to be movement participants. The organizations people are affiliated with, e.g., labor unions or business associations, may officially voice a grievance. Those in the organization participate in a movement as an extension of their group affiliation.

For example, mountaintop mining, a highly contentious method of extracting coal in West Virginia and elsewhere in Appalachia, involves dynamiting the tops of mountains to expose and extract coal. The unwanted rubble is disposed of in nearby valleys and creek beds. Both energy companies and the United Mine Workers oppose legislation that would significantly restrict this practice. Miners and their families are among the most active protesters against the legislation, identifying themselves as community citizens, working people, and union members. Social networks increasingly are constructed in cyberspace and putatively through exposure to the grievance-voiced media, bringing people into a common orbit of communication and providing opportunities to sign a petition, have a message sent, and jot a note of support or opposition, rather than donating money and engaging in more costly and time-consuming forms of participation.

Second, persons who are already politically engaged are more likely to be movement participants, and those persons who have been active in a movement are more likely to be active at a later time. Social movement participation tends to create a greater commitment to the movement for some that translates into a sense of personal efficacy and willingness to participate again. In a similar vein, persons whose parents have been movement participants and have been socialized to lend support to social movement activity are more likely to be participants themselves. This is what Snow and Soule (2010: 126) call "intergenerational transmission [of] activist values."

---

[5]Edward Walker's (2009; 2010) analysis of corporations and industry groups that provide financing for activities such as telephone campaigns and public rallies in support of corporate interests, giving the appearance of their being spontaneously organized citizen responses to legislation, applies to aspects of this social movement. Corporate planning and sponsorship compares to authentic "grassroots" organizations, hence the term "Astroturf" (see also Mayer 2010).

The third reason is ecological, a person's location in social time and space. Someone's everyday situation can obstruct or contribute to participation. Work and family obligations, proximity to social movement activities, and the involvement of others in the immediate vicinity all contribute—some negatively, some positively—to the likelihood of participating. A "stay-at-home mom" is possibly more likely than many other people to have the flexibility with her time. She may regularly watch cable news and hear talk radio personalities voicing a call to arms that tell her when and where to go to express her grievance. A good predictor is that she has friends or family who would be with her if they could.

College students across the globe have for decades been in a position of "ecological proximity" to social movements. They are often exposed to grievance issues. Many students are incredibly busy with work and school, but many also have flexibility in their schedules and limited responsibilities to others. They can devote time to movement activities and find themselves among activists more readily than most. No wonder so many social movements originate or find a friendly locale on college campuses.

Resource Mobilization

Vital to the chance that a social movement will be effective is its success at resource mobilization, i.e., acquiring for its use those things necessary to function as an organization and carry out social movement activities. As enumerated by Snow and Soule (2010: 91), the first things that come to mind are financing and material support. Indeed, being able to pay for materials, office space, salaries, travel, advertisements, communication, and even to post bail money can be critical.

Resource mobilization also includes human capability. Movements must be able to create and maintain a website, address the public, draft documents, organize contacts and ways to quickly communicate to participants, provide legal representation if necessary, and exert leadership. These require a range of capabilities of people who can get necessary things done as well as inspire participants to work for a cause.

Moral resources, Snow and Soule's third category for resource mobilization, include the good opinions of widely recognized and admired people who can present themselves on behalf of the social movement. Moral resources also include a favorable public opinion, at least by a supportive minority of the population, and the widespread view that the movement is serious and has the right to advocate for its cause. Around the world, Women in White protest outside government offices, quietly demanding information about their politically active sons, daughters, and husbands who were taken into police custody, never to be seen again.

When social movements use disruptive tactics, the credibility with the public of those who disrupt is critically important. Priests who douse the street with red paint to symbolize the blood spilled in an unpopular war may find more sympathy with the public than if hooded youth do it. When a pious Buddhist monk or a despairing college graduate whose only available work is to sell produce on the street sets himself on fire at a busy intersection, this conveys the message he seeks to spread more widely than someone who could be characterized as deranged and suicidal.

Finally, social organizational and cultural resources are valuable. The former include networks of association between the movement and potential allies, "free space" in which to operate, venues to gather publicly, a division of tasks and responsibility, and channels of communication. Writing songs, designing logos, composing slogans, adopting distinctive dress or costumes, and even modes of comportment can be valuable cultural resources.

The dignity and fearlessness with which Malcolm X, Dr. Martin Luther King, Jr., and other civil rights leaders carried themselves was a deliberate expression of a valuable cultural asset: pride. Whites who had little firsthand experience with Blacks and often held demeaning stereotypes of Blacks as shuffling, docile, and willfully subservient saw something far different. In contrast, the Guerilla Girls' attire, complete with gorilla masks, made their sudden, unscheduled appearance at art galleries and cultural events both shocking and enlivening. They created a "brand" that gave them notoriety and popular support. Provocative in appearance and actions, they communicated their disrespect for established authorities of art and indelibly thrust their cause—the promotion of women artists and criticism of the paucity of women's work in galleries—into the public consciousness.

Social Movement Framing

Social movements are forms of "contentious interaction" (McAdam et al. 2001: 16–18) that question not only existing relationships and accepted practices but may challenge their very legitimacy, morality, utility, or viability. Movements do this, in part, by insisting that participants and target audiences see things differently, what sociologists call the definition of the situation.

One of the most interesting aspects of social movements is the process by which they define situations. They "frame" situations in new but comprehensible and compelling ways. Their grievances, actions, and goals begin to "make sense" of the challenge to the taken-for-granted and commonly accepted view of the status quo. An image of same-sex marriage, an important issue for the gay rights movement, shows two middle-age, middle-class, simply dressed women sitting at opposite ends of a couch in their living room. One is a nurse, the other works in sales. The frame's message: for goodness sakes, why should this couple be denied the rights of married couples!

Doug McAdam's study of the U.S. civil rights movement pays particularly close attention to how the pursuit of equality of opportunity and an end to discrimination and prejudice framed the issues by challenging racial prejudices and long-accepted patterns of interaction as contrary to the founding ideals of the nation. Social movement frames for target audiences are preceded by less clearly developed but fundamental changes in worldviews of participants themselves. Social movement participants invariably "reject institutionalized routines and taken for granted assumptions about the world and . . . fashion new world views" (McAdam 1999a: xxi).

Though discrimination in public transportation was illegal after the 1947 Supreme Court decision in *Morgan v. Virginia*, Blacks in the U.S. South continued to be expected, when approached by Whites, to give up preferred seats near the front door of a bus and move to a seat in the back. In a highly symbolic act, Rosa Parks and others who had lived with racial segregation expressed their rejection of second-class citizenship by refusing to give up their seats to white riders, in many cases able-bodied white males, and move to the back of public buses.[6] They challenged what had been taken for granted by Whites for generations by framing the denial of civil rights in a circumstance that struck the sensibilities of someone like Iris Summers, far removed but able to identify with Rosa Parks and see clearly the indignities Jim Crow foisted upon her.

---

[6]Ending Jim Crow in public transportation began with Irene Morgan who in 1944 refused to give up her seat on an interstate bus. The court's decision in her favor, *Morgan v. Virginia*, was met with new state and local laws throughout the South to circumvent the decision even in interstate travel, the province of the federal statuses enforceable by the Interstate Commerce Commission. The strength of the Court's decision was tested by sixteen activists, including Bayard Rustin. Their Journey of Reconciliation in 1947, organized by the Fellowship for Reconciliation and the Congress of Racial Equality, was a precursor to the 1961 Freedom Riders (Meier and Rudwick 1973: 33–41). In 1952, Sarah Keys, a WAC (Women's Army Corps soldier) stationed at Ft. Dix, New Jersey, similarly refused to give up her seat to a white Marine. Three years later, in *Keys v. Carolina Coach Company*, in tandem with another NAACP suit, the U.S. Supreme Court ruled that laws designed to maintain Jim Crow practices on interstate transportation were unconstitutional. This decision preceded by two weeks Rosa Park's test of the enforcement of the *Keys* decision. Her refusal set off the Montgomery Bus Boycott and brought international attention to the civil rights movement.

Diagnostic, Prognostic, and Motivation Frames. Robert Benford (1993) describes three types of frames in his analysis of the nuclear disarmament movement. The first, *diagnostic* framing, implicitly explains how a situation is problematic. Things are not supposed to be this way. The movement provides reasons for this and puts forth a need for change. For example, the social movement among American Indians seeks to rid sports of team mascots that demean Indians by using pejorative or inappropriate designations such as *warriors*, *redskins*, *braves*, and *chiefs*. To put the issue in a new light, the movement developed a poster of team pennants with such names as the Kansas City Jews, Pittsburgh Negroes, and San Diego Caucasians. Seeing these fictitious pennants, it is difficult not to think that something is amiss.

A sense of victimhood can be a powerful part of diagnostic framing. People often live with injustice, deprivation, unequal life chances, and impotence without complaining, but if they begin to see this as unfair and unnecessary, they may cast themselves as victims. Indeed, they may *be* victims and experience victimization in ways both large and small. When they see themselves as victims, however, the world begins to look different, and their role becomes less passive and less accepting of their fate. They are more receptive to solutions that will ease their lot, elevate their status, and reduce or end the conditions they now see as abusive.

The second, *prognostic* framing, establishes what is needed to change a situation. In Benford's study, disputes about strategies and tactics—what should be done and how it should be done—divided the antinuclear movement. One strategic solution to the threat posed by nuclear weapons is a nuclear-free world. All nations possessing nuclear weapons should disassemble them and pledge never to have nuclear weapons. Brazil, South Africa, and a few other countries have done this, but major powers (the United States, China, United Kingdom, France, and Russia) and minor powers (North Korea, Israel, India, and Pakistan) are loath to dismantle their nuclear arsenals. Some antinuclear groups advocated "direct action" tactics by demonstrating outside nuclear weapons sites, blockading entry, and using other forms of civil disobedience. Less confrontational tactics were proposed by more moderate groups who organized workshops and sought to convince the public ("middle America") about the danger of nuclear weapons accidents and the huge financial costs that could be used for more important public services.

Finally, *motivational* framing or "frame resonance" is designed to enhance social movement participants' involvement and commitment to working on behalf of a movement. "[T]he question is not what is or ought to be, but rather how reality should be presented so as to maximize mobilization" (Benford 1993: 691). Appeals to become involved are what Mills

(1940) called the "vocabularies of motives," e.g., *the cause is just, history is on our side, time is slipping away, the issue is a matter of life and death, no one else cares, we have suffered enough.* Motivational frames help participants and those who would join the movement identify the reasons for personal sacrifices and the importance of making efforts on behalf of the movement.

Motivation can also be enhanced by the possibility of tangible benefits, either material, spiritual, personal, or collective, to be gained by the movement's success. Earlier these were discussed as collective goods. Higher retirement benefits or better health care for veterans, an end to discriminatory lending, the abolition of quotas restricting the number of women, Blacks, and Jews in higher education, lower taxes, better schools, stronger job security, a healthier environment, and access to affordable health care are just some of the many benefits that could come to participants, as well as being collective goods. Motivational frames may promise these outcomes as long overdue, just, or impossible to secure without an active effort on behalf of the movement. Because they are collective benefits, movement participants are able to see themselves as benefactors to others, itself an attractive status claim and benefit.

Social Movement Tactics

Social movements are characterized by "transgressive contention," especially in their early phases, operating outside the normal range of interactions with established organizations. When the annual property tax assessment comes in the mail, it is not unusual for homeowners to balk and call the tax assessment office to complain. They write letters and provide documentation, ask for an interview, or even threaten to sue the city or county. This is quite normal. When a group of citizens begins communicating with one another about their property tax assessments, are joined by several local businesses, circulates petitions, insists on meeting with the city or county commissioners, organizes a letter-writing campaign, and protests on the courthouse steps, they are acting outside the norm, as an ad hoc social movement. They are engaged in a social movement "tactical repertoire" (Soule 2009: 153).

Following the contested reelection of Mohammad Ahmadinejad as Iran's president in the summer of 2009, mass street protests ensued in Tehran and other cities. These included silent marches of hundreds of thousands that became violent clashes when marchers were attacked by the police and the

Republican Guard.[7] Facebook and other social networking sites, text messaging, and e-mails were used for planning and distributing news among dissidents, despite the government-controlled media. Among the many other elements of their "tactical repertoire" was a campaign to write messages of protest on millions of Iranian banknotes. The government issued warnings about defacing the country's currency and banks were ordered to confiscate defaced bills, but the messages were read for months throughout the country.

Blockades by environmentalists of logging trucks and whaling ships, sit-ins at university presidents' offices, workers' sit-down strikes to protest the sale and impending closure of their employer's company, disrupting meetings, and occupying empty buildings draw attention to a committed group's opposition. Even when rallies, vigils, public protests, marches, speeches, parades, demonstrations, petition drives, and prayer services fail to get the public's attention, they can help solidify participants' commitment to a cause.

Tactics in Western countries have become "more self-consciously performative" in the last half of the twentieth century (Snow and Soule 2010: 180). The need to "cut through the clutter" of news that would otherwise crowd out the social movement's actions has increased the likelihood of using dramatic and flamboyant staging to get the public's attention. In his study of the 1960s antiwar movement, *The Whole World Is Watching*, Todd Gitlin (1980) elaborated on repertoires of contention that were often highly provocative (e.g., those of the Yippies and Abby Hoffman's colleagues) but increasingly became part of the standard repertoires of groups of all political persuasions.

Unexpected and seemingly out-of place theatrics and generally harmless antics are a standard part of some movements' tactical repertoires. These include street or guerilla theater that tells a compelling story of outrage, absurdity, loss, abuse, or suffering, e.g., lying on the ground naked to spell out a word, number, or message and selling calendars with monthly photographs of disrobed wilderness or clean-water enthusiasts. More arresting and likely to be deemed newsworthy are cross burnings in front of homes; protest camps of tents and shacks assembled in a highly visible public space; blocking the entrance to a health clinic, corporate office, or research laboratory; as well as self-immolation and mass suicide.

---

[7]The Republican Guard is "a hard-line branch of the country's military that was founded in 1979 to defend the Islamic Revolution" and is currently under the president's jurisdiction (Fathi 2010: 13).

Not all repertoires of contention are intended to gain public attention or spark the energy of social movement participants. Lawsuits and litigation have been used to challenge laws, expose failures to enforce the law, delay the actions of corporations and governments, and expose personal and institutional malfeasance. Trials themselves, however, can be venues for social movement actions, in defense of an accused person or to visibly rally support for the prosecution of those whose actions are either supported or opposed by movement participants. At the trial of popular Palestinian leader Marwin Barguthi, daily demonstrations by Israel's Jewish settlers from the occupied territories protested not only against the defendant but against the Palestinian Authority and Palestinian resistance more generally. The trial of Scott Roeder for the murder of Dr. George Tiller was accompanied by antiabortion demonstrators from across the nation who supported his right to kill the doctor in order to "protect the children" (Davey 2010).

Political Opportunity for Social Movements

Not all societies are equally friendly venues for social movements. When hundreds of Buddhist monks marched peacefully against the government in Myanmar (Burma) in the summer of 2007, nighttime arrests quickly depleted their numbers and made it clear there would be no tolerance of their actions. When the state controls what is broadcast, can monitor, filter, and suspend personal use of the Internet, and acts with impunity to arrest, imprison, shoot, and otherwise silence any opposition, it is difficult to sustain social movement activity.

Authoritarian and Democratic Societies. Such coercive responses are common under authoritarian regimes and may include killing demonstrators, their torture and death in police custody, destruction and seizure of personal property, threats to the families of protesters and damage of their homes and businesses, deportation, and long prison sentences or execution. Repressive measures by the state can attract unwanted attention, however, and a sophisticated repressive response will seek as much as possible to limit public and international attention, lest this earn sympathy for the movement and its participants.

In more open and democratic societies, the rule of law is intended to protect unpopular and unofficial views. Legal and social prescriptions guide those who seek to influence the political process "by other means." Democracies are expected to tolerate and protect dissenting views and the right to express them. Political opportunities, including sympathy within the political system, are important for social movement success. Groups can obtain permits and otherwise use public space to hold rallies, marches, and demonstrations. Contested elections, popular suffrage, campaign contributions, and multiple political parties give social movements an opportunity to influence party platforms, candidates, and issues important to voters. Circulating petitions that require voter referendums is an even more direct way for social movements to have a grievance considered. In the United States, aggrieved groups can even circulate unfounded rumors, scathing accusations, and untruths, so long as they do not intentionally libel an individual or organization.

Elite Competition as Opportunity. Elite fissures in democratic and authoritarian states provide political opportunities for social movements to find advocates and sympathy for their cause. A strong authoritarian state system characterized by unified elites in government, business, and the military that can control the media is barely threatened by public opinion and can effectively crush a social movement. However, unity can break down in authoritarian systems when elites are competing among themselves. In democratic societies, a stable unity among elites is even more elusive.

An example is the ongoing contest for state support of synthetic fuels and other renewable energy sources. For many environmentalists, this provides political opportunities and potential pitfalls for their social movement. The current vast production of corn in the United States means a low market price and a constant search for new uses of corn. This, in turn, has led corporate agriculture elites as well as many farmers to pursue production of biofuel made from corn (ethanol) as a way to raise the price of corn. They are joined by agricultural seed and chemical corporations who try to develop the means for ever-larger harvests (Big Farm). Big Farm and ethanol producers have successfully won supportive legislation and regulations, subsidies, and tax benefits. The political domain to retain and expand these is contested terrain.

On the other side in the contestation are energy corporations (Big Oil), food processors and food retail corporations (Big Food), and large corporations who dominate grain trade and transportation. For them, ethanol production translates into less and more expensive corn to trade and move. Predictably, Big Oil opposes expenditures for, and inflates public skepticism about, the capacity of renewable energy to help meet demand. Big Oil objects to legislation mandating that more than minimal levels of ethanol be added to petroleum-based gasoline. Big Food benefits from subsidies to corn producers, which translate into high volumes of inexpensive corn for high-fructose corn syrup, corn oil, and other products that go into a large share of processed foods they make and sell, as well as fast food. Higher corn prices are not in their interest.

When corporate elites are split over an issue like ethanol, environmental activists seeking an expansion of renewable energy find political opportunities. Ethanol has been presented by its advocates as an alternative to nonrenewable carbon-based fuels. The environmental movement supports renewable energy, including biofuels, and for some groups this includes ethanol extracted from corn.[8] Those in the environmental movement advocating for renewable energy, including ethanol, can seek to advance their cause by aligning themselves with Big Farm, but this is an uneasy alliance and a tenuous political opportunity. Environmental activists advocate land-use planning over unfettered property rights. They often openly criticize industrial farming practices that rely on large inputs of petrochemicals for fear they will wash into and pollute waterways, damage the natural soil biomass, and contribute to the production of unhealthy food. Big Farm sees these environmentalists as antiagriculture but will grudgingly hear their concerns if it is politically necessary to gain support for their own issues.

Resistance to Social Change

Social Movements Opposing the Direction of Social Change

You might have taken note in Snow and Soule's conceptualization of social movements, quoted earlier in the chapter, that concluded with, "or resisting change in such systems." The world never stops changing. It is important to remind ourselves that institutions are constantly in flux: passing new laws; generating new consumer goods and services; adopting new ways to deliver education and public services; focusing on health and scientific challenges in new ways; generating popular culture and broadcasting it to new and established markets; creating, redesigning, and eliminating jobs. Social change is ubiquitous.

The phrase "resisting change" implies bringing things to a halt. In reality some social movements resist change—current and impending—they see as going in an unfavorable or threatening direction. Sometimes called reactive or reactionary social movements, such efforts seek to reverse course or channel social change away from what is seen as confusing, threatening, untried, untested, or in conflict with material interests or core values and beliefs.

---

[8]Many environmental groups challenge the claims of energy savings provided by ethanol. Groups working to alleviate hunger in poor countries object to using food for fuel and fear the effects of higher grain prices.

Following the 1973 passage of the U.S. Supreme Court's *Roe v. Wade* decision on abortion, public attitudes shifted more favorably toward abortion as a woman's choice (Tedrow and Mahoney 1979). Within two years of *Roe v. Wade*, a significant antiabortion movement, initially led by the Catholic Church and later joined by Christian fundamentalists, grew into an effective counter to the court's liberalization decision. For more than thirty-five years, the battle has raged to reverse the social changes legal abortion made possible, to in effect change course.

Most organizations identified with the libertarian movement in the United States are very pro-capitalism and decry the "progressivism" of liberal social movements that have sought measures to limit corporate discretion. In response to most social and regulatory legislation, libertarian social movement organizations reject the interstate commerce clause of the U.S. Constitution and call for the end to regulations and statutes over food and workplace safety. They oppose the power of the Federal Reserve to regulate the monetary system and supervise banks and other financial institutions. By opposing unionization, the U.S. mail service, civil rights laws, public libraries, national parks, and public education, and in seeking to roll back corporate and personal income, business, and estate taxes, they seek to return the United States to an earlier era of global economic dominance. Perhaps more importantly, by nullifying the policies espoused by more than a century of liberal social movements, they believe that a kind of personal freedom and individual initiative of an earlier era can be restored.

The antiglobalization movement includes many social movement organizations opposed to the reduction in trade tariffs and regulation of the global flow of manufactured goods, raw materials, and capital. The direction of change in the global economy and labor market is viewed with alarm. Antiglobalization participants see it as driving down the wages of workers in more affluent countries, shifting pollution-generating industries to countries with lax regulations, and flooding distant locales with Western popular and commercial culture designed to sell mass-produced commodities to people whose basic needs have yet to be met. While pro-globalization has been less a social movement than the evolution of global capitalism in search of markets and materials, antiglobalization has become a very visible and diverse social movement allied to existing organizations such as international labor unions and environmental advocacy groups.

State Resistance to Social Movements as Agents of Change

In the summer of 2007, more than a thousand Buddhist monks seemed suddenly to mobilize in opposition to the dictatorial control of the Myanmar

(Burma) state. They took their grievances to the streets for nonviolent protest and were joined by thousands of others. Initially the monks and other protesters were bullied and bludgeoned by the police and military, attracting the attention of the international media. The government of Myanmar quickly recognized that it was a tactical mistake to confront the saffron-robed monks in plain view of the news cameras and journalists covering the story. Taking a lead from other authoritarian governments with restive populations, a series of mass arrests, usually at night, led to the imprisonment of untold numbers of monks, including those leading the protests. Within days the protests had stopped.

In June 2009, Iran held presidential elections. Hard-line conservative incumbent Mohmoud Ahmadinejad was threatened with defeat by opposition political groups unhappy with the country's economic problems and the government's increasingly repressive and undemocratic practices. They rallied around a single candidate who, if polls were correct, was poised to win against the unpopular president. Within hours of the polls' closing, the courts declared Ahmadinejad the winner, setting off charges of voting fraud and street protests.

Thousands of people marched in Tehran, Iran's capital, and other cities. There were street skirmishes, mass arrests, interrogations, and killings by the authorities. As the size of protests grew to the hundreds of thousands, the police, the voluntary Basij militia, and Republican Guards escalated their use of force. International journalists were forced to leave when their visas were withdrawn, cell phone service in Tehran went down, and quiet arrests of leaders and participants increased. The government-controlled media, including state-run TV, condemned the demonstrations and blamed the United States, Britain, and Israel for inciting the social movement.

As many as three dozen of those arrested died in police custody, and the government brought hundreds of people to trial. It began executing members of the movement, the first being young students, and effectively used other less visible but effective means of coercion and suppression. For several months, pro-democracy activists continued to stage marches in which people silently protested the government's actions. By 2010, however, the movement had lost strength and appeared to have lost its contest with the state.

Responses to State Suppression. In many cases, the tactics of suppression help determine the tactics of social movements. State violence often sparks a violent response by movement participants. Paul Almeida (2003) examined social movements in El Salvador between 1962 and 1981, a time when political opportunities of a relatively open society gave way to intense threats against any protests critical of the government. His analysis shows that threats to livelihood and rights can generate or escalate social movement activity in an authoritarian state system. In an authoritarian context, a social movement is also likely to become more radical and even violent. In turn, when violence is used by social movements, even if provoked, it is convenient for the state to justify more repressive violence as an obligation to keep public order and safety.

Mao Zedong famously said, "Drain the pond to catch the fish," recognizing that the best way to defeat one's enemies was to deny them the resources they needed. This takes a page out of the theory of social movements' resource mobilization. When movement leaders are silenced, new leaders may adopt disguises, go underground, or the movement may fracture into cells, dispersing leadership across the movement and making it less hierarchic and more horizontally integrated. Being less coordinated from the top, social movement cells can chart their own course, some taking on more aggressive or violent tactics while others extract resources using coercion, blackmail, and extortion.

Today social movements are supported by Internet sites that provide evidence, expertise, and action plans as well as a platform for making and communicating decisions. A common state response to social movements is to kill or imprison journalists whose writings challenge the authority's frame.[9] The Internet, however, is proving to be one of the greatest challenges to authoritarian states. China's periodic crackdown on dissidents is accompanied by increasing state control of the Internet, including monitoring and blocking web traffic and restrictions on who can register and operate a website. These steps interrupt but rarely stop the flow of information and communication. Activists today hack into government and corporate sites, leak secrets, and, in other ways, use the technology to their advantage. Social network sites and streaming news over the Internet were apparently very significant in activists' communication and mobilization of thousands in support of regime change in Tunisia, Egypt, and across the Arab world in the first months of 2011.

---

[9]In 2009, twenty-four journalists were in prison in China, and twenty-three journalists were in prison in Iran, according to the Committee to Protect Journalists. That year seventy-one journalists were killed, most while covering conflict situation, but some were assassinated for criticism of their government's actions and public officials. Updated information is available at http://www.cpj.org.

CROWDS, SOCIAL MOVEMENTS, AND POPULAR DEMOCRACY

Just as, all too often,

some huge crowd is seized by a vast uprising,

the rabble runs amok, all slaves to passion,

rocks, firebrands flying. Rage finds the arms.

—Virgil *The Aeneid* (I.174–177)

To see a demonstration or public protest is to see people acting in ways unexpected, even unusual, from everyday life. Shouting, waving plac-ards, singing, and bringing attention to themselves, acting like friends with people they hardly know, there is a strong sense of camaraderie among participants. Voices are adamant and insistent about the way things are and the way things should be, repeatedly singing and chant-ing a simplistic rhyme or sentiment: "Hey, hey, LBJ! How many kids did you kill today?" "No blood for oil!" "The people, united, they cannot be defeated." "Drill, baby, drill."

And sometimes crowds turn violent, as in this description of students, angry over rising college tuition. Protesting in London in December 2010, they confronted Prince Charles and his wife as their car passed through the crowd.

About 50 protesters, some in full-face balaclavas and shouting "Tory scum!" broke through a cordon of police officers . . . protesters beat on the side of their armored, chauffeur-driven Rolls-Royce with sticks and bottles, smashing a side window, denting a rear panel and splashing the car with white paint. A Jaguar tailing the car and carrying a palace security detail was so battered that the police ended up using its doors as shields . . . Other violence across the city continued into the night, with demonstrators trying to smash their way into the Treasury building at the heart of the Whitehall government district with makeshift rams made from steel crowd barriers, shouting "We want our money back!" The protesters set small fires and clashed with riot police officers and mounted cordons outside govern-ment buildings. BBC reporters at the scene wore helmets as the rioters threw shattered blocks of steel-reinforced concrete. (Burns 2010)

This is pretty extreme, but how much different is this from overly enthusiastic rowdies at a soccer match? To a lesser degree, but equally quixotic, is fairly typical die-hard fan behavior, with painted faces and bodies, headgear and apparel in tribal colors, screaming for their players to smash the opponents. If you have no familiarity with sports, such behavior looks outlandish and threatening.

Even a nonviolent street protest can look like a crowd of boisterous, angry people. The potential volatility of their emotions makes them unpredictable. A charismatic leader or a powerful speaker can ignite their ire. A sudden threat or a show of force by authorities can lead to confrontational shouting, pushing, fighting, and serious violence. This is the image of social movements held by many, and not just those who are threatened by the protests. But like the proverbial iceberg, there is much more below the surface.

Mobs and angry crowds have often been associated with popular revolutions of agitated people who have dropped their everyday routines to voice their grievances. This is what was seen from the balconies of the genteel and privileged class in the tumult of nineteenth-century France. Gustav Le Bon (1841–1931) wrote of crowds as a "social realist," (1903) unearthing what he thought was an unconscious crowd psychology. An unreasoning herd mentality enveloped crowds, subverting individual identity, and destroying the accumulation of individual consciences. He described crowds, but he was also metaphorically describing democracy. Le Bon's real dislike was widespread political participation, where the "popular classes" become the "governing classes," pushing out hereditary elites or members of the educated aristocracy who had been groomed to rule. Crowds for Le Bon typified the modern era that was increasingly chaotic, bordering on anarchy in politics, cultural mores, and intellectual thought, in a word, democratic.

Elias Canetti, winner of the Nobel Prize for literature in 1981, similarly examines crowds as a stand-in for his view of modern society. Drawing unflattering observations about leaders and followers, his *Crowds and Power* (1984) is not a work of social science but has a great deal to say about his view of society. Canetti uses historical illustrations and anecdotes to describe a troubling state of current affairs. Unscrupulous leaders (power) direct the people (crowds) wherever it is convenient, in order to remain in power. Crowds like to follow, especially when they think they are taking the lead. Canetti, like Le Bon, offers various types of crowds, e.g., rhythmic, stagnating, feasting, prohibiting, that seek to dissuade any reader from trusting participatory democracy.

Soon after World War II, Eric Hoffer (1951), a very likable longshoreman and autodidact, found popularity with *The True Believer*, a primer about the people who find solace and camaraderie in mass movements.

(Continued)

(Continued)

They are the gullible, the weak, the poor, the bored, the sinners, the cynics; but most importantly, they are persons eager to follow a leader. His insights did not pretend to reference history or any other disciplinary research to support his bromides and platitudes. Offering a combination of pop psychology ("Hatred springs more from self-contempt than a legitimate grievance") and pop sociology ("The mass movement aims to infect people with a malady and then offer the movement as a cure"), Hoffer was often interviewed and endlessly quotable by anyone wanting to dismiss social movement activities that might upset the apple cart.

It wasn't until social and behavioral scientists began looking at social movements as social organizations, nascent and sometimes transitory, that the outlines of a much more complex and interesting social phenomenon became apparent. With more research, especially through participant observation, social movement crowds—protests, pickets, demonstrations, rallies, and riots—began to make sense, and the very rational and deliberate role they play as a tactic became obvious. Not surprisingly, most successful social movements carefully script their public displays of emotion. Even civil disobedience and property destruction can, on occasion, prove to be a rational and effective tool in furthering movement goals.[10]

---

[10]Try to see a film of the modern civil rights movement, e.g., *Freedom Riders*. The participants were dedicated to nonviolence, and they learned to publicly express their grievances with stoic determination and apparent fearlessness. It is painful to watch the white crowds and mobs that attacked them.

Social Movements as Drivers of Social Change

Linking Social Movements to Social Change

> [T]he sixties was without doubt the most portentous decade in the twentieth century from the perspective of grass-roots social change . . . Our behavior and our values [today] bear the imprint of these movements.
>
> —Robert Putnam (2000: 152)

Despite what Robert Putnam and many others contend, it is not readily obvious how successful those and any social movements are in creating social change. This may seem like a peculiar thing to say, in light of the earlier discussion, "How Social Movements Matter." There are many forces influencing the direction of change. The impact of social movements on social change cannot be assumed or taken for granted. How, in fact, do social movements create or influence change? Edwin Amenta (2006: 6) poses this issue very well: "The empirical challenge comes down to demonstrating that important changes would not have occurred, or not in the way they did, in the absence of the challenger or the actions it took."[11]

The worldwide antiwar movement of the 1960s and early 1970s that sought to end the U.S. invasion and occupation of Vietnam and the bombing of Cambodia was extremely active, visible, and had hundreds of thousands of participants in marches, rallies, teach-ins, protests, and acts of civil disobedience. It seems certain that the end of military conscription (the draft) in the United States was the result of antiwar activity and public opposition. It is much less clear, however, that the course the U.S. government pursued in the war was greatly altered, and the end of the war hastened, by the antiwar movement.[12]

For several decades, social movement organizations have opposed nuclear power as an unsafe way to generate electricity. They cite the risk of a nuclear accident such as the meltdown of a reactor at Chernobyl, Ukraine, in 1986 that resulted in radiation spreading as far as Ireland and the resettling of more than a third of a million people. The difficulties of disposing of spent nuclear material and other nuclear waste make it a serious environmental threat. In Europe, the antinuclear power movement appears to have had significant success after the Chernobyl catastrophe, but Denmark, Luxembourg, and Norway had already dropped nuclear energy by 1986. Social movement activity was not alone in influencing the move away from nuclear power throughout much of Europe (Kolb 2007: Chapter 12). The part it did play is difficult to gauge.

Anthony Obershall, Timothy Garton Ash, and many others have written admiringly about the democracy movements that precipitated the collapse of the Soviet Union and communist-led governments in Eastern Europe between 1989 and 1991. Others are not so sure. Stephen Kotkin's *Uncivil Society*

---

[11]"Social scientists . . . all seem to agree that the study of the effects of social movements has largely been neglected" (Giugni 1998: 373; see Soule 2009: 35). Largely, but not entirely. William Gamson's 1975 seminal work, *The Strategy of Social Protest*, examined social movement consequences of fifty-three "challenge groups" throughout America's history, and other researchers have followed his lead.

[12]Doug McAdam and Yang Su (2002) provide several tests of this, with decidedly mixed findings.

(2009) describes the implosion of the Soviet Union and how the regimes in Romania, East Germany, and Poland essentially collapsed as a consequence. Certainly the bravery and creativity of Poland's *Solidarność* movement and the Velvet Revolution in Czechoslovakia are indisputable, but would things have been otherwise if they hadn't happened? That is the question.

There is sometimes a fine line between social movement formation in response to social change and social movements that engage in and direct change in otherwise quotidian circumstances. Doug McAdam's study of the U.S. civil rights movement observes that "most shifts in the [political opportunity structure] are themselves responses to broader social change" (McAdam 1999a: xviii), i.e., shifts that open and close doors for social movements. Social changes resulting from technological innovations, corporate decisions, war, and government action can lead to grievances and create conditions affecting the possible success or failure of social movements. Social movements, in turn, may become participants in directing the changes to which they are responding.

Untangling the Causal Processes. There are two approaches to understanding how social movements create social change. The first approach analyzes social movements as "transgressive contenders": their internal dynamics and the societal conditions in which they operate, e.g., the political opportunity structure. The other approach sorts out movement goals, actions, and claims of success, distinguishing them from other causal factors, and draws empirical connections between what a movement does and what happens as a consequence.

Charles Tilly provides a useful way to think about social movements as change agents by distinguishing between social movement claims, the effects empirically attributable to social movements, and effects resulting from other events and actions. These often overlap, as Figure 5.1 shows.

Social Movements as Effective Change Agents

As Marco Giugni (1998: 373) succinctly says, "The ultimate end of movements is to bring about change." Social change is the consequence of how movements are organized and operate, i.e., what they are. It is also the consequence of movement activities, i.e., what they do. Movement success depends on social context and how they exploit opportunities in the larger society. Finally, social change can go in unintended directions, sometimes as a result of the movement's efforts.

To Be Effective: What Social Movements Are. Successful social movements persist by developing some of the efficiencies of bureaucracies,

Figure 5.1 Identifying social movement outcomes

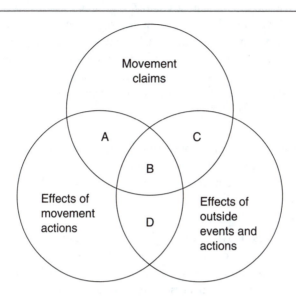

A = Effects of movement actions (but not of outside influences) that bear directly on movement claims
B = Joint effects of movement actions and outside influences that bear directly on movement claims
C = Effects of outside influences (but not of movement actions) that bear directly on movement claims
D = Joint effects of movement actions and outside influences that *don't* bear on movement claims

Source: Tilly 1999: 269

including established channels of communication and authority, a division of labor among groups of participants, and a deliberate strategy for obtaining and using resources.[13] Framing becomes linked to the collective

---

[13]The pursuit of resources, however, can distract or detour a movement from effecting social change. Few of those who provide resources to social movements give without some sense of reciprocity, i.e., that they be heard and heeded. Benefactors, allies, sponsors, and others who support a social movement may try to direct the movement's actions or burnish its image in ways that limit it and so may dilute the effectiveness of the social movement's efforts.

identity of the movement that is shared to some degree by participants and the target audience.

Successful social movements are cohesive. Sometimes factions occur over basic ideology, with groups disagreeing in terms of morality or fundamental principles. Factions may hold differing "prognostic frames," with fights over what actions to take. Other times factions are more cultural, breaking along divisions of age, social class, ethnicity, and occasionally gender. When factions form, successful social movements find ways to retain a common purpose and effort.

To Be Effective: Tactical Repertoires. The tactical repertoires of social movements explicitly target desired effects. When a corporation or the state responds as the movement intended, it is fair to credit the movement with effecting change.

Economic Impacts. Having a material impact is especially effective when a grievance is focused on a profit-seeking organization. The state is also vulnerable to this, as tax resisters have found. Sarah Soule (2009) provides an accounting of successes of several social movements, from university divestment campaigns against South Africa's apartheid government to the student antisweatshop movement against labor conditions in Third World sport apparel factories and the campaign to protect dolphins from tuna fishing practices. King and Soule (2007) document the "abnormal stock price returns," i.e., a fall in stock prices attributable to social protests of products or practices of three dozen corporations. In all of these cases there was a real or threatened loss to a corporation caused by a boycott of products. The losses were mitigated by the corporations responding to social movement grievances.

Violence and Disruption. There are movements that have benefited from highly disruptive activities, including the use of violence. Frances Fox Piven and Richard Cloward's 1979 study, *Poor People's Movements*, explains how rioting in the late 1960s in U.S. cities, much decried by most Americans, was a successful tactic in getting legislators to take action.[14] Gaining recognition for a union and negotiating a collectively bargained contract are highly improbable outcomes without the threat or actual use of disruptive

---

[14]The success may have been short-lived and, though supported by other research (see also Button 1978), not all researchers agree with their conclusions (e.g., Schumaker 1975).

tactics (e.g., strikes, sit-ins, picket lines) toward those who will be required to share profits in ways they otherwise would not.

THE MOVEMENT TO WIN COLLECTIVE BARGAINING

By the time Iris Summers was a mother, most of her siblings and cousins had left the farms on which they grew up. They moved to towns and cities, many to work in factories. They had little recourse but to quit a job when they objected to the low pay or working conditions; if they objected openly, they were fired. During the era, the idea of workers forming associations that could bargain collectively for wages, benefits, and improved working conditions was gaining public acceptance, despite decades of antiunion violence, the hostility of newspaper editors and owners, and no shortage of antiunion politicians.

Craft guilds that protected the knowledge and practice of crafts date from the Middle Ages in Europe. In the United States, guildlike unions were established soon after the Revolutionary War. They were seen as conspiracies, however, and usually held their meetings out of public view (Hodson and Sullivan 2008: 17). Efforts to unionize workers across crafts and regions in the United States date from the first year after the end of the Civil War, 1866. The National Labor Union, the Knights of Labor, and other industrywide unions were largely unsuccessful in sustaining large organizations between 1866 and the mid-1930s, however, due to a combination of employer hostility and government opposition (Griffin, Wallace, and Rubin 1986).

In the late nineteenth century, craft workers formed unions of artisans and mechanics in select trades. They joined together in 1886 as the American Federation of Labor (AFL), led by Samuel Gompers, and achieved a membership of more than two million by 1914. The AFL limited its scope to work-related issues and posed no threat to free-market practices per se, rarely entering into the realm of political give-and-take by endorsing candidates or forming a political party.

Without legislation that gave unions legal status to represent associations of workers, they could be suppressed with whatever means of coercion the state had at its disposal, including using troops and police to break up rallies and marches and the imprisonment and deportation of labor leaders. Newspapers regularly attacked efforts to unionize workers, and union activists were often viewed as anarchists, violent

(Continued)

(Continued)

bomb throwers, assassins, and enemies of private property (Brecher 1977).[15]

A hundred years ago, union organizing of workers across many skills, work sites, and states was the program of the Industrial Workers of the World (IWW) or Wobblies. The Wobblies were frequently involved in violent confrontations with the police and government troops. The killing by Colorado National Guard troops of twenty people, including eleven children, living in tents during a strike in Ludlow, Colorado, was followed by violent retribution by miners. Allied with the Western Federation of Miners, the Wobblies reached a membership of perhaps 100,000, mostly in the Rocky Mountain States and the Northwest. After the First World War, repression of the IWW was fierce and largely successful. Hundreds of Wobblies were imprisoned and deported, and many of their leaders were lynched while in police custody. Joe Hill, executed by the State of Utah, and WWI veteran Wesley Everest became martyrs to the cause. By 1930, membership in the IWW had dwindled to about ten thousand men and women.

In 1935, a group of unions broke with the AFL and began the Congress of Industrial Organization (CIO) in order to pursue the IWW goal of industrywide unionization. Led by the fiery orator and leader of the United Mine Workers, John L. Lewis, the general strike (stopping all work and public services for an entire city) and other aggressive actions were the tactical repertoire for this new labor movement. The doorway to widespread unionization, however, was the front gate of an automobile factory, not the opening to a mine shaft.

With the onset of the Great Depression and the high unemployment and poverty that struck the nation and the world in the early 1930s, calls for changes in corporate behavior and the role of the state were widespread. As Irving Bernstein (1966) writes in *The Turbulent Years*, popular doubt was growing about industrial capitalism's ability to solve the nation's problems without both state action and a change in the

---

[15]In fact, some were. Among the repertoires of contention of anarchists in the early labor movement was the use of violence and a willingness to fight back against police violence. Beverly Gage's *The Day Wall Street Exploded* (2009) describes violent anarchist cells as well as police and company strategies of falsely convicting labor leaders of crimes they didn't commit. See also Gary Gerstle's (2008) fascinating history of anti-immigrant movements that targeted labor activists.

operations of the economic system. After twelve years of Republican presidents, there was a more favorable opinion about industrial unions in the presidential administration of Franklin Delano Roosevelt. His administration created the National Labor Relations Board, and in 1935 the federal government passed the Wagner Act. This made legal the formation and operation of industrial unions and required companies to negotiate the terms of work with union representatives.

A critical episode involved the United Auto Workers' (UAW) effort to gain union recognition by the major automobile companies. In light of the corporations' opposition to unions, the Flint, Michigan, local of the UAW, led by 24-year-old Victor Reuther, organized a bloody but success-ful sit-down strike at the General Motors' die-making plant in Flint, Michigan. Because of this and other events—some of them involving police, worker, and company violence—unions became a mainstay for millions of workers.

General Motors and the other automobile companies had tried to stop the union from signing up members and refused to recognize the UAW as representing their workers. On December 20, 1936, after find-ing out that GM was planning to move die making to another plant, auto workers locked themselves in the plant, shutting down production. Clearly an illegal occupation of the company's property, this brought out the police and National Guard. The workers inside, however, insisted the company recognize the UAW and enter into bargaining over wages and working conditions. They were backed by their families and friends, along with political and civic leaders who supported the formation of unions and collective bargaining rights.[16]

The attempt to drive the workers from the Ford plant lasted more than a month and on several occasions became violent. The graphic photographs of workers being beaten and tear gassed made front pages across the nation. These, and the framing of events in terms of workers fighting for a better life during an economic depression created by financial elites, drew the attention of the entire country and evoked widespread sympathy. On February 11, 1937, GM agreed to collective bargaining, and Chrysler Corporation followed shortly thereafter.

---

[16]The events of the Flint sit-down strike are captured very sympathetically in inter-views with striking workers' wives and historical footage in the film *With Babies and Banners*.

(Continued)

(Continued)

In 1937, nearly three out of four Americans approved of labor unions, and in that year the UAW grew from 30,000 to 500,000 members. Between 1939 and 1978, public support for unions never dropped below 60 percent. Curiously, based on Gallup public opinion surveys done since 1952, "the percentage of the workforce belonging to a union was at an all-time high in the 1950s ... [But] public sympathy for unions was actually higher in 2005" (Panagopoulos and Francia 2008: 137). The strong positive public approval does not translate into widespread union membership, as it has declined in every decade since the 1950s.

Litigation and Rights. The environmental movement in the United States uses the tactic of litigation far more effectively than many movements, petitioning the courts to delay and block the issuance of permits needed by coal mining and petroleum corporations to access energy resources and by logging companies to cut timber on public lands. Both the gay rights movement and its opponents have used the banner of legal rights to frame their cause (Werum and Winders 2001). Proponents of gay rights, however, are much more likely to seek social change through the courts, relying on the U.S. Constitution and legal precedent. Their opponents use the language of rights as well, but to mobilize public opinion in order to win referendums and ballot initiatives, a tactic normally less favorable to gay rights.

Tracing the Effects of Social Movement Actions

Social movements drive social change in at least four ways.

- They change public opinion by inserting at least part of their diagnostic and prognostic frames into popular discourse.
- They help create or alter policy, both in the way policies are made and in the effects of policies on state administrations, public organizations such as universities and churches, and corporations.
- Social movements impact the culture: music and dress, the way individuals interact with one another, the language they use, and the values, beliefs, and attitudes that unite and divide them.
- Finally, social movements have a lasting effect on many of those who participate in them: their identity, life choices, and level of public involvement.

Public Opinion and Movement Success. Researchers differ among themselves about the effect of public opinion on the political process.[17] A conventional politics model (Kolb 2007: 156–158) sees public opinion as important in deciding what elected representatives do, along with lobbying, constituent pressure (e.g., letters and community meetings), and electoral strength, including campaign contributions.[18]

Mobilizing public opinion may be most effective when the target is a corporation. For example, in 1977 the Infant Formula Action Coalition (INFACT) brought to the public's attention the downside in poor countries of feeding infants baby formula rather than having babies suckle their mother's breast milk. The coalition, joined by thirty-seven other groups, singled out one company, the Nestlé Corporation, and publicized Nestlé's billboards and other ads encouraging women to use infant formula. INFACT marshaled statistics on malnutrition resulting from poor mothers watering down the formula, as well as Nestlé's earnings from infant formula sold in poor countries. INFACT was successful in organizing a worldwide boycott of Nestlé's products (Solomon 1981). After seven years, Nestlé's changed its marketing practices to conform to International Health Organization guidelines, and in 1984 the boycott ended. All of the features of a successful social movement were present in INFACT's campaign. It is empirically valid to credit it with success (Soule 2009: 109, 148).

---

[17]Paul Bernstein observes how some scholars believe that social movements' efforts to sway the democratic state through public opinion "seldom have much impact because democracy works so poorly, while others argue that they have little impact because democracy works so well" (Bernstein 1999: 3). On the one side is the view that democracy is highly imperfect; the political system caters to elites and only pretends to be doing the public's business. The other side believes that democracy is an open political system and responsive to the public; "elected officials know what the public wants and respond to its demands . . . they would be foolish to respond [to social movements] . . . rather than the majority, because doing so could cost them reelection" (Bernstein 1999: 4).

[18]For some researchers, this is an overly optimistic view of politics. They stress the importance of power and the ability to apply pressure not only on politicians but on corporations and other institutions. Public opinion can easily be ignored by politicians and manipulated by elites. When the middle class is aroused, organizations may take note, but the poor and powerless have little voice and gain minimal attention. Only when they are effectively organized for action and grievances are framed in ways that can challenge authority, will they be addressed.

Public opinion can be powerful when it recognizes movement participants as victims of injustice. McAdam's research on the U.S. civil rights movement, and especially its dramatic victories from 1960 to 1965, charts its exposure of the ongoing injustice of racism (racial victimization). The movement invoked powerful scenes of discrimination and injustice in the confrontations between nonviolent protesters and a violent white public and law enforcement (see also Lee 2002).

The sit-ins at lunch counters by four black students at North Carolina-Greensboro were quickly followed with sit-ins by students elsewhere (Morris 1984). These presented searing images of White's hostility and violence against young people asking for nothing more than a cup of coffee. Starting in 1961, arrested sit-in protesters began refusing to post bail, creating jail-ins that threatened to overwhelm law enforcement resources. Freedom riders, black and white civil rights activists, boarded interstate buses to challenge state laws allowing segregation of public transportation. Many were brutally beaten by white mobs. Three hundred of them filled the Mississippi State Prison in Parchman. Beatings, imprisonment, and a burning bus made for unavoidably disturbing images in newspapers around the country and the world and forced a reluctant federal government, the Kennedy administration, to intercede.

White racial supremacists reacted to nonviolent protest actions exactly the way the civil rights leaders expected. Attacks on marchers by police, sometimes using dogs and blasts from high-pressure water hoses, as well the murders of civil rights workers engaged in voter registration, finally brought the federal government into the equation, leading to landmark legislation.

On March 9 [1965] state troopers attacked and brutally beat some 525 persons who were attempting to begin a protest march to Montgomery. Later that same day, the Reverend James Reeb, a march participant, was beaten to death by a group of whites. Finally, on March 25 . . . a white volunteer, Mrs. Viola Luizzo, was shot and killed while transporting marchers back to Selma from the state capital. In response to this consistent breakdown of public order, the federal government was once again forced to intervene in support of black interests. On March 15, President Johnson addressed a joint session of Congress to deliver his famous "We Shall Overcome" speech. Two days later he submitted to Congress a tough voting rights bill . . . The bill passed by overwhelming margins in both the Senate and House and was signed into law on August 6 of the same year. (McAdam 1999a: 179)

ABORTION AND THE BATTLE FOR PUBLIC OPINION

A woman's access to abortion in the United States illustrates the role of public opinion in social policy. Looking at public opinion in the past half century and the way social movements for and against access to abortion have sought to advance their movements' goals, public opinion stands out as a very significant part of social change through social movement activity, especially in determining the limits imposed on women who seek to terminate their pregnancy.

In the 1960s, there was widespread public discussion, and public opinion was not particularly favorable toward abortion, but it was also not an issue that tapped widespread and deep public concern. Studying Gallup polls between 1962 and 1969 and the 1965 National Fertility Study, Judith Blake found that approval of abortion varied greatly with the reasons for terminating a pregnancy. Most adults approved of abortion to save a mother's life, but between two-thirds and three-quarters disapproved otherwise, including "if the family does not have enough money to support another child." Even more people disapproved of terminating a pregnancy "if the parents simply have all the children they want" (Blake 1971: 541).[19]

Abortion was restricted in most states in the early 1970s. In the most restrictive states, it was legal only in the case of incest or rape or to save the woman's life. In 1973, the U.S. Supreme Court heard *Roe v. Wade*, a case pitting a woman against the state of Texas, whose statute prevented her from legally terminating her pregnancy. The Court ruled against the state statute, saying that any law creating a barrier to access to abortion in the first three months of pregnancy was unconstitutional. They based their ruling on the equal protection clause of the Fourteenth Amendment of the Constitution and the opinion that the Constitution implicitly protects the right to privacy.

Examining NORC General Social Survey data, Lucky Tedrow and E. R. Mahoney (1979) show increasingly favorable public opinion with regard

---

[19]Concern for a woman's health was spurred by the public's response to women's deaths as a result of efforts to abort the fetus themselves or having a nonprofessional terminate the pregnancy, the so-called "back-alley abortion."

(Continued)

(Continued)

to abortion between 1972 and 1976. This was also the beginning of a polarization of attitudes and public opinion that is very much a part of the pro-choice/pro-abortion and pro-life/antiabortion social movements today. Overall, a majority of the population agreed with the *Roe v. Wade* ruling, believed abortion should be legal and available, and thought abortion was largely a matter to be decided by a woman and her doctor. A dissenting opinion was equally clear in its opposition to the court's majority and was willing to disallow women the personal decision of rejecting or choosing abortion. So began the modern pro-choice and pro-life movements.

Pro-choice groups were initially successful with judicial decisions and have continued to litigate on behalf of a women's right to "control her reproductive health." Their major repertoires of contention have been large rallies and marches in prominent venues, strong public spokeswomen such as U.S. congresswoman Bella Abzug, funding research on the personal consequences of unplanned pregnancies, and garnering support for funding of family planning that includes abortion counseling. Alliances with the medical and public health communities, groups concerned about overpopulation, and the Democratic Party have also benefited the movement.

Pro-life groups have also used the judiciary to their advantage, especially regarding restrictions on teenagers' access to abortion. They have been very visible during confirmation hearings for U.S. Supreme Court justices and have been able to make abortion restrictions a central plank in the Republican Party and with regard to party candidates. Direct actions, such as blocking access to medical clinics where abortions are performed and identifying doctors who perform abortions, have maintained the visibility of the antiabortion movement. It has strong allies among Christian conservative groups, popular right-wing radio and television personalities, and the Catholic Church hierarchy.

A majority of the U.S. population has approved of abortion throughout the past forty years. There have been some fluctuations, but support for the decision of *Roe v. Wade* has never dipped below 50 percent of the population. Similarly, when given the choice of self-identification, more people consistently identify as pro-choice, with a minority identifying as pro-life. Greg Shaw concludes from his analysis of public opinion polls in the 1980s and 1990s that "general support for abortion...remains, more often than not, the majority position" (Shaw 2003: 408) despite decades of vociferous and well-organized efforts by both the pro-choice and pro-life social movements. Framing the issue as one of privacy (as Justice Harry Blackman did in writing the *Roe v. Wade* majority opinion) and personal control resonates with many people, including conservative libertarians, but the argument that abortion is murder is equally convincing for those opposed to abortion.

Public opinion on abortion became a highly contentious political issue in the 1970s. In time, it became an indicator of a constellation of opinions, often referred to as "family values." Those agreeing that abortion should never be permitted have consistently remained less than a quarter, and often less than one in five, of the population. Those believing abortion is entirely a matter of personal choice are about a quarter of those surveyed. The remaining 50 percent or more believe abortion should be legal, but restrictions should be applied (Shaw 2003). Opinions have changed surprisingly little in recent decades, despite the ongoing public relations war.

Abortion is legal and available, though state and national legislation and legal opinions have created restrictions for some procedures and reduced access to abortion, especially for teens, military personnel, rural and small-town women, the poor, and for late-term abortions. As the political winds have shifted, public funding for international family planning services has sometimes been available and sometimes been suspended, depending on who is in the White House. Nominees to the Supreme Court continue to be asked about their views of *Roe v. Wade*. The major parties and candidates continue to stake out their pro-choice or pro-life positions. Because policy on this issue seems strongly driven by public opinion, until one social movement or the other can radically shift the public's views, and barring a Supreme Court opinion reversing *Roe v. Wade*, policy is likely to remain much as it is today.

Political Process and Policy Changes. One way to see how social change is driven by the successful actions of social movements is to apply Gamson's (1975) concepts of *acceptance* and *advantage*—what might be called access to and benefits of political action by the state. Getting into the halls of power, being heard, and finding a place at the table is part of access and acceptance. Favorable changes in legislation, court rulings, desired enforcement, tax breaks and expenditures, research support, and regulations are among the advantages or benefits conferred by states as measures of social movement success.

Similar but more encompassing are Felix Kolb's (2007) five tactics to affect political institutions and governmental power: disruption, changing public preference, gaining political access, acquiring judicial favor, and using international politics. Each results in a type of political and policy change.

Disruption mechanisms involve the actions that slow down or halt an ongoing state action (enactment and enforcement of laws) to gain the attention of both political officials and the public at large. Attention to disruptions allows members of a movement to present their grievances and advance public preference mechanisms, including changing public opinion in favor of a social movement's proposals for change.

Political access mechanisms are usually conventional: e.g., energizing voters and exhibiting electoral power, gaining seats on committees and boards, and creating organizations to raise money, lobby, and offer ideas and analyses for the solution to problems. David Vogel's (1978; 1996) study of the public interest movement that began in the 1960s shows how a coalition of consumer protection, environmental, and other groups very systematically worked to change government oversight of corporate activities and expand the state's efforts to plan urban growth, reduce poverty, improve housing, and generally use the state to represent citizen interests against corporate power.

Judicial mechanisms such as legal suits against actions by public officials and others (e.g., importers, doctors, polluters) seek redress for a wrongdoing. If successful, they create legal precedent for what the movement believes are rights and wrongs. Finally, international relations move nations to take action as participants or signatories to international conventions and treaties. Domestic and international nongovernmental organizations like Human Rights Watch, Amnesty International, and World Wildlife Fund are active in urging countries to sign and ratify treaties that favor their causes.

Largely confined to the "causal dynamics" of social movements for political change, when states alter budgets, change modes of administration, adopt policies, grant legal statuses, and shift public resources in response to, and acknowledgement of social movement efforts, the movements have been successful change agents.

Cultural Impacts of Social Movements. The construction of culture and changes in cultural practices are a significant outcome of social movements. Some social movements—like the Guerilla Girls—are directly focused on the arts, what might be called "high culture." Others promote changes in outlooks, personal practices, uses of technology, and ways of living. Religious social movements, as well as secular social movements in which religion or religious groups play an organizational part—for example the civil rights movement—have the expressed purpose of changing personal beliefs, practices, attitudes, and values beyond participants in the movement itself. Movement participants are among the first to adopt these, and when shared by many and promoted intentionally or by example, they become available to a wider public. This is what Mohandas Gandhi meant when he said, "If you want to change the world, be that change."

When movements, including those espousing personal change, seek a wider audience, the intention is to advance social change. Cultural markers such as music, dialect, dress, leisure pursuits, literature, art, and the arrangement of living space express a group identity. Individuals only mildly sympathetic to the movement may be attracted to its music, dress, or style and incorporate these into their personal lives. Sensing popular appeal, the culture items, in turn, can be co-opted by commercial interests that profit in their sale and disseminate them widely, though the original meanings may be entirely lost or abandoned (Earl 2004).

The Hari Krishna movement is more than proselytizing the Hindu faith. It encouraged changes in the way participants and potential converts view the world, relate to other people, live day to day, and focus their life. The proliferation of cultural products of this movement, and related cultural trends, extend beyond Hari Krishna adherents to include vegetarian and vegan cuisine, various forms of yoga, a preference for light, layered, loose-fitting, natural-fabric clothing, and households and yards adorned with religious figurines, shrines, and prayer flags.

The "beat culture," actually a subculture movement in the early post-World War II era and captured in Norman Mailer's *The White Negro* (1957), was culturally oppositional by design, rejecting mainstream lifestyles it characterized as conformist, conventional, and "square." Howard Becker (1951) explains this in his classic essay, "The Professional Jazz Musician and His Audience," showing how language, patterns of interaction, and especially music characterize a social movement that was increasingly adopted by a wider population. Jack Kerouac, Lawrence Ferlinghetti, Alan Ginsberg, James Baldwin, and other writers popularized a style for young people—especially in the rejection of racial and sexual barriers and middle-class lifestyle—that presaged the culture and social change of the 1960s counterculture.

In much of the United States and Europe, the environmental movement's stress on reducing carbon emissions stimulates a preference for bicycle transportation. This, in turn, has spawned a new bicycle culture that can be very political but is sometimes completely focused on changing cultural practices and building a subculture around bicycles (Mapes 2009). Significant efforts by alternative-transportation groups seek to influence policy, especially working for street and highway construction that incorporates bicycle traffic, enhances bicycle safety, and reduces the marginalization of bicyclists on city streets. The bicycle movement for many participants, however, centers on bicycle events and "bike fun" that draws in a wider audience who share elements of bicycle culture and, in so doing, selectively share the collective cultural identity conveyed by the movement.

Similarly, the cultural products of the emerging "food movement" go beyond changing public opinion and legislation. Movement leaders like writer Michael Pollan, restaurateur Alice Waters, and former Texas agriculture commissioner Jim Hightower have effectively advanced both diagnostic and prognostic frames of the movement for "slow food" and healthier eating, largely by contrasting their agenda with those of fast-food retailers, university programs in agriculture and food science, and industrial agriculture corporations. They challenge the authority of corporations and governments, especially government support for major grain commodities, by stressing today's unhealthy diets and environmentally costly farming and livestock practices. The core of the food movement, however, is personal choice and the alteration of peoples' behavior: what to eat, how to grow or otherwise get it, how to prepare it, and the consequences of doing otherwise. Beyond changing public opinion, the food movement actively promotes a cultural identity and values that are having significant social change consequences (Pollan 2010).

Personal Change as a Consequence of Social Movement Participation

Insular social movements focus their energy on participants themselves, and a few "escapist movements" completely shelter their adherents from the outside world. Insular movements promote "personal transformation as the key to societal transformation" (Snow and Soule 2010: 208). They foster and support lifestyle practices: e.g., becoming a vegan or vegetarian, devoting time to prayer, buying fewer commodities, recycling household waste, avoiding alcohol, tobacco, and caffeine, or practicing marital fidelity or premarital chastity. The personal changes are paramount. They are the limit of what the movement seeks to accomplish.

As agents of social change, most social movements focus less strictly on individual change and have wider goals of changing institutional and organizational practices. In the process, however, the participants themselves undergo changes that cumulatively become socially significant (Giugni 2004). George Mayes was made homeless in the Great Recession of 2007–2010 and participated in the January 20, 2010, rally in San Francisco for housing and support for the homeless. This is his reflection on the experience.

It was the first time I was ever in a rally or march in my life. I was looking at all the people on the side of the streets, looking at us and thinking, that used to be me. It's made me think a lot and reflect. The hollering and screaming and the camaraderie from the entire group of people from all over the country. The energy was high . . . It has pumped me up to do some more and I'm looking forward. It's important to advocate for the homeless. I feel like I'm a part of something. I've always tried to help, and since I found myself on the streets, I've learned what it means to advocate not only for myself, but for others . . . All of this has changed me and I'm still processing it, to be honest . . . It's hard to describe. (Bayer 2010: 3)

There is "a powerful and enduring effect of participation on the later lives of activists" according to Doug McAdam (1999b). He and others who have studied movement participation found that it is a formative event in peoples' lives and identities and continues to have salience for them (e.g., Griffin 2004, Whittier, 1997). Movement activists subsequently have higher rates of participation than others, and participants consistently recognize changes in themselves as a consequence of participating.

As discussed in Chapter 2, involvement is one of the mechanisms that contribute to activist political generations. In the formative young-adult years, social movement participation establishes, for a portion of an age cohort, an orientation to their social milieu that has a residual effect or staying power. It can be most significant years later, when the generation occupies positions of power and authority in social, economic, political, cultural, and educational organizations.

Social movement participation influences people in ways not directly related to movement grievances. For example, Hasso's (2001) study of Palestinian women attributes a greater sense of gender equality as well as greater personal efficacy and empowerment to their having participated in building Palestinian nationhood and opposing Israeli occupation of Palestinian territory. Taylor and Whittier's (1992) study of lesbian feminists and other studies of the women's movement emphasize the personal changes that became part of a broad orientation toward others and influenced a range of life choices.

Period Effects of Personal Change. Beyond the effect on the participants themselves and the impact of innovative groups like the Hari Krishna movement on cultural styles, much less is known about how social movement participation results in structural change. What little research has been done points to strong "links between social movement processes and aggregate-level changes in the life-course" (McAdam 1999b: 136). The question goes beyond the participants to ask, Does the experience of participating in social movements also cause those who do not participate to alter what would otherwise be their own life's trajectory? Is there a kind of spillover effect of personal change that becomes so widespread as to have a period effect of social change?

Doug McAdam's (1999b) ingenious study of movement participants in the 1960s (specifically the civil rights movement, opposition to the war in Vietnam, and the women's movement) begins with an analysis of early and later Baby Boomers' life-course trajectories and compares these to nonparticipants. He was able to statistically "hold constant" the influence of factors that differed between participants and nonparticipants in order to isolate and measure the influence of the social movements on three features of their adult life: having children, having lived with a sexual partner before marrying, and whether or not the person ever married. Widespread adoption of these practices constitutes structural change in society.

Other things being equal, movement participants were less likely to marry and have children and more likely to have cohabited, to the degree that he concludes, "Movement participation [was] a force shaping individual life-course choices" (McAdam 1999b: 128). This is biographical change. Was there also social structural change?

McAdam found that nonparticipants, especially those slightly younger than the participants, were individuals most likely to be aware of the alternative lifestyle options adopted by many of the activists. They also were more likely to not marry, to not have children, and were more likely to cohabit. That is, those participants whose biographies were altered by social movement participation created an "atmosphere of choice" that was responded to by those who came after them. Together, their lives were different from their elders and expressed social change that was important and measurable.

The research reveals "the clear imprint of those struggles" for civil rights, to end the Vietnam War, and for women's equality in the 1960s "in the current structure of the American life-course" (McAdam 1999b: 143). Over the next decades the demographic changes pioneered by movement participants became widespread. Unlike fifty years ago, it is not unusual for someone to live with another person before marriage, to never marry, or to choose not to have children. "The emergence of this broader set of life-course options," according to McAdam, was at least partly the result of the personal experiences of those who participated in the social movements of the 1960s.

As discussed in Chapter 2, this becomes a period effect that, along with cohort and age effects, helps explain the life course of a generation. But moreover, by having deep personal consequences, social movements can also generate choices for others who come later and may take for granted ways of living unavailable to or inappropriate for an earlier generation. For them, the political content of the choices (rights, war, equality) is less important or not even recognized, but the lifestyle options remain.

Topics for Discussion and Activities for Further Study

Topics for Discussion

1. Who in the class has been involved with a social movement? Is anyone a social movement activist? Describe your experiences. How well do these match what you have read in this chapter? Talk about how you came to be involved. What personal consequences, if any, did you experience? Will you be more likely to become involved again?

2. A popular tactic increasingly used in the age of digital, omnipresent cameras is to visually record an individual or group acting inappropriately, post this on the web, and embarrass them, possibly discrediting them entirely. Conservative videographers have found success in posing as disreputable clients and eliciting seemingly incriminating evidence of organizational wrongdoing, e.g., in the antipoverty organization Acorn and Planned Parenthood. Liberal opponents have used bogus identities to contact and record conversations with their opponents, and animal protection groups are fond of getting footage of concentrated animal feeding operations (CAFOs) while posing as employees. What do you think of such tactics? What are the possible pitfalls and unethical dimensions of these activities?

3. What social movements are most in the news today? How much do they appear to be grassroots social movements? How much are they more "Astroturf," as Edward Walker (referenced in Footnote 5) describes some social movements? Does it matter if it is grassroots or Astroturf? Why would an Astroturf movement want to appear to be a grassroots movement?

4. Discuss Figure 5.1. Try to think of examples that illustrate each of the spaces in the diagram. Same-sex marriage advocates seem to be gaining in pursuit of their goal. So are those who support the legalization of marijuana. Can you locate their claims and tactics and the social changes going on that are both an outcome of, and are occurring despite, what their advocates are doing?

5. As a class, discuss the three frames Benford conceptualized: diagnostic, prognostic, and motivational. What frame is most likely to demonize the government, corporations, or an adversary's social movement? What frames have been crafted by those who deny that global climate change is being caused by human activity?

Activities for Further Study

1. Select a social movement from the past that interests you. It might be the 1930s Townsend Plan studied by Edwin Amenta, the suffrage movement or the pre-Civil War abolition movement. Or you might pick a more contemporary movement such as the public interest movement David Vogel studied, the antiabortion movement, the Tea Party, or the antinuclear movement. Find out about its grievances, its framing, its participants, and its repertoires of contention. What impacts has it had on social change?

2. Spend some time interviewing an older social movement activist. In almost every community there are people identified as activists, but you may need to ask around in yours. You might want to read about interviewing techniques before doing your interview, in order to be prepared with questions, the best way to take notes, and what you do to analyze the information you collect. Talk to this person about the movement, the context in which the movement operated, and their experience as a social movement participant. Compare what you find to what you have read in this chapter.

3. How does a political campaign differ from a social movement? Why do social movements often attach themselves to campaigns? How do campaigns reach out to social movements? Find an example of this from a recent past election and analyze the interactions between campaigns and movements.

4. Many corporations would like to stay off the radar screen of any social movement, but it is sometimes difficult. How do corporations respond to social movements? Look for writing that advises corporations on how to respond. (There is a huge literature on how to beat back challenges from labor movements, but also from social movements more generally.) Find a case where a corporation has had to defend itself, responded to the social movement, and took steps to get back off the radar.

5. The recent popular protests in many countries in the Middle East have been described as democracy movements. Much has been written about how information technology, especially social networking and the Internet, has been integral to their actions. There has also been some analysis of how these same technologies have been used against the democracy movements. In research groups, look at several sides to this phenomenon, including the idea that information technology has been involved but has not had the significant impact some attribute to it.

6

War, Revolution, and Social Change

Political Violence and Structured Coercion

"No one is ignorant enough to prefer war to peace," admonished the Greek historian Herodotus (485–425 BCE). "In peace sons bury their fathers, and in war fathers bury their sons."

Abhorring war seems the most natural sentiment. Yet war is and has been an omnipresent part of contemporary human existence. The destructive power of modern war takes an enormous psychological and economic toll, and the deaths resulting from wars in the past century are in the tens of millions. War is a powerful force, capable of altering national borders, dramatically changing the ethnic composition of people over a large area, ushering in as well as destroying political, religious, and economic systems, altering cultural practices and deeply held beliefs, realigning social classes, national groups, and gender relations, and both destroying and redistributing wealth. Social change is driven by war.

It would be comforting to think that war is an interruption in the normally pacific affairs of human endeavor, but this is not the case. War may not be, as Winston Churchill remarked, "the story of the human race," but its presence is unremarkable. United States troops have been actively fighting somewhere in the world for forty-eight of the past one hundred years. Wars in the last century were fought so often and on such a massive scale, they cannot be thought of as aberrations. War influences the path and scope change takes, depending on the precipitating circumstances, mode of conduct, and settling of accounts in its aftermath. War is a part of the continual process of social change characterizing the human condition.

War as an Instrument of Social Change

Significant scholarly effort has gone into understanding the causes of war. The horrific personal and social experiences of war and the capacity for total annihilation made possible by nuclear weapons makes this imperative. Because wars are intended to benefit victors and penalize the defeated, understanding the conduct of war can increase a nation's chances of being the former and not the latter. How to conclude wars and, in their wake, take steps to reconcile and rebuild—both on humanitarian grounds and to reduce the likelihood of future wars—challenges victors and vanquished alike. War, however, is only occasionally examined as a driver of social change.

Given the intimate relationship between state power and organized, lethal coercion of war, it would be possible to treat war making as one of the ways states effect social change, the topic of Chapter 8. That would consider war as a means to an end, much the way legislation providing for universal education is a means to enhancing a nation's human capital and increasing its overall well-being. War is much more than this. As Michael Mann (1988) convincingly argues, modern war, even when motivated by a desire to establish or maintain peace,[1] is an ongoing and major force influencing and structuring social, political, and economic life.

Though war itself, protracted and expensive in dollars and lives, is a force that effects social change, those who conduct war may intend for it to do just the opposite: to maintain a way of life or a position of security or domination. Rarely, if ever, do things remain the same when a war is underway, as this chapter shows. When it is over, indeed, a great deal has

---

[1] *Si vis pacem, para bellum.* If you want peace, prepare for war, according to Roman historian Vegetius. This is, in one version or another, a sentiment widely shared, as reflected in public opinion polls and the willingness to spend large sums to support a modern military. People in many nations believe with great sincerity that war's purpose is to ensure peace. Some people would destroy others' peaceful way of life without the threat of war. Occasionally, those using hostile aggression must be vanquished through war. Wars are fought to win and guarantee peace and freedoms. These views, strongly held, are not disputed in this chapter. They are largely beyond the scope of the author's ability to empirically confirm or refute.

changed. The many ways war directs the scope and depth of social change is the subject of the chapter.

The topic of war and social change includes the resistance to war as a mechanism of social change. Social movements and nongovernmental organizations work for peace and to heal the wounds of war. As discussed later in the chapter, peacemaking is not confined to the national state. International diplomacy and international organizations work to prevent war and assume responsibility in its aftermath. Building the foundations for peace, as well, is part of the process of making social change.

Versions of War as Coercive Politics

War is hostile action, including the use of lethal violence, carried out by political groups (e.g., national states and insurgents) of a duration and magnitude to require sacrifices and resources of a wide population, in whose name the conflict is conducted. In designating a conflict as war, it is common to set the magnitude at fifty thousand combatants. War by definition assumes some degree of symmetry or equality of capability between warring groups. War is neither irrational nor a consequence of human nature. It is a tool of contention used by nations against nations and against stateless insurgents. Both for reasons of custom and law, wars are expected to be fought according to rules that limit how they are carried out. When rules of war[2] are breached, the result may be prosecutions of war crimes. In order to understand how war drives social change, it is useful to distinguish types of state-sponsored wars as well as organized lethal conflicts that are not by definition war.

Total and Limited Wars. The world wars of the twentieth century were total wars. They required society-wide mobilization of resources, extensive

---

[2]The rules or laws of war are an evolving collection of treaties, protocols, conventions, and agreements signed by national states that codify acceptable and unacceptable actions in war. Most well known are the Geneva Conventions and Protocols, signed by 194 nations, covering the treatment of prisoners, the wounded, and civilians in war situations. The Hague Conventions of 1899 and 1907 have been built upon as new weapons have been developed. For example, the Geneva Protocol of 1925 prohibited the use of poison gas following World War I, and the Nuremberg Principles emerged from the war crimes trials against the most inhumane actions of the Nazis. It is perhaps easier to agree on how prisoners and civilians should be treated—for instance, the Hague Convention of 2000 outlawing child soldiers—than the use of violence more generally. Laws of war covering conventional and unconventional weapons urge combatants to use violence only when militarily necessary, a very vague guideline.

government control of industry and communication, and conscription of civilians into the armed forces. Total war aligns a group of nations in a commitment and effort to block the war-making actions of another group of nations. In the final analysis, national states use war to further or protect national interests or the interests of dominant national elites.

Limited wars are more likely to involve fewer nations and are more modest in their goals. States may be restricted in fighting a wider war because of economic, geostrategic,[3] and public support constraints, or they may simply lack the ability to rally allies needed to engage in a more concerted and protracted conflict. As will be seen, limited wars are not limited in their destructiveness, having been some of the most lethal wars in the past century.

Symmetrical and Asymmetrical Conflict. When nations attack one another, it is expected that they will contest each other's military actions with similar technologies and equipment, strategies, and capabilities. Something other than this is usually considered an asymmetrical conflict and given a name like pacification campaign, police action, intervention, clandestine war, or quelling an insurgency, rebellion, or insurrection. In many warlike cases, the term "war" is probably misapplied, such as Algeria's struggle for independence against the French that cost a million lives, the nation-building conflict between Israel and the Palestinians, and the violent resistance to Russian rule in the Russian republic of Chechnya. In a situation of occupation, where a powerful military fights to control a foreign site, the initial conflict may destroy the conventional military capacity of the occupied or invaded country, or the country may have been too poor to have developed a security capability in the first place.

Such asymmetry should benefit the invader, but this has not always been the case. The history of British (1789–1799), Soviet (1979–1989), and U.S. (2001 to the present) invasions of Afghanistan, and the U.S. invasions of Vietnam (1959–1973) and Iraq (2003–2011), illustrate how indigenous forces organize widespread guerilla movements with strong public support. In a terrain known best by the local fighters supported by the local population and using a combination of direct engagement and hidden-weapons

---

[3]Geostrategic most generally applies to considerations of international relations. National states have many agendas, and their war-making efforts may conflict with what they are doing elsewhere. For example, at the time of the U.S. invasion and occupation of South Vietnam, the Soviet Union was North Vietnam's ally and China's opponent, creating a tightrope for U.S. relations with both countries on matters unrelated to the war, thus limiting the scope of the U.S. conduct of the war.

tactics, they can prove to be formidable and sometimes successful opponents in seemingly asymmetrical conflicts (Arreguin-Toft 2005).

Third parties may join in, bolstering the indigenous forces and moving the conflict toward symmetry (Regan 2002). If, in the course of resisting the more powerful force, invaded nations or militant groups are able to amass arms and organize a fighting force, and a protracted conflict ensues, it may become a more symmetrical conflict, approximating or becoming war.

The Cold War and Proxy Wars. The threat of war, rather than war itself, has often been sufficient for one nation to gain advantages over another without resorting to open conflict. The development of nuclear weapons during the Second World War extended the threat of war to the possibility of global annihilation. This and strong military alliances—e.g., the Warsaw Pact and the North Atlantic Treaty Organization—created a Cold War between previous allies, the United States and the Soviet Union. The threat and preparedness are credited with both the lack of war between major national states and a number of "proxy wars" over the next forty-five years.

After an initial demobilization following World War II, military budgets began to increase worldwide, and the portion of persons in the armed forces also began to grow. As Ruth Sivard (1988) shows, this was more the case in poorer countries, especially after 1960, but the United States and Soviet Union spent huge sums contesting each country's efforts to limit the sphere of influence and military capacity of the other. The United States "built a chain of military bases . . . in effect ringing the Soviet Union and China with literally thousands of overseas military installations" (Johnson 2000: 36). The Soviet Union created a strong military presence or capability in all the nations contiguous to it and along its border with China. The United States had nearly half a million troops abroad in 1987 and the Soviet Union had almost three-quarters of a million (Sivard 1988). Both nations had ships and submarines, including ones armed with nuclear weapons, trolling the seas, and each had arsenals of thousands of nuclear weapons poised to strike the other.

Following the failed Bay of Pigs invasion by the United States in 1961, the Soviet Union prepared to place nuclear weapons in Cuba, sparking a confrontation between the superpowers that came very close to setting off a nuclear war. The almost inconceivable destruction that would have followed reinforced the sense that countries possessing nuclear weapons and their close allies could not engage in direct conflict. The price of nuclear conflict was too high.

This did not prevent the intrusion into, and in some case the abetting of, wars in other locales by the Cold War adversaries. The rivalry took place in proxy wars where the United States and the USSR fought one another through their support of smaller, often very lethal wars. All the Cold War proxy wars during the last quarter of the twentieth century, such as Mozambique's civil war and in Afghanistan during the Soviet occupation, took place in poor countries. Angola's war is a good example of how proxy wars were fought.

The former colony of Angola on the coast of southwest Africa was a Portuguese colony in territory originally claimed by Portugal in the sixteenth century. Angola's war of independence, begun in 1961, was fought by three rebel factions. Two were primarily socialist and had Soviet support. The third had support from South Africa and the large African country, Zaire, an ally to the United States and a reliable conduit for U.S. military aid to proxy wars in Africa. After years of fighting, an agreement was reached in 1975 that would reconcile the three Angolan factions and allow joint participation in the new government. When the agreement fell apart, the strongest group, with Soviet backing, negotiated independence from Portugal and installed the first government of Angola. The United States joined South Africa, then an apartheid regime, in continuing to back its rivals, and the war soon resumed.

Fighting continued for years, with 300,000 more civilians dying in the conflict[4] and more than a quarter of the Angolan people driven from their homes. The United States provided money and munitions, and South Africa sent troops to support the main rebel group, while the Soviets gave funding and arms to the Angolan government. Its ally, Cuba, provided soldiers, teachers, doctors, and other professionals. Only in 1988—on the eve of the collapse of the Soviet Union and the end of the apartheid regime in South Africa—was agreement reached for a South African and Cuban withdrawal. Though the United States recognized the Angolan government in 1993, fighting continued until 2002 when the U.S.-backed rebel leader, Jonas Savimbi, died.

The collapse of Somalia is the exception that proves the rule. A civil war in 1991, after the Soviet Union's implosion, produced a terrible famine, with hundreds of thousands of peasants dying when roving bands of militia seized their stored food. A UN military intervention to stop the fighting and end the famine resulted in casualties to Pakistani and U.S. troops, and the intervention was cancelled. It was considered a disaster, but in fact it did end the famine and so could be considered a success. The Cold War had ended, however, and there was no longer an interest in using the

---

[4]The best estimate is that, between 1961 and 2002, 90 percent of the war's deaths, 750,000 people, were civilians (Sivard 1996: 19, 71).

intervention forces to install or gain favor with a new government that would be a U.S. ally against the USSR or vice versa. This would have stopped the factional fighting among Somali clans, but, because it did not happen, Somalia has been without a functioning government for two decades.

Civil Wars and Wars for Independence. War is defined in part by geography. War between nations, as usually conceptualized, is fought outside the territory of at least one of the warring nations. In contrast, civil wars are violent contestations within the nation or region of both warring parties, though other nations may provide munitions, financial support, and safe haven or reconnaissance sites to belligerents. Civil wars can involve an insurgent group or militant party taking up arms against the state in order to alter a region's relationship to the central government, to achieve greater independence and autonomy,[5] or to secede.

In recent decades, civil wars have proved to be tremendously costly in terms of human life. Three examples, in Sri Lanka, East Timor, and Chechnya, illustrate this all too well. Tension and violence between Hindu Tamils and Buddhist Sinhalese began in 1975 and erupted into terrible violence in 1983. The war, from which emerged the first systematic use of suicide bombers, continued until January 2009 when the Sri Lankan government annihilated the Tamil independence movement's army and leadership, killing thousands of civilians in the process. The war took more than 100,000 lives, at least half of whom were noncombatants, on an island three-quarters the size of Ireland.

In 1975, East Timorese declared independence from the former Portuguese colony of Timor but were immediately and forcibly annexed by Indonesia. The war of independence raged for more than twenty-five years, with possibly 15 to 25 percent of East Timor's population dying. In Chechnya, possibly 200,000 people were killed—as many as 80 percent of whom were civilians—in two wars (1996–98; 1999–2007) between Chechen insurgents and Russian soldiers fighting to suppress the independence movement.

Grievances motivating civil insurgencies range from economic disparities to political abuses and social discrimination. Civil wars may be framed as being defensive on the part of a minority when the intention is to provide

---

[5] Autonomy usually refers to some measure of self-governance, including raising and spending tax revenues, carrying out judicial proceedings, and determining the content and language of education. In the case of Kurds in Iraq and American Indians, it also involves control of natural resources (e.g., oil, timber, minerals) and revenues from leasing or extracting these.

greater security or autonomy for an aggrieved region or group within a country (Harff and Gurr 2004). Irredentist conflicts seek to redraw national or regional boundaries to better incorporate a population of common ancestry or national identity by reclaiming real or imagined lost territory.[6] They may begin as civil wars (one region against the national state) and escalate into interstate war when a contiguous state is threatened with refugees or shares the insurgents' ethnic identity.

Throughout the twentieth century in parts of the world controlled by distant nations—usually with occupying armies and colonial settlements—nationalist movements initiated wars of independence, setting off protracted conflict. Because the colonial powers had made alliances with indigenous groups and provided the groups arms to protect colonial rule, the conflicts between them and the independence movements had at least the appearance of a civil war.

In the twentieth century, civil wars of independence have followed a centuries-old script of ending foreign domination and expelling the agents of imperial rule. Because European colonialism drew national borders that reflected their interests in extracting the things of value in the regions they possessed, the borders were largely arbitrary with regard to tribal, linguistic, and historical circumstances. In most cases, especially in Africa, these seemingly inappropriate national borders have been very resilient and by and large remain the dividing lines for national sovereignty today. Nonetheless, civil wars have been common as ethnic groups have fought to acquire greater autonomy (e.g., the Ibo in Nigeria, unsuccessfully) or independence (Eritrea in Ethiopia, successfully).

ETHNIC CONFLICT AND CIVIL WARS

It is not unusual for groups of citizens to be treated inequitably. Economic opportunities and jobs, access to education, legal status and treatment before the courts, the right to live in a particular area, and social status and recognition are made unequally available. Sometimes the reasons for inequality and disparate rights are justified in terms of

---

[6]Irredentist claims can be based on ancient historical myth, a felt loss of status, or actual experiences. They are often bolstered by a religious claim, e.g., Our God intended this land for us (Smith 1992).

talent and hard work. Sometimes heredity—the luck (and bad luck) of birth—is accepted as a natural right. Inheritance of wealth, property, and social standing as an entitlement is often supported by law and custom. In nations that extol particular cultural practices, a religion, or ascribed identity—e.g., nationality, ethnicity—doors may be opened for those possessing the desirable features and closed for others. When doors are closed or only slightly ajar, conflict may be the result.

Barbara Harff and Ted Gurr's *Ethnic Conflict in World Politics* provides a useful conceptualization of four types of ethnic contenders whose grievances can result in conflict and even civil war: (1) ethnonationalists, (2) indigenous peoples, (3) ethnoclasses, and (4) communal contenders. "The first two types are peoples who once led a separate political existence and want independence or autonomy from the states that rule them today." Kurds, Tibetans, Tamils, and Palestinians "want to (re)establish their own states." (Harff and Gurr 2004: 19). States block their independence or sovereignty, sometimes violently, and civil war ensues.

Indigenous people like the Miskito Indians in Nicaragua and other native peoples of North and South America, the Karen of Myanmar (Burma), the Papua of New Guinea, and indigenous people in Bangladesh, Finland, and Canada" are mainly concerned with protecting their traditional lands, resources, and culture within existing states." Like ethnonationalists, they seek independence when autonomy or protective legislation is not forthcoming (Harff and Gurr 2004: 19).

The third type, ethnoclasses, such as Turks in modern-day Germany and Palestinians within Israel, seek full citizenship rights and an end to discrimination, without abandoning their cultural distinctness and ethnic identity. Their tactics are usually peaceful and within the political process. This is not always the case. Naxalites, with initial support from the Santhal and other tribal peoples who are among the poorest of India's landless people, have engaged in guerrilla struggle since 1967 to secure land rights for tillers against absentee landlords. They are now demanding far-reaching changes in India's economy and have mobilized in large areas of India with a Maoist ideology that attacks globalization and neoliberal capitalism (Ramakrishnan 2005). Deaths in this ongoing battle increase each year.

Communal contenders can be found in the political arrangement of countries such as Lebanon where Druze, Maronite Christians, Shi'i Muslims,

(Continued)

(Continued)

and Sunni Muslims compete for power. Political affiliation and voting are along ethnic lines, and power is balanced between contending ethnic political parties. Nigeria and many other African nations have constitutions and electoral laws recognizing that all major ethnic groups need to have political representation and so restrict the power of one group, even if it is a majority of the population. When the political balance is upset, civil war can break out, as it has in Lebanon, Sri Lanka, and Nigeria.

Resource Wars. The numbers are eye opening: a quarter of the earth's known petroleum reserves are in one country, Saudi Arabia. A third of the earth's natural wealth was used up between 1970 and 1995. Only 3 percent of the planet's water is fit for human consumption, and half of it is already appropriated for human use (Klare 2001: 18–19). Given increasing population and rapidly expanding economies in the world's most populous countries, it is not difficult to be frightened about a growing scarcity and competition for resources. In turn, the possibilities for conflict over resources seem to grow exponentially. Some scholars have predicted that the twenty-first century will be the century of resource wars, led by water wars but followed by wars for petroleum reserves and scarce minerals vital to information technologies.

Water has for thousands of years been contested. It is linked to many conflicts, most recently the genocide in Sudan and Israeli settlements in the West Bank of Palestine. But to date, no war has been declared by any nation with the stated goal of claiming or reclaiming water (Economist 2010b). In the case of minerals and timber, according to Michael Klare, "it is not uncommon for competing elites or power blocks to feud over control of valuable commodities, often producing prolonged civil wars . . . States with significant 'lootable resources' are 'four times more likely to experience war than a country without primary commodities'" (Klare 2001: 13, quoting a 1999 World Bank report). Petroleum is, however, the prize, as Daniel Yergin (1991) calls it in his history of the search and struggle for control of petroleum in the twentieth century.

The U.S. military is deployed around the world to protect economic interests and the movement of goods. Klare views this as a return to the pre-Cold War use of structured coercion, before the need to "contain the communist threat" became the dominant military and diplomatic preoccupation. Today, maintaining resource security is a critical task of the military, especially the protection of oil fields and shipping routes. Extending and strengthening the U.S. global reach into Turkmenistan, Uzbekistan, and Kazakhstan—the Caspian Basin—where a fifth of the earth's petroleum reserves lie, and maintaining a friendly Middle East will possibly prevent a resource war in coming decades. At least, that's the plan.

The New Wars and Terrorism. Much of the previous discussion of the twentieth century's wars may seem anachronistic today. Nuclear weapons have shifted the theater of war away from open clashes among affluent nations that possess the means to annihilate a foe and, possibly, life on earth. Proxy wars are less common today, replaced by resource wars and massive international military engagements against both potentially threatening regimes and nonstate militants and terrorists. These are the wars, and threats of war, that keep large portions of national budgets committed to military might.

The changes in warfare have been going on for several years. They include the military's recruitment and training of highly skilled individuals and the shift to high-tech defense industries. War itself may no longer be a historically episodic event with a beginning and end, a set of clear goals, and a plan to achieve them through organized violence. It may instead become a steady state of preparedness and deployment across the globe. It may be about building national infrastructure and protecting populations from threats within their own country. There may be less concern for violations of the sovereignty of nations whose undesirable leaders come under outside attack or where nonstate groups train and launch terrorist operations intended to destabilize political and economic systems rather than to defeat armies (Barnett 2004).

For people in the theater of war today, destruction and the personal and social costs of living with violent coercion are very real. For others, including civilians in countries at war but one fought in a distant land and with technologies that reduce one's own casualties, war can be largely ignored, as if nothing is amiss. Charles Tilly expresses this well: "Most powerful states enjoy a partial exemption from war on their own terrains and therefore, perhaps, become less sensitive to the horrors of war" (Tilly 1992: 68).

Ironically, from the ability to be sheltered from war emerges another form of violent coercion that goads states into war: terrorism.[7]

Doku Umarov was a seventeen-year-old pan-Caucasus separatist from the Russian republic of Dagestan. Following her suicide bombing that killed forty people in a Moscow subway, a message she left read, in part, "I promise you the war will come to your streets, and you will feel it in your own lives and on your own skin" (Barry 2010). Asymmetrical wars fought between a modern, conventional military and groups without sophisticated weapons, conventional training, or even uniforms, often unprotected by any national state, have become more common today. For such groups, terrorism is a tactic of contention of the most lethal kind.

In Chechnya, Sri Lanka, Indonesia, the Philippines, Iraq, Afghanistan, and elsewhere, this violent tactic of contention has altered the ways war is fought and the very idea that war has a delineated battlefield. The ramifications of terrorism for civil liberties, criminal justice, and judicial process (e.g., the rights of *habeas corpus* and personal privacy, interrogations and gathering evidence, military tribunals for civilians, extrajudicial assassinations) are great and only gradually being understood. The impact of terrorism is changing the use of war to pursue national security and global national interests.

War and the State

In an oft-cited passage from his essay "Politics as a Vocation," Max Weber explained the essence of state power.

> Force is a means specific to the state. Today, the relation between the state and violence is an especially intimate one . . . [A] state is a human community that (successfully) claims the *monopoly of the legitimate use of physical force* within a given territory . . . [T]he right to use physical force is ascribed to other institutions or to individuals only to the extent to which the state permits it. The state is considered the sole source of the "right" to use violence. (Weber 1946: 77. Italics in original.)

---

[7]Terrorism is variously defined. The U.S. Army Manual's definition of June 14, 2001, reiterated in the Defense Department's *Dictionary of Military Terms* is "the calculated use of unlawful violence or threat of unlawful violence to inculcate fear. It is intended to coerce or intimidate governments or societies in the pursuit of goals that are generally political, religious, or ideological." Note that attacks on armed forces and random acts of violence with no coercive intentions are not terrorism, regardless of how frightening they may be or who carries them out.

Weber was referring to the state's police powers to stop, detain, and arrest; judicial decisions that impose penalties, fines, jail sentences, and executions; and its other uses of coercion, including war making. Disputes cannot be legally settled with a personal duel. Retribution and vengeance are not "mine" but must be approved and apportioned by the state. The Second Amendment to the U.S. Constitution may guarantee an individual's right to own, carry, and use guns, and private property laws allow armed militia groups to not only target practice but conduct drills and exercises that look like paramilitaries preparing for war. The First Amendment protects these militias to speak ill against ethnic, national, and religious groups as well as politicians and the government. The line is drawn, however, when they conspire to or actually act with organized violence. At that point the state, having a monopoly on the use of coercion, intervenes.

The national state structure that pervades the political landscape today has its origins in, and owes much to, the monopolization of organized coercion and lethal violence. In Europe, where the national state originated, this began ten centuries ago. "From AD 990 onward, major mobilizations for war provided the chief occasions on which states expanded, consolidated, and created new forms of political organization . . . Large-scale warmaking and the negotiations of large-scale peacemaking drove all European states toward a new organizational form: the national state" (Tilly 1992: 70, 195).

When successful in preparing for and engaging in wars of conquest, powerful warlords consolidated land and power, enriching themselves and those loyal to them, and so began the nation-building process. The necessity to extract resources and borrow funds in order to conduct war led to the creation of nascent bureaucracies and the organization of executive political bodies that evolved into more comprehensive state systems.

Creating Nationalism and Patriotism

Through war, requiring the sacrifice of its citizens, the modern state has gained in public support and loyalty. Michael Mann (1988) describes the curious paradox of "citizen warfare" that involves mass mobilization and a nation's commitment to war. He explains the paradox as part of the "dialectic of nationalism" that not only built the national state as the primary modern political unit, often through war, but elicits popular support for war, despite its staggering costs. The key to the paradox is the sense of belonging to a nation, i.e., nationalism. It became a vital source of identity and the basis for a contemporary sense of community, of belonging to something larger than oneself.

An illustration of this is found in the U.S. Civil War. While the abolition of slavery was a central motive of many supporters in the North, the primary reason for the war, as President Lincoln often said, was the preservation of the Union, barely "four score and seven years" old. The world's first constitutional democracy was threatened with dissolution, and the state waged—with widespread popular support in the North—a tremendously costly war of attrition. In large part the war, painful and prolonged as it was, could be carried on by the state because it tapped into the sense of national and human progress that was threatened by secession. It and later "citizen wars" solidified, for millions of people who came from across the oceans, a personal identification with and loyalty to the larger community of the nation, i.e., patriotism.

Power and Coercion

The essence of power is the ability to get another party to do something that they would otherwise not do. This can be done according to a set of rules. A college wrestler exerts rule-governed physical power, just as does a professor, a police officer, and an office supervisor. Competition is the rule-governed use of power in interaction between parties who want the same thing. In international oil trading, companies compete for buyers and sellers and are expected to do this within an established legal framework. Labor unions and employers compete for the profits of a corporation within the guidelines established by collective bargaining rules. Competition is a form of conflict, but stipulations limit what can and cannot be done, i.e., how power can and cannot be used.

Most organizational power is predicated on the willingness of those toward whom it is used to accept it as legitimate, what Max Weber defined as authority (Weber 1964: 324 ff.). Police carry lethal weapons, but few officers ever use or threaten to use a gun. A routine traffic citation is not accepted and paid because the officer can shoot the violator. Rather, officers have the authority to enforce traffic laws, and this is accepted by offenders as legitimate, gun or no gun.

Power does not need to be rule governed, of course, and the ability to exert power randomly or arbitrarily can itself enhance power. The sources of power—e.g., money, bombs, public shame—may not actually be used if the mere threat of their use can gain compliance. Violence also may or may not be rule governed. It may be a threat but when used causes damage, destruction, and loss of life. Torture is a form of violence, as are acts of terrorism that are carried out against innocents in order to gain compliance from authorities with whom the terrorists have enmity. Even when those possessing the means of violence are recognized as legitimate authorities, violence, torture, and terrorism are recognized as extreme forms of coercion.

War as Structured Coercion. War is an act of power, most visibly in its use of violence to achieve desired ends. It contains elements of competition, as in the competition to develop nuclear weapons and the competition between adversarial nations for allies. It includes threats. The mid-twentieth-century strategy of international "brinkmanship" brought adversaries toe-to-toe, exchanging threats to use nuclear weapons. The threat sequence is well known. If disputes cannot be resolved with rule-governed competition, the response may be to threaten, and finally to attack militarily, using violence to force an adversary to accede.

Through war, state power in its most extreme form compels behavior that would not otherwise be forthcoming. "Why did wars occur at all? The central, tragic fact is simple: Coercion *works*" (Tilly 1992: 70). It certainly doesn't work all the time and doesn't work for solving many, if not most, of the problems that plague humankind. It is, hopefully, not the only solution to even seemingly intractable problems that threaten peoples' lives and livelihoods. But it is and has been a fact of life for millennia and has played an enormous role in making the world what it is, for better and for worse. It is a fundamental fact of social change.

PURIFICATION OF SPACE

One version of the modern nation, sometimes called organic nationalism, insists that a single dominant culture be practiced by all those within a geographic boundary. Protected by the national state, the culture can thrive, lending a deep sense of a shared community and identity to the citizenry. Under threatening circumstances, personal security and mutual trust are predicated on familiarity with those who bear visible cultural affinities, including language and religion, of one's own nationality.

The dark side of organic nationalism is what Istvan Deàk (2002) calls "the crime of the century," i.e., mass expulsions, including systematic killing, to rid an area of persons identified as "the other." Military forces order and supervise expulsions and killing, and states authorize, sanction, and pay for the effort. In the past century, however, ordinary people, armed with crude weapons and handguns, often carried out the tasks of driving neighbors from their homes, destroying and seizing their property, incarcerating them in concentration camps, and committing mass murder in order to eliminate a group, i.e., genocide.

(Continued)

(Continued)

The purification of space (Sibley 1988) is a surprisingly new phenomenon. While there have been mass expulsions, pogroms, and the destruction of entire tribes for millennia, genocide requires a fairly articulate and popularized theory of race,[8] as well as a militant group's or a national state's capability for structured coercion, to carry out the "purification of space" on the scale seen in the last hundred years. Racial theories are only somewhat more recent inventions than are the powers of coercion developed by national states.

The impending collapse of the Ottoman Empire in the late nineteenth century, finalized in World War I, gave rise to Turkish nationalism and the creation of a state dominated by Turkic-speaking people. Among the first engagements of the "Young Turks," bent on building a nation out of the Ottoman ruins, was the 1915 expulsion of Armenians and the killing of possibly a million people. Threatened by Russia and fearing disloyalty among Armenian nationalists who saw the war as an opportunity to advance their cause, Turkish authorities, with the participation of non-Turkic people, committed the century's first of many genocides. In forging a national state, a second "purification of space" was carried out soon after the war[9] when a million and a half Greeks were deported from, and a third of a million Greek Turks were sent to, Turkey.

---

[8]Racial theories are pseudoscientific concoctions that try to find demarcations and establish categories among the welter of people in the world. Skin color, hair texture, cranial features, cultural practices, intelligence, physical and sexual prowess, hardiness and environmental adaptability, attractiveness, and historical experiences are combined into a mash and distilled into beakers of racial purity. Interbreeding is racial pollution. Intermingling of cultures (acculturation) dilutes and blemishes a superior way of life, way of speaking, and way of believing. The historically recent efforts of various disingenuous racial theories are captured well in Stephen Jay Gould's *The Mismeasure of Man* (1981) and Michael Banton's *The Idea of Race* (1978).

[9]In their desire to unify all Turkic peoples, the Young Turks rejected the Treaty of Sevres that divided the land of the Ottoman Empire. When Greece sent troops ostensibly to protect ethnically (but not culturally) Greek people in the new nation, they were routed, and an agreement was reached—the Treaty of Lausanne—to "purify" each other's space. As Istvan Deák describes it, "Hundreds of thousands of Greeks, Bulgarians, and Turks were 'exchanged'... into each other's country. Many of the so-called Greeks either spoke not a word of Greek or spoke a dialect that proved to be incomprehensible in the home country. Many of the so-called Turks did not know any Turkish because they were Muslim Greeks, Gypsies [Roma], Slavs, and Vlachs" (Deák 2002: 50).

Mauthausen-Gusen concentration camp prisoners, 1945. Ebensee, Austria. Unknown photographer.

Source: http://commons.wikimedia.org/wiki/File:Ebensee_concentration_camp_prisoners_1945.jpg

Mass expulsions, what became known as "ethnic cleansing" during the wars that dismembered Yugoslavia twenty years ago, are a radical form of demographic change, altering both the population composition and the life chances and experiences of a nation's people. The creation of India and Pakistan in 1949 was a terrible bloodletting that cost perhaps a million lives and the forced migration of nearly 100 million people. The war that ensued after Israel declared independence in 1948 and Arab nations attacked Israel made refugees of three-quarters of a million Palestinians, followed in the 1967 war by another quarter million refugees, most of whom were barred by Israel from returning to their homes. Twentieth-century ethnic cleansing included the forced removal from Central European nations of German-speaking people near the end of and following World War II. As many as seven million people—the Expellees—were forcefully emigrated to the Federal Republic (West Germany) and the Democratic Republic (East Germany) by

(Continued)

(Continued)

both the advancing Soviet army and newly created democratic governments (Naimark 2001).

The purification of space includes both mass expulsion and genocide. The most terrible genocide, the killing of millions of Jews by the Nazi regime, had its origin when the fascists first took power, with the sterilization and killing of homosexuals, the mentally ill, the crippled, and other Germans considered socially undesirable. Such heinous crimes against humanity are supported by a sense of superiority in both race and purpose. Behind the psychology, however, are often rational calculations—at least by the leaders—of potential gain. In Rwanda, Tutsi attacks on Hutus in the early 1990s, in retribution for earlier attacks by Hutus on Tutsis, swelled into genocide. The rhetoric of "kill the cockroaches" could not hide the fact of intense economic competition. Genocide became a terrible means of redistributing land in one of the most densely populated parts of the world (Gourevitch 1994).

War is inextricably part of the purification of space in modern times. Stalin's deportation of six hundred thousand people from Chechnya and Ingushetia and the Crimean Tartars (literally depopulating Chechnya) during World War II was ostensibly to prevent Nazi collaboration within the borders of the Soviet Union. The war in Bosnia pitted ethnically based armies and paramilitaries against one another, but people from all ethnicities became victims of the "purification of space" that left 102,000 dead and created perhaps a million refugees (Power 2002).

As Barbara Harff and Ted Gurr (2004) explain, far from being spontaneous byproducts of an emotionally aroused citizenry, such events are part of deliberate policies of nation building and the maintenance of national boundaries. Often with little forethought or an ability to prevent atrocities, the goal is to enclose a peaceful and compliant population of like-minded and patriotic people, bonded by a common culture and identity, in a more pure, if not completely purified, space.

War as a Driver of Social Change

Lessons From Twentieth-Century World Wars

Arthur Marwick's (1974) study of the consequences of the First and Second World Wars provides three ways war acts as an agent of social change: war's destruction, the test of war, and participation in war. Given the impact of these total wars and the intense scrutiny they have received in the years since, World Wars I and II provide a good starting point in understanding war as a mechanism of social change.

War's Destruction. War kills and injures people and animals, damages and destroys homes, businesses, highways, power plants, schools, and sewage systems, ruins cropland and animal habitat, and pollutes the environment (Hooks and Smith 2005). The strategic destruction of cities and factories, mines, harbors, railroads, and oil fields is well known. It bankrupts national economies and erases long-standing trade relationships. Understanding the social changes of war includes the ways war's destruction alters how people go about their daily lives.

In situations of total war and war fought among civilians, death is inflicted on noncombatants in huge numbers. In the German air raids on London during World War II, more than fifty thousand civilians were killed. This number pales in comparison to the civilian death toll as a result of U.S. and British bombing of Germany. The firebombing of Hamburg killed a quarter of a million people in one week, probably 70 percent dying of carbon monoxide inhalation while seeking safety in the deep tunnels under the city.[10] In 1945, with the fate of the war nearly sealed, the Allies killed as many as 140,000 people in bombing and fire-bombing Dresden. Allied bombing of Japanese cities during the war killed 400,000 to 600,000 civilians prior to the United States dropping atomic bombs on Hiroshima and Nagasaki that killed 320,000 people, 90 percent of whom were civilians.

Germany's war in the East, particularly in Russia, was even more horrendous. As many as seventy thousand villages were entirely destroyed; their inhabitants killed or made refugees. The siege of Leningrad by the German army and the siege of Moscow that followed were brutal and protracted, resulting in hundreds of thousands of people starving to death. The Soviet Union suffered at least twenty million deaths in World War II, the majority of whom were civilians. In nations that have not felt the impact of invasion or defeat, the destructiveness of war is sometimes hard to grasp.

The U.S. war in Vietnam and Cambodia claimed 2.5 million people, nearly half of them civilians. The war destabilized Cambodia and led to the Pol Pot regime of terror (the Killing Fields) that claimed another million people. The war between Iran and Iraq from 1980 to 1988 resulted in nearly one million casualties. The U.S. invasion and war in Iraq led to at

---

[10]Kurt Vonnegut's *Slaughterhouse Five* captures this in vivid, if fictionalized, form.

least fifty thousand, and possibly more than a hundred thousand, civilian casualties, more than 4,400 U.S. military deaths, and more than thirty thousand Americans seriously wounded.[11]

Melvin Small and J. David Singer (1982) compiled the Wages of War Database, the updated version of which is their book, *Resort to Arms*. It provides a vivid picture of the terrible destruction of war. These and dozens of other quantitative compilations, many other books and essays that tap historical memory and personal experiences, and the multitude of novels and poems about war would fill a massive library. Movies and music, too, capture the destructive experience of war, not only as it is fought but as it affects those far removed and long after it is over.[12]

Much like the catastrophic effects of a natural disaster, personal security, work, education, travel, health, commerce, leisure, communication, and relationships are all altered, especially for those who live in the immediate site of war. Preparing for war, including expansive military expenditures as a prelude to war, and making reparations and repairs in war's aftermath are similarly tied to war's destructive power.

MIGRATION AND WAR REFUGEES

Since World War II, war and roving militias have become the greatest force for creating the millions of people who are uprooted, displaced, and made refugees worldwide. The United Nations High Commissioner for Refugees' (UNHCR) most recent count was forty-two million people not living in their homes due to war, civil unrest, fear of violence, environmental problems, and natural disasters. Most are not classified as refugees but rather internally displaced persons and uprooted people. They come from war-torn and violent situations in countries such as

---

[11]There is much contention about civilian deaths in Iraq, but icasualties.org figures for civilian deaths actually reported and military casualties are considered accurate by the mainstream media. See also Sivard (1996).

[12]*Invisible Wounds of War* (Taniellian and Jaycox 2008) is an excellent collection of studies of the long-term consequences of recent U.S. wars, especially for U.S. troops who fought in Iraq and Afghanistan.

Sudan, Afghanistan, Colombia, Sri Lanka, Somalia, and the Democratic Republic of Congo. Palestinians displaced by war and their families make up 4.7 million of the total (UNHCR 2009).[13]

Natural disasters and environmental crises have created millions of refugees, but flights from violence, expulsions following military conquest, and the inability to resume a productive and secure life because of war's destruction have created millions more, many never to return to their homes. They fear for their safety, have lost their possessions and means of livelihood, are often targeted by the military and militias for having sympathized or associated with one or another faction, or simply have the wrong name, religion, or tribal identity.

Internally displaced persons are refugees who leave their homes but do not cross international borders. Sudan has more than five million internally displaced persons who have fled both the North/South war that killed upward of two million people and the genocide in Darfur. To migrate internationally requires resources many people do not have. To do so risks being stateless or prosecuted as an illegal immigrant.

Many countries confine refugees to camps and settlements, segregating them from the local population, sometimes for years. Eritreans are living in confined areas in Ethiopia; Somalis are in camps in Kenya; displaced Burundians are in Tanzania; Tibetans are in India; Myanmarese (Burmese) of various ethnicities have fled to Malaysia, Bangladesh, India, and Thailand; Afghans are living in Russia and India; and three generations of Palestinians are in refugee camps in Lebanon, Jordan, and other Middle Eastern countries, many living stateless and without citizenship rights (UNHCR 2010). Less-affluent countries host four out of five of all refugees worldwide, putting tremendous pressure on their nation's resources and political stability.

---

[13]The number of refugees worldwide is an inexact estimate. The Office of the United Nations High Commissioner for Refugees (UNHCR) identifies nearly ten and a half million people as refugees or in refugee-like situations, with millions more seeking asylum status (UNHCR 2010). According to the United States Committee for Refugees and Immigrants (USCRI), more than half of the fourteen million persons who they officially classify as refugees or asylum seekers—a slightly higher figure than UNHCR's—are victims of wars, civil and international (USCRI 2010). Not surprisingly, the largest numbers of refugees today are persons from Iraq, Sudan, Myanmar, and Afghanistan.

The Test of War. War puts enormous stress on institutions and organizations. In some cases, states are not able to cope with demands placed on them, and systemic breakdown proves pivotal to the war effort. Occasionally, entire nations collapse, as happened during World War I in Tsarist Russia. Among other factors, the strain of a long and costly war against the Mujahideen in Afghanistan contributed to the collapse of the Soviet Union (Kagarlitsky 2008).

When tested and found wanting, political and economic practices can be altered, and significant change can occur in order to meet the challenge of engaging in war. At its best, organizational practices become more efficient, doing more with fewer resources. Hidebound practices may be shaken up, and new leadership comes forth. Challenge stimulates a search for technological solutions. In the case of Britain during twentieth-century wars, "one can detect change towards management of the economy, towards a more science-conscious society" (Marwick 1974: 12).

Not all challenges are met with either collapse, more planning, or improvements in efficiency. The state may initiate draconian measures that violate constitutional law and legal precedence in order to protect against possible subversion: e.g., imprisonment without due process, summary deportations, press censorship, domestic spying, suppression of dissenting opinions, increased state secrecy, and even dictatorial power for the state's executive. At the conclusion of a war, precedent has been set by these measures, providing a legal avenue for their later return. For example, after the 9/11 attacks on the World Trade Center and the Pentagon, the Bush administration lawyers cited Abraham Lincoln's suspension of *habeas corpus* during the U.S. Civil War as precedent to justify imprisonment without trial for terrorism suspects, including U.S. citizens.

Participation in War. "War participation must mean civic [i.e., public] participation," according to Michael Mann (1988: 175). In many cases, the state seeks to convince the nation's people, especially the lower or less-privileged classes, that war is necessary or desirable. In the political bargaining, previously excluded or shortchanged groups may be courted with promises of changes in political and economic rights in order to elicit their support for going to war.

Rosie the Riveter, the iconic figure of women's participation in industry during World War II,[14] was not a new phenomenon. Women's labor force participation during World War I was a contributing factor in the passage of

---

[14]The poster adopted by feminists in the early 1980s depicting a muscular woman boasting, "We can do it!" was designed originally for the Westinghouse Company's "campaign against strikes and absenteeism" (Williams 2010).

the Nineteenth Amendment to the U.S. Constitution that gave women the right to vote. It was publicly touted as recognition of the vital role women had played in the war effort and in filling the labor void created by the mobilization of men for the war. Nearly a century and a half earlier those who wrote the U.S. Constitution were able to convincingly argue that the sacrifices of the landless in fighting for the American Revolution justified extending the vote beyond property owners, thus enfranchising most of the white males in the original thirteen colonies (Hooks and Rice 2005: 576).

Abraham Lincoln's admiration for the 100,000 Blacks who fought for the Union in the Civil War shifted his thinking, causing him to support universal enfranchisement of all persons of African descent (Foner 2010). In both the First and Second World Wars, African American males were able to acquire reasonably improved treatment in the military and, prior to World War II, increased job opportunities that included reducing discrimination in industries and government service. The attacks on Blacks who served in World War I[15] were not repeated after 1945. The armed forces began full integration with President Truman's 1948 executive order (Dalfiume 1969). As Charles Moskos observed, within four years, largely due to mobilization for the Korean War, racial "integration was a fact of life. At a time when blacks were still arguing for their educational rights before the Supreme Court . . . the Army accomplished integration with little outcry" (Moskos 1986: 66).

The modern civil rights movement, gaining force in the 1930s, continued apace in seeking political opportunities to advance the cause following World War II. Civil rights leaders took the United States at its word when its propaganda contrasted the free world against what Winston Churchill dubbed the Iron Curtain. They had only to ask, Free for whom? The civil rights movement gained an important international relations advantage from this Cold War contrast, recognized by President Truman in his steps to begin dismantling Jim Crow.[16]

---

[15]J. H. Franklin (1956), in *From Slavery to Freedom*, cites 167 verified lynchings of returned African American World War I veterans from 1917 to 1919.

[16]President Truman established the Committee on Civil Rights that expressed the sentiments of many as a consequence of the war: "The record shows that the members of several minorities, fighting and dying for the survival of the nation in which they met bitter prejudice, found that there was discrimination against them even as they fell in battle. Prejudice in any area is an ugly, undemocratic phenomenon; in the armed services, where all men run the risk of death, it is particularly repugnant" (quoted in Marwick 1974: 177–179). Richard Polenberg (1972) offers an excellent summary of circumstances, contending groups, and actions that advanced civil rights during and after WWII in Chapter 4 of his *War and Society*.

The migration of African Americans out of the South to work in industry during and after the war changed the political balance in many congressional districts in the North and West. Their greater electoral opportunities made the black vote critical for presidential elections (Lemann 1991; McAdam 1999a) as Blacks grew into a political force that subsequently moved the Democratic Party to embrace civil rights as a major issue of social change (Kryder 2000).

Clearly demarcated wartime battle lines and timelines seem a thing of the past. Wars often are endless, with continuing military occupation (peace-keeping) rather than victory. Preparations for war and the threatened prosecution of war are ongoing. War's demands transform the fundamentals of how the state operates. Examining these things—the psychological impacts of war, the economic impacts of a state's military preparations, and state planning required for the prosecution of wars—shows how few people today can escape the personal and social consequences of war and the way it shapes social change.

Psychology of War

The psychological impact of war is seen in the perceptions and feelings, the attribution of blame, and wartime experiences of Iris Summers. She was a young wife during World War II, involved only as an unemployed civilian in supporting the Allies' efforts to defeat Germany and Japan. Her husband was working in a strategic industry. Her younger brother was drafted and flew supplies to troops, and a brother-in-law was shipped overseas to play a supporting rather than combat role. Despite the losses and hardships, it seemed to her that it was, as Studs Terkel (1984) described it, a "good war."[17] It made her proud and drew her closer to the idea that her nation was special and somehow better than the rest.

The other side of Mann's curious paradox of "citizen warfare" is that indeed, war can build patriotism and loyalty, as members of veterans groups proudly display on public occasions. It can as well create doubts, regrets, cynicism, and disappointment that erode not only the state's legitimacy but its capacity to mobilize public support for any agenda of social change, the topic of Chapter 8.

---

[17]Studs Terkel's conversations, interesting as they are, exhibit some of the "recall" problems described in Chapter 2. In contrast to Terkel, Paul Fussell's *Wartime* (1989) takes issue with the "good war" and "greatest generation" ideas associated with World War II as does Richard Polenberg (1992) in his essay, "The Good War?"

Despite her experiences and opinion of World War II and her formative years during the Great Depression, including the government's activism to restore the economy, Iris' trust declined for many things the state did as she grew older. How she saw her government, i.e., the test of war on the state's ability to claim legitimacy for its actions, was weakened by its defeat in Vietnam and questionable wars elsewhere. This also may have been an "age effect" inasmuch as older people tend to be less enthusiastic about their country going to war than are middle-aged and younger people (Mueller 1973).

CONSTRUCTING MENTALITIES FOR WAR

In the modernization of Japan, the traditional authority of the Emperor system was the dominant political institution: [i.e.] the preservation of the traditional family and village communities as social institutions and the inculcation of the traditional values of loyalty to the Emperor and filial piety to one's own parents.

—Kazuko Tsurumi (1970: 5)

The Meiji Restoration in 1867–1868 marks the beginning of modern Japan. From the outset, the new leaders promoted a very aggressive, militaristic national revival, led by Japan's preparations for war. In 1873, conscription was made universal. Few males were exempted from spending six months to two years in military service, with compulsory military training that involved severe discipline, regimentation, and the use of often arbitrary violence as punishment for any reference to one's own needs or thinking. This drove home "the necessity of unquestioning obedience" (Tsurumi 1970: 116). Soon after the Restoration, the country declared its intentions on the Asian mainland and carried out an Asian war with Russia (1904–1905). Japan invaded China in 1937. Within a generation, the military experience had become formative for all men in Japan. When extended to the schools by means of military drill, dormitories designed like army barracks, and military discipline, militarism became the official norm for the entire society.

Prior to modern Japan and its adoption of military conscription, only the men of samurai families, a social class of perhaps 7 percent of the population, were allowed to have weapons (though not firearms). In ending this privilege and amassing a huge army, the samurai ethic of *bushidō*, the Way of the Warrior, was established society-wide. "The feudal samurai ethic of dying honorably for the sake of one's lord was transformed into a universal ethic for the entire population of Japan,

(Continued)

(Continued)

through the requirement of honorable death for the sake of the Emperor" (Tsurumi 1970: 81). All Japanese were the children of the Emperor and were to provide unquestioned loyalty to him, to the extent of giving up their own sense of individuality and personal responsibility. The military dominated the nation and the marshal virtues dominated the society, including the virtues of courage, selflessness, and simplicity.

Period Effects. In Chapter 2, social change through period effects was described in terms of changes in the values, personal orientations, norms of behavior, and attitudes of one or more generations of individuals as a consequence of new circumstances. Preparations for war, the conduct of war, and war's aftermath are among the most dramatic circumstances that create period effects. The social changes in Japan attributed to preparations for war provide a strong illustration of this. Not a single generation but the entire society experienced the shift in Japan's social and political organization as the country sought to expel the West from Asia and establish Japanese hegemony throughout the Pacific, but for the young the effect was most formative.

The Emperor system so effectively transformed the society that Japan could sustain incredible destruction and loss of life, yet continue to pursue the war. Even before the Japanese attack on the assembled U.S. naval fleet at Pearl Harbor in 1941, Japan had lost nearly 200,000 soldiers, with a third of a million unaccounted for or wounded. A million and a half more Japanese soldiers died by war's end in 1945, compared to approximately half a million U.S. military casualties in both the European and Pacific theatres of war.

In her analysis of the statements, diaries, poems, and letters of more than two-thirds of the 1,068 leaders, officers, and soldiers sentenced to be executed by the International War Tribunal following the war,[18] Kazuko Tsurumi concludes that the psychological effect of the Emperor system was indelible. There was very little change of heart after the war among the 701 individuals whose writings she studied. Only three wavered in their belief in the ideology of imperial Japan, though some criticized the Emperor himself for not having committed suicide as an act of honor in accepting defeat. A few expressed doubts about what they personally had done, but a sense of guilt was almost entirely

---

[18]The Tribunal categorized the offenses of those arrested as (1) having planned, prepared for, initiated, and waged a war in violation of international laws and treaties, (2) having ordered and directed atrocities, and (3) having carried out atrocities.

directed at not having served the Emperor and their country more successfully. The preparations for war, as much as the war itself, had had a profound impact on the psychology of the Japanese people.

In-Group Solidarity and Out-Group Hostility. Wars have the power to strengthen in-group feelings and sow hostility to out-groups. Former U.S. Supreme Court justice John Paul Stevens' reflection of World War II evidences a typical response to war, especially at the outset and when people feel endangered yet confident they can repel the threat.

> It really was a unique period of time in the sense that the total country, with a few exceptions, was really united. We were all on the same team, wanting the same result. You don't like to think of war as having anything good about it, but it is something that was a positive experience. (quoted in Liptak 2010: 3)

Conflicts between social classes and local animosities are papered over. It is unpopular, to the point of being unpatriotic, to express concerns about social inequities and even social injustices when everyone is expected to be focused on the need for unity to face the difficulties of war. In situations of austerity, when personal sacrifice of the civilian population is sought, the majority of people enthusiastically embrace the opportunity to participate. Buying war bonds, planting victory gardens, cooperating in rationing, and conserving and reducing the use of resources that are needed by the military help to support a sense of in-group cohesion and belonging.

At the same time, fear, hostility, anxiety, distrust, and even hatred and loathing are not unusual feelings toward out-groups. Stories of wartime atrocities[19] committed by the enemy combine with a portrayal of foreign people who care little about their own deaths or that of loved ones, of fanatics bent on creating a perverted world order, people who lack basic decency and common sense and who are best characterized as animals or insects.

---

[19]Early in the Vietnam War, then Secretary of Defense Robert McNamara ordered a history of the war to be compiled. It was kept top secret throughout the war. One of the first things McNamara asked Daniel Ellsberg, a Pentagon employee who helped craft the U.S. conduct of the Vietnam War, to do was find an incident of a Viet Min or North Vietnamese atrocity committed against American troops in order to help convince President Johnson to escalate the war. Ellsberg later provided members of Congress and major newspapers with copies of the secret history of U.S. planning and discussions during the war's buildup, what became known as the Pentagon Papers (Gravel 1971–72).

There is a tendency to dehumanize enemy combatants and everyone who lives in the countries of one's adversaries, using slurs and accepting accounts that often have no veracity. Racist and jingoistic opinions are tolerated, and media portrayals both build and reinforce the view that one's adversaries are fundamentally inferior to oneself. The intemperate characterization of Japanese Americans by the wartime governor of Idaho, Chase Clark, is all too common: "Send them all back to Japan, then sink the island. They live like rats, breed like rats, and act like rats" (quoted in Polenberg 1972: 65).

Euphoria and Disillusionment. The euphoria and sense of victory that greeted the conclusion of hostilities in World War II has rarely been repeated since.[20] By the end of World War I, deep disillusionment with the war, especially the massive and seemingly endless and purposeless deaths of stalemated trench warfare, colored an entire generation of Europeans, typified in the label "lost generation" discussed in Chapter 2. In most wars since World War II, public support has declined well before hostilities ceased.

The Vietnam War was the first televised war, and Iris Summers watched it with horror. She couldn't quite accept the reasons why the United States invaded this far-away Southeast Asian country of peasant farmers that had been colonized by France. When one of her sons was drafted and sent to Vietnam, she found a personal reason to support the war. As the fighting dragged on and weekly "body counts" became the indicator of progress, rather than battle lines showing where people were being liberated from tyrants (her image of World War II, the Good War), Iris, like a majority of Americans, came to oppose the war, though never publicly. Its conclusion was a relief, and she couldn't find it in herself to support another war during her lifetime.

As public polling shows, the longer a war proceeds, the less inclined a population is to support it (Mueller 1973). There is a decline in hopefulness, the sense of common purpose, and the belief that it is succeeding as planned.

---

[20]Many times it is unclear when, in fact, a war is ended. A U.S. war has not been officially declared since World War II, despite the Constitution's requirement that the Senate issue a declaration of war before committing the military to a major sustained foreign engagement. This adds to the difficulty of knowing the exact beginning and ending of a war. Though the commitment of "combat forces" following the Gulf of Tonkin resolution is considered the Vietnam War's beginning, the incremental buildup of forces in the early 1960s gave it an ambiguous starting point. The invasion of Afghanistan in 2001 and Iraq in 2003 had clear starting points, but what will constitute their conclusions is unclear. The Korean War was called a U.N. police action and remains a cold war sixty years later, with tens of thousands of U.S. and Allied troops still stationed along the border between North and South Korea.

As casualties mount and innocent people in the theater of war are killed ("collateral damage"), doubts are raised. States spend a great deal of their political capital as well as their budgets in order to continue what they began, trying to keep up public support and citing their international credibility. When it goes badly, as it did for the Soviet Union in Afghanistan, it can significantly weaken a state, both financially and in terms of citizens' sense of the state's legitimacy.

The Social Economy of War

What war brings is not just work for undertakers. Many fortunes have been made from wars, both through conquest and the financing of military expeditions that is repaid handsomely. To defense-contracted researchers, arms manufacturers, suppliers of contract workers, and arms dealers, war is business. Some communities, regions, and countries owe their economic vitality to military installations and industries providing the range of products and services the military requires. The post-World War II economic success stories, the so-called Asian Tigers—Taiwan, South Korea, Hong Kong, Singapore, and even Japan—owe much to the U.S. war in Vietnam and Cold War hostility toward China that funneled billions of dollars their way for military support facilities (Johnson 2000).[21]

Despite the destructiveness of war and the damage it can do to a nation's economy, there is widespread belief that preparations for war, in the form of defense spending, are an economic boon. Not everyone agrees. Most economic research finds that military spending distorts the nation's economy by investing in less-productive sectors. It generates debt that raises interest rates and the ability to obtain credit for productive investments (c.f. Ram 1986; Chan 1985).

A Permanent War Economy. War as a force for social change involves the preparation for war as well as the actual use of arms. The allocation of resources to carry out war, whether defensive or offensive, drives social change by focusing both state and private corporate endeavors in one direction rather than another. This has become chronic rather than occasional in the past half century. With the development of the nuclear bomb, according to Herrera (2006: 186), "the difference between wartime and peacetime becomes largely indistinguishable" in terms of state resources directed

---

[21]Chalmers Johnson's three other books on the U.S. military's global presence include *The Sorrows of Empire* (2004), *Nemesis* (2006), and *Dismantling the Empire* (2010), all well-regarded warnings of the wider implications of U.S. military reach.

toward war making. The growth of a wartime state is difficult to reverse when peace comes, especially when a wide range of interests benefit from military spending, including industries, unions, communities near military installations, and the bureaucracies created to pursue war.

Seymour Melman (1985) and others called this the "permanent war economy," involving huge state expenditures for war, even during peacetime. The military's demand for goods and services engages a large portion of a nation's productive capacity. Millions of people find direct employment in the permanent war economy, and tens of millions of jobs are devoted to research, technology, and providing goods and services for military endeavors. President Dwight Eisenhower, former general and commander of Allied forces in Europe during World War II, in his often-quoted remarks upon leaving office in 1961, found this justified, but worrisome.

> Until the latest of our world conflicts, the United States had no armaments industry. American makers of plowshares could, with time and as required, make swords as well. But now we can no longer risk emergency improvisation of national defense; we have been compelled to create a permanent armaments industry of vast proportions. Added to this, three and a half million men and women are directly engaged in the defense establishment. We annually spend on military security more than the net income of all United States corporations . . .
>
> We must guard against the acquisition of unwarranted influence, whether sought or unsought, by the military-industrial complex . . . Only an alert and knowledgeable citizenry can compel the proper meshing of the huge industrial and military machinery of defense with our peaceful methods and goals, so that security and liberty may prosper together. (Eisenhower 1960)

Direct Costs. World War I cost the United States $334 billion, while WWII cost $4.1 trillion. The Korean War cost $341 billion, and the invasion and occupation of Vietnam cost the United States $738 billion. As of mid-2010, the war in Iraq had direct costs of $784 billion, and Afghanistan added another $321 billion (Bumiller 2010).[22] Nobel Prize–winning economist Joseph Stiglitz and Linda Bilmes (2008) estimate the Iraq War, begun in 2003, will ultimately cost three trillion dollars when all costs—including paying the debt for the borrowed money and veterans' long-term medical care—are paid. With the end of the Cold War and the implosion of the USSR, military budgets of the former Soviet Union and Eastern Europe shrank by more than half. After the United States initially reduced defense spending as a portion of GDP in the 1990s, military spending increased

---

[22]All figures are in 2010 dollars, i.e., adjusted for inflation to reflect 2010 dollar values.

dramatically to fight the Persian Gulf War, two long-term wars, and maintain a strong military presence around the world. Since the 9/11 attacks on the World Trade Center and Pentagon, U.S. military spending has doubled.

Today, the U.S. defense budget is nearly equivalent to the military expenditures of all other nations combined. It is an expensive proposition to be the world's greatest military power and to engage in wars. The "base budget" for 2010 was $664 billion dollars, but this did not include "overseas contingency operations" and defense-related funding for other agencies of the government, such as NASA, Veterans Affairs, Homeland Security, FBI and CIA counterterrorism, and interest on the debt of previous and current wars. This probably raises the annual military budget to more than $1 trillion, nearly a quarter of all U.S. federal spending.[23]

During the Cold War, between a third and a quarter of U.S. military expenditures went to nuclear arms (Sivard 1996: 26). Worldwide since 1945, $8 trillion dollars has been spent on nuclear weapons, half of that (at least $4 trillion) by the United States and half by the other four major nuclear powers (China, Britain, France, and the USSR/Russia). While nuclear weapons stockpiles are being reduced today, continued nuclear weapons research, maintenance, and modernization costs tens of billions of dollars annually.[24]

In 2010, the annual training and support for one U.S. soldier was about $100,000. To support that soldier in war increases the cost by ten times, i.e., as much as one million dollars annually (Bumiller 2010). In fighting "new wars" and nonstate combatants, discussed earlier, there is an increasing shift in tactics, replacing men and women with machines as a way to reduce casualties. It costs much more, however, to use drones and other highly sophisticated technological alternatives as a substitute for troops on the ground.

---

[23]Computing military spending is no easy task, but data are available at http://www.usgovernmentspending.com and on the various federal government agency websites, including the Congressional Research Service.

[24]At the height of the Cold War, the United States and USSR deployed (had ready for launching) a combined total of 19,000 nuclear warheads. These were reduced with the 1991 Strategic Arms Reduction Treaty and in 2010 numbered 2,100 and 2,600, respectively. Thousands of warheads remain in reserve. The arms reduction treaty signed by President Obama and Dmitry Medvedev, president of Russia, will reduce deployed nuclear weapons by about two-thirds over several years (Baker 2010). Mr. Obama pledged $85 billion for continued nuclear weapons "modernization: to congressional critics of the treaty in order to gain their support for the New Start Treaty of 2010.

Economic Growth. The United States produced the lion's share of war material for itself and its allies in WWII: planes, ships, and munitions. As a consequence, in 1942 U.S. manufacturing was twice what it had been in 1940. By war's end, U.S. industry was far and away the most dynamic and robust in the world. The federal government borrowed heavily during the war, in an amount exceeding the annual national GDP, but the debt would be repaid with taxes in a booming economy of the postwar years.

The shortage of jobs that inevitably occurs in the economic contraction soon after a war was addressed in the United States by laying off about a third of the women who had taken up jobs vacated by men and women going to war. The economic slack was also addressed by the Serviceman's Readjustment Act of 1944, i.e., the GI Bill of Rights. It authorized unemployment insurance for the nearly five million returning GIs, provided home and business loans, and made available education grants that allowed war veterans to continue their education rather than immediately seeking jobs. The development of human capital through higher education paid a bonus as the economy expanded and loans through the newly formed International Monetary Fund helped rebuild markets in a shattered Europe (McMichael 2004). These efforts by the state, rather than direct funding for war and the military, contributed greatly to the economic rebound that ushered in more than two decades of unprecedented economic growth that helped pay off the national debt of the war.

Paying for wars, however, is not the same thing as a permanent war economy. As happened in the United States in the early 1940s, and in Nazi Germany the decade before, heavy military expenditures provide a short-term boost to the economy. In countries with growing economies that are moving away from primary agriculture and natural resource extraction—what are often called developing economies—there is some evidence that military spending as an annual budget expenditure correlates with economic growth. This is unusual, however, and happens in very few of the 103 cases Greg Hooks and James Rice (2005) studied.

There is zero long-term economic benefit from military spending for most countries. Iran, Iraq, and a few other nations are exceptions, though most of these—including Taiwan and Israel—for years "have received large amounts of military aid" primarily from the United States, thus reducing the burden of military spending on their economies (Mintz and Stevenson 1995: 299). Overall, and especially in countries with more mature industrial and service economies, military spending either does not stimulate economic growth or may have the opposite effect (Hooks and Rice 2005: 272).

Opportunity Costs. Economists describe the things that are not done, not invested in, and not purchased as opportunity costs. Opportunities are lost because time, effort, and funds go elsewhere. At this very moment you could be traveling the world and being totally carefree, but instead you are taking two, four, or more years out of your youth to go to college, study hard, work at a low-paying job, and worry about making ends meet. The things delayed or forgone, experiences that will perhaps never be retrievable, youthful exuberance smothered by classroom decorum and library fatigue—these are opportunity costs. The cost of preparing for and engaging in war has its own opportunity costs. A "peace dividend," e.g., the amelioration of social problems, is neglected, and resources that could have gone into labor-saving or empowering technologies are expended instead on the technologies of military security and war making.

Critics of military spending lament the inability of governments to spend the money on other things instead of war: e.g., health care, social services, scientific research, education, space exploration, parks and public land, and diplomacy. At the same time, military planning and execution divert human resources—e.g., highly educated professionals, especially in technological fields—away from what they might otherwise be doing to enhance the well-being of people and the nation. Sometimes referred to as the "crowding-out effect," public investment in other things could be greater and private investment could be going elsewhere.

This is a sentiment of neoisolationism that has usually been associated with conservatism, though this view is more likely to be held today by liberals and others on the left of the political spectrum. Conversely, despite the conservative shibboleth of spending money prudently and limited government borrowing, otherwise fiscally conservative and pro-business political groups tend to support larger defense budgets. Arguing on the basis of national security and the economic stimulus provided by this spending, most moderates and conservatives in the United States back large military expenditures for corporations as well as small businesses near military bases in their districts.

Labor's Gains. Wars are good for employees, or so it seems in the short run. When conscription takes men into the military and companies find themselves with new orders for munitions, ships, equipment, contractors, and technology, workers are in demand. Market principles apply: greater demand for workers becomes an opportunity for higher wages and better benefits.[25]

---

[25]Sandra Halperin's *War and Social Change in Modern Europe* provides a more radical but well-researched interpretation. The state's preparations and conduct of war "brought the working class to power" (Halperin 2004: 29) and, she concludes, transformed Europe into a social democracy in the mid-twentieth century, making possible Europe's advanced industrialization.

Workers in the United States and France won the eight-hour day on the strength of labor's importance during World War I. World War II provided irreversible job gains for women and minorities. Gender and racial discrimination were not eliminated, but positive changes that were underway when the war began accelerated. There would be no more sacking of women when they married (the marriage bar), the dual-wage system (higher wages for Whites) was made illegal, and the call for adequate day care for children of working women became louder. Though some unions continued discriminatory treatment or remained segregated (Polenberg 1972: 115), the modern civil rights movement was given a strong boost by changes in minority employment as well as the experience of Blacks and other minorities who fought in the war.

Women found a growing demand for their labor in both Britain and the United States and entered the workforce in large numbers. In the United States, women's labor force participation went from 25 to 36 percent between 1940 and 1945 (12 to 16 million women). The War Labor Board authorized equal pay for women and men, though women on average continued to be paid 40 percent less than men. Two million women were laid off in 1946, but others kept their jobs, despite a new popular image in advertising that placed women in the center of a newly revived consumer culture revolving around home, husband, and children (Friedan 1963).

Just as women migrated to the shipping yards on the East and West Coast and to other industrial areas away from home, black migration from the South increased significantly with job opportunities the war created. Recognizing the inevitable need for more black workers, A. Philip Randolph, leader of the Railroad Porter's Union, planned a march on Washington in 1941 to protest racial barriers to employment. This prompted President Roosevelt to issue an executive order forbidding discrimination in defense industries, a major precedent for later civil rights efforts. Many African Americans and Latinos, newly hired in industries in the West and North, were let go at the conclusion of World War I and II, both because of racial prejudices and in light of the recessions that followed the wars. The experience of having worked at a skilled job in industry, however, and having served capably in the military emboldened and empowered many Blacks and Latinos to demand greater equality of opportunity (Polenberg 1972).

It may be a situation that will rarely be repeated, but World War II brought a bounty to workers. Unions were gaining ground in the late 1930s, and union membership continued to climb after the United States declared war in 1941. In 1942, there were 1.5 million more union members than there had been in 1939. A 1941 no-strike, no-lockout agreement among unions, corporations, and the government did not prevent railroad and coal industry workers from striking for better wages, but overall it was a period of labor peace. To get and keep good workers, however, wages moved upward, and the average weekly wage doubled between 1939 and 1945. In addition, companies began offering pension plans and other benefits in order to retain their best workers, especially white-collar professional staff. Benefit packages proliferated and expanded after the war.

The end of war in Europe in 1945 and the occupation of Germany by Allied forces resulted in a new constitution for the German Federal Republic that sought to prevent a repeat of fascism. The victorious Allies insisted that German industry share corporate governance, i.e., codetermination, with newly established labor unions in order to minimize the likelihood of another powerful, corporate-dominated German state bent on military conquest. Since the war, codetermination has been a powerful force in the creation of universal public health and other social welfare and unemployment provisions that are far more generous than those found in the United States (Hodson and Sullivan 2008: 398).

Military Research and Development

To a large extent, military planners created "big science" in the Manhattan Project that produced the first nuclear weapon. Since World War II, science and technology have been harnessed and in important respects controlled by national security planners (Hooks and Rice 2005: 575). More generally, the laboratories and facilities at the heart of the research industry in the United States have been built, in part, by billions of dollars of military funding and are thus organized to respond to lucrative opportunities offered by the Pentagon and military-related agencies. Because the best research is done in universities, this is where much of the money is channeled. Many universities depend on military-related research contracts to support their educational and service missions and would be unable to balance their budgets without the "overhead" funding for doing this work. They, and research institutes with direct ties to the Pentagon, e.g., the RAND Corporation, benefit from the development of military research that finds civilian use, in the way of patents and copyrights for technology derived from military-related research.

Worldwide, about half of all publicly sponsored research and development in the last half century has been in support of military and defense activities. In the United States, 60 percent of government-sponsored research and development is military oriented (Sivard 1996: 5). As discussed in Chapter 4, the U.S. Defense Advanced Research Project Agency (DARPA) funds billions of dollars in research annually. The

Defense Department's fiscal year 2010 expenditures allocated $79 billion or 11.5 percent of its $664 billion base budget to research, development, testing, and evaluation. Military research and development is not only in the Pentagon but in the Department of Energy and in the budget of the National Aeronautics and Space Administration (NASA), the former for nuclear research and the latter for the military use of space. In the 2011 budget request to Congress, defense-related expenditures were approximately 50 percent of NASA's budget.

Technological Spillovers. The direction of social change has been influenced by various technologies initially created for military purposes. The technological innovations that find civilian use, paid for with military spending, are wide ranging, from medical to materials research, from ballistics to cyberpiracy, and from bio- and nanotechnology to optics.

In the flurry of concern about math and science education and the commitment to government research following the successful launching of an orbiting satellite by the Soviet Union in 1957 (*Sputnik*), DARPA was created and tasked with devising a way to coordinate the U.S. armed forces in the event of a nuclear attack that destroyed conventional communication channels. RAND's Paul Baran recommended a network of computers (very rudimentary, expensive, and large) through which digitally transmitted information, "datagrams," would be sent and received. Collaboration between Stanford University researchers and DARPA developed the network, now known as the Internet.

Jet engines, helicopters, submarines, penicillin, the design of airliners, prosthetics, and artificial intelligence are only some of the more visible technologies that have found civilian use. No civilian applications were considered when, in the 1970s, the Navy developed Navstar as a means of orienting and locating objects on earth by way of coordinates drawn from satellites twelve thousand miles in space. Only when Korean Air Flight 007 strayed from its flight path and was shot down by a Soviet fighter jet did President Ronald Reagan authorize Navstar's use as a guidance system for civilian airliners. Today we know Navstar, much expanded, as the global positioning system (GPS) that is in hand-held mobile communication devices and cars today. Whether GPS would have been developed without the effort to prepare for war, as well as the political and economic pressure exerted by interested corporations and communities that benefit from military spending, is probably impossible to know.

Health technologies owe a great deal to war. The story of Florence Nightingale and the birth of modern nursing during the Crimean War are in every child's history book. Developments in surgery, immunology, emergency medicine, and prosthetics owe much to the demands placed on medicine by war. The challenges of long-term care and recovery for those injured in today's wars are greater than ever (Taniellian and Jaycox 2008). Fortunately for persons injured in automobile accidents and in other ways traumatized, what is gained by medical research and in clinical settings on account of war can make a huge difference in their chances of survival and recovery.

State and Corporate Planning

Preparations for war and the conduct of wars have, since their inception, necessitated that nations plan accordingly. The sometimes desperate commitment to garner resources and allocate them with a clear objective has given an advantage to states that do this on the battlefield. The lessons of preparation and planning remain vivid when the guns fall silent: rational, deliberate, calculated steps can be effective in efficiently using resources, meeting challenges, and solving large and complex problems.

As has been alluded to several times in previous chapters, many of the functions of the modern state, including its administrative bureaucracies, had their origin in the conduct of war. Over the centuries, war has extended the state's reach and power. For the victors, the effectiveness of mobilizing the power of the state offers new possibilities, especially for agendas that are set aside but are kept alive during war. As economist John Kenneth Galbraith (1967) extolled in his book, *The New Industrial State*, wars provide a "best practices" model for state planning.

During the Second World War, John Maynard Keynes returned to the British Treasury to create a policy for using government expenditures to avoid mass unemployment after a war (Marwick 1974: 162). Keynes' *General Theory of Employment, Interest and Money* (1936) had set out a bold new way of thinking about economics, challenging previous theories and practices across a range of topics. His focus on avoiding economic recessions found an eager audience in light of the recurring financial crises that plague modern capitalism and the failure of neoliberal (i.e., laissez faire) economics to right the disaster of the Great Depression. Keynes convinced mainstream economists that recessions and high unemployment can be addressed by governments willing to increase the money supply and target spending to build enough private demand that businesses can operate and grow in an environment of low inflation. Until productive capacity is in better balance with the demand for goods, public spending is an effective way to lessen or soften the damages to workers, businesses, and public programs caused by recessions. This is popularly called Keynesian economics.

In Britain, the outlines of the postwar welfare state, laid down in the 1930s, were revived as the war drew to a close. As well, the war exposed inadequacies in public health infrastructure and the ability to respond to widespread public need. Plans for the welfare state, including a national health service—universal health care—were already underway in 1945 (Marwick 1974) and implemented by the Labour government elected at the end of the war. In Italy, Germany, France, and Britain advisory and planning boards were created to provide closer cooperation and collaboration between corporations and governments. "The war and postwar emergencies accelerated" the synthesis of private enterprise and collaborative planning. "In each case, they led to important technological adaptations, to improved engineering and marketing methods, and to the spread of industrial planning" (Hogan 1987: 435).

The White Plan drawn up by the U.S. government for postwar planning focused on financial issues and moderate government programs to convert wartime industry to peacetime use. The Council of Economic Advisors was established in 1946 to provide guidance to the president about spending and coordinating government and corporate trends. It retains this role today.

The Serviceman's Adjustment Act of 1944 or GI Bill, described earlier, was among the most significant plans designed to ease the transition of returning GIs to civilian life. It authorized unemployment insurance for returning service personnel. It established programs for home and business loans for men and women who served in the armed forces. Its education grants delayed the onrush of GIs into a contracting economy. Its loan programs gave a huge boost to home construction and the subsequent growth of suburbs. The GI Bill provided a useful coincidence of rewarding participation in the war with increased planning of the national economy.

Discussions of state planning for social progress in the United States during and following World War II were met with strong opposition from many quarters, predictably businesses and corporations but also labor unions that wanted to keep issues like health care as part of union benefits, to be negotiated with employers (Edwards 1979). Irrespective of this opposition, and as will be seen in the next chapter, corporations also learned from the war by recognizing the value of planning. They were eager to have clear ideas of both labor costs—established through collective bargaining with unions—and multiyear government spending plans, policies, and regulations that provided opportunities for expansion and investment safeguards.

Internationally, the war made the United States much more eager to plan for future international economic relations. A design for international banking and postwar global monetary policies was adopted at the meeting of

Western allies in 1944 at Bretton Woods, New Hampshire. International agencies were created, nominally to be part of the new United Nations. They included the International Monetary Fund to stimulate international trade and address currency problems, along with the World Bank for Reconstruction and Development. The World Bank pooled capital for lending, primarily to poor and war-damaged countries for the purchase of technology from abroad (McMichael 2004: 43–44). The Bretton Woods agreements pegged national currencies to the U.S. dollar. For the next twenty-five years, this provided a stable currency exchange system for international transactions that greatly benefited the United States.

Revolution and Social Transformation

You say you want a revolution, Well, you know, we all want to change the world.

—John Lennon and Paul McCartney,
"Revolution"[26]

Ways of Understanding Revolution

The French Revolution (1789–1799) has long been the most popular subject for the study of revolution. The conditions for the revolution were auspicious. The press of modernizing economic forces was obstructed by the privileges of a profligate monarchy and a decaying rural gentry. Joblessness was high. Class antagonisms were strong, and a restive urban population voiced deep grievances against the state's impotence and disinterest in addressing their plight. Ideas of social change and human possibility filled the space once occupied by clerical authority.

The increasingly violent attacks on the Ancient Regime precipitated the takeover of the French state by revolutionaries and radicals. Followed by

---

[26]The Beatles recorded "Revolution" in May 1968, soon after returning from India. Lennon and the other band members had avoided any references in songs and public statements about the social upheavals of the time. Months before, the Tet Offensive of the Vietnam War galvanized the antiwar movement. Dr. Martin Luther King was assassinated on April 4, 1968, and urban riots ensued. The lyrics—rejecting violence and questioning such New Left icons as Mao Zedong—were criticized by those who wanted a clear statement of support for radical social change. Nineteen years later "Revolution" became the first Lennon and McCartney song to be featured in a commercial advertisement, by Nike.

reactionary bloodletting, a new political system was established. It combined much of the old order with novel and progressive experiments in governance and social relations, and launched France as a modern global power. The revolution's example showed the way for revolutionary changes elsewhere and provides a rich illustration for understanding social change to this day.

Four historical transformations became the subject of Crane Brinton's classic 1938 study of comparative history, *Anatomy of Revolution*: the English (1640–1660), American (1774–1783), French (1789–1799), and Russian (1917–1918) revolutions. His "natural history approach" focused on the antagonisms and battles between social classes, the inability of the state to govern effectively, the detachment of key intellectuals from the ruling class, and the support they provided in framing the need (the diagnostic frame) for radical change.

Fidel Castro, Cuban revolutionary and president, 1959–2006, in New York City, 1959.

Source: Public Domain. Library of Congress Prints and Photographs Division Washington, D.C LC-U9-2315-6. digital ID ppmsc.03256

Importantly, Brinton recognized that, despite the clamoring of the common people for economic security and a higher standard of living, these four were not societies in decline. If anything, rising expectations fueled fervor for a better future.[27] In the Russian case, war was a critical part of the revolutionary transformation of society. In the others, wars contributed to economic distress and had weakened the state's power to the point that it was vulnerable to the defection of important elites.

Anticolonial Revolutions. After World War II, decolonization began in much of Africa, South Asia, and Southeast Asia. England, France, Portugal, and other European colonial powers gave most of their attention to rebuilding their countries rather than administering colonies. The participation in World War II of troops from colonial holdings and the education of a small elite offered a sense of possibility in the colonies, providing a basis for making claims of independence. Their nationhood was in some cases accomplished by wars of independence and in some cases was accompanied by civil wars between groups vying to control the new states.

The length, severity, and costs of these wars had a major impact on the kind of new nations they became. In regions that were colonial domains controlled by foreign powers, there have been many genuinely revolutionary transformations to self-governing national states. In some cases, colonial elites were replaced by local elites who introduced radical changes but ensconced themselves into a lucrative dictatorship.

In other cases, leaders responded to popular grievances by initiating radical change addressed to solving the problems of the poorest people in their new countries. When peasants were the strongest force for social transformation, they demanded land redistribution, but rarely were they able to obtain economic policies to the long-term benefit of rural farmers. New regimes' fear of counterrevolution and a desire to focus the state's resources on building a strong national economy encouraged the adoption of one-party rule. This often had the same effect, i.e., creating an authoritarian state.

The revolution in China that brought the Chinese Communist Party to power, discussed in Chapter 8, exemplifies Mao Zedong's famous *Little Red Book* maxim that political power grows out of the barrel of a gun. The Japanese invasion of China, as Crane Brinton would have predicted, added to the weakness of the ruling Kuomintang Party and provided the destabilization and ineffective-state conditions required for successful revolutions.

---

[27]As pointed out in Chapter 5, James Davies (1969) and Ted Gurr (1970) developed the social-psychological model for rising aspirations as a force for change, contrasting the disparity between what people want and what is actually the case. When the distance between these is widening, or is wide and not closing, regardless of what the actual living conditions may be, there is potential for a revolt of the masses.

Equally important was the success of the communists in organizing a mass army of distressed peasants (after having failed to do this in the 1920s) and leading a civil war that began before and continued through World War II.

A Model of Revolutionary Change. Theda Skocpol's *States and Social Revolutions* (1979), perhaps the most compelling analysis of revolutionary change, drew parallels between revolutionary transformations of China, Russia, and France. She questioned the model of successful revolutions as being the result of uprisings of aggrieved and angry citizens, with skilled leaders and a just cause, attacking an unresponsive government. Skocpol analyzed in detail the structural conditions that make for successful revolutions, playing down the leadership and defection of intellectuals of earlier, often Marxist, interpretations. In Tilly's summary of Skocpol's theses, the old regimes were

> hampered by existing relationships between their central authorities and their agrarian economies, weakened in the course of failed responses to foreign pressures, dissolved through challenges from more powerful states and/or their own landed upper classes, became vulnerable to assaults from below, and succumbed when those assaults actually materialized. (Tilly 1984: 113)

The Iranian revolution (1979–80) exhibits most of the elements in Skocpol's model. Mass uprising against the Western-oriented Shah of Iran replaced the Shah's regime with Shi'i Muslim clerical rule and a subordinate civil government of elected officials. Military expenditures—rather than state support for the urban poor and rural agriculture—and an unpopular alliance with the United States had weakened popular support for the Shah. Popular grievances were abetted by anti-American rhetoric of the revolutionary Ayatollah Khomeini who cultivated a large, loyal following while exiled in France. Restive groups mobilized around Iran's universities and mosques, drew material resources from the bazaar economy, generated fissures in the army, and delivered a quick, lethal blow "from below" to a secular dictatorship.

The transformation not only of the state but of society, i.e., strict adherence to a fundamentalist version of Shi'i Muslim beliefs and practices, met initial resistance that was ruthlessly suppressed. For the past thirty years, the religious elites have successfully blocked efforts to significantly alter the state, with continued though occasionally contested support from moderate clerics and nearly unanimous backing of Iran's rural population.

War-Weakened States and Defecting Militaries

There have been hundreds if not thousands of rebellions in human history, but few have matured into revolutions. Despite the visibility of revolutions in the historical memory, most states have successfully suppressed efforts by those who would overthrow them. Powerful, unified states do not succumb to revolution. They subdue revolts and rebellions through intimidation, cooptation, coercion, and violence. States that cannot solve problems or address grievances, however, are vulnerable to economic depressions that create high unemployment, rising prices, especially for food, and population pressures for arable land.

A costly war—made more expensive if munitions and military hardware must be purchased abroad—contributes to economic failure. State legitimacy suffers if the war ends in stalemate or defeat. War and ongoing support for a large military make enormous demands on a nation and its people in terms of resources and personal commitment. It is no wonder that wars, especially when unsuccessful, create situations with revolutionary potential.[28]

National states need armies to accomplish their goals. Paradoxically, a powerful military creates the conditions for the state's possible overthrow. A weak state that fails to control and satisfy its military leaders and troops makes itself vulnerable to an internal military coup. A powerful military, when disillusioned by civilian rule or when ambitious officers seek personal opportunities, may initiate what Kay Trimberger calls a "revolution from above." She analyzed the military takeovers in Japan (1868), Turkey (1923), Egypt (1952), and Peru (1968) that "destroyed the economic and political power of the dominant social group of the old regime." The army or a military faction leads an "extralegal takeover of power . . . with little or no mass participation in the revolutionary takeover" (Trimberger 1978: 2, 3).

D. E. H. Russell (1974) studied fourteen twentieth-century rebellions, half ending in revolutionary changes, half ending in the rebellion being crushed. It is nearly axiomatic that a state's ability to maintain its rule depends on the loyalty and cohesion of its military. When the military joins

---

[28]Jack Goldstone's (1986, 17) excellent essay outlining the study and theories of revolution qualifies this important point. Citing many cases of defeat in war, he concludes, "The broad relationship between defeat in war and revolution is virtually nil." Defeat in war is certainly not a necessary condition for rebellion or revolution, but as a force that both weakens and shows the weakness of a state, it is often a significant contributing factor.

one of the contending classes or elite groups, a revolution is much more likely. As discussed earlier in the chapter, state power is backed by coercion. To cite a well-known observation, "Inside the velvet glove of the state is always the iron fist." If a state loses its monopoly on coercion to a breakaway faction of the army, it may well lose its ability to rule.

Revolutionary Outcomes: Political Change and Social Change

Barrington Moore's classic *Social Origins of Dictatorship and Democracy* (1966) offers a set of general principles about revolutionary outcomes in seeking to explain the perplexing question of the last century: why are national states governed in the ways they are: fascist, socialist, or capitalist democracy? His focus on state capacity, international challenges, the unity or defection of elites, class competition and conflict, and popular uprisings by urban workers and rural farmers or peasants comes down to this, again summarized by Charles Tilly:

> [C]apitalist democracy resulted from bourgeois revolutions that transformed or liquidated the old landed classes . . . [F]ascism grew from the development of capitalism with a relatively weak bourgeoisie and without a liquidation of the old landed classes . . . [S]ocialism developed from the stifling of commercial and industrial growth by an agrarian bureaucracy which ultimately succumbed to a peasant rebellion. (Tilly 1984: 121)

Outcomes are more complex than this, of course, and it is necessary to add some details. When elites—including those in the military—defect from the dominant state, they need popular support in order to govern effectively. To gain support, they may introduce social innovations (e.g., ethnic group autonomy, increased social benefits, gender equity), reduce material inequality, expand economic opportunities for previously disadvantaged classes or castes, abolish traditional privileges, redistribute land, and seek new international allies, actions, and policies that move the country in a new direction.

Revolutions and Social Change. Seizing state power confers an obligation to use the state for social change. Greater resources, expanded administration, and state planning often follow a revolution, with mixed results. Active popular participation in revolutionary change entitles a wide portion of the nation to benefit in the way of changes that address their interests (i.e., provide democratic outcomes). This can result in the spread of popular democracy and political influence (i.e., provide for democratic process), though this may conflict with a centralized state's intentions of carrying out a set of planned changes, to say nothing of corrupt elites posing as revolutionaries who seized power only to enrich themselves.

Does revolutionary change result in changes in culture and social relationships? Changes in intimate forms of everyday life may be facilitated by structural change or made difficult by legal prohibitions, and in time these can become widespread. For example, the Iranian Revolution restricted women's public appearances and how men and women could interact outside their homes, encouraging greater patriarchal practices in family and public life. Well-established cultural practices and beliefs are, however, less amenable to political persuasion and the rhetoric of revolutionary possibility. The persistence of habits of thought, including religious beliefs, gender relations, caste and class identities, differences in social status, tastes and aesthetic preferences, forms of leisure, and orientation to time usually change slowly and only through a complex process of personal experiences and social support.

Revolutions, even those that radically alter political and economic systems, are frustrated by this cultural entropy. In response, new regimes may expend energy on fruitless efforts, often using disquieting tactics, to no lasting avail. Language and clothing regulations, prescriptions and proscriptions that defy the common-sense or deeply held values of a people, may use all the tools of modern technology and image creation to accomplish superficial adoptions and fads. Significantly altering cultural practices, however, is a much more laborious process.

Beware, Mao Zedong cautioned, of "sweeping the floor out the door" or "cutting the chives off at the roots." Pushing too deeply into the core values of a people can destroy the revolutionary efforts of political and social change.[29] Post-Soviet Russia and post-Mao China—as will be seen in Chapter 8—give some indication that change on the level of culture and social relationships may take more than a revolution and the forty years Moses kept the Israelites in the Sinai Desert to rid them of the habits developed during generations of slavery in Egypt.

War and Resistance to Social Change

War!

What is it good for?

---

[29]As will be shown in Chapter 8, his notion (borrowed from Leon Trotsky) of continual revolution that would keep the revolutionary spirit alive and block retreats into prerevolutionary social and economic relations led to the Cultural Revolution (1966–76) that ended the Mao era.

> Absolutely nothing!
>
> War, it ain't nothing but a heartbreaker.
>
> War, friend only to the undertaker.
>
> —Norman Whitfield and Barrett Strong

"War," a soul/rock antiwar song, was first popular in 1970 during the Vietnam War. It has become one of the best known modern protest songs, revived in 1986 by Bruce Springsteen and banned on Clear Channel following the 9/11 attacks in Washington, D.C., and New York. Its sentiment is not unusual nor is it new. War has always had its opponents and resisters.

Today, trying to solve problems using structured violence is almost universally condemned, publicly regretted, and grudgingly accepted only as a last resort. The reasons are obvious. Wars are destructive, costly, and have unpredictable outcomes. They require people to do many things they otherwise would never do. The innocent all too often suffer the brunt of the violence. Tactics of economic coercion and diplomatic isolation are much preferred when issues between contending nations cannot be solved with discussion or legal procedures. The tools of diplomacy are designed to prevent war through nonviolent engagement, even between enemies. International relations efforts seek to configure a balance of interests and nonlethal power that prevent war's outbreak.

War in Opposition to Social Change

States can try to use their coercive power to prevent any change in the status quo. By waging (civil) war, political arrangements can remain as they are, not as rebellious others would have them. As the Arab Spring of 2011 vividly illustrates, reactionary states violently defend the regime and dominant elites against would-be usurpers, competing elites, or democratic idealists who challenge existing arrangements and would make major changes that jeopardize those in positions of power and privilege.

States have often gone to great lengths to maintain themselves against independence and autonomy for regions within their jurisdiction. Soon-to-be president of China, Xi Jinping, spoke for many states in praising the Communist Party secretary of Xinjiang, a region in western China occasionally experiencing ethnic strife. "He has firmly erected the idea that stability overrides everything, unswervingly safeguarded national unity, and struggled with a clear-cut stand against the forces of ethnic separatism" (Wong and Ansfield 2010: 10).

There are several reasons for this opposition. Breakaway regions are often rich in petroleum, minerals and ores, or other natural resources the larger nation does not want to give up. Nations sometimes have such ethnically mixed populations that allowing one to succeed will signal to others that they, too, can break away. The restive region's location or affinity to a neighbor might, if it was given more independence, present a security threat to the nation. Wars to maintain national unification are largely respected by the international community. Many of them could also be vulnerable to similar challenges. The example shown by successful independence movements is antithetical to their own interests.

Because "good wars" have the potential to unify a nation, an unpopular regime or one threatened by elite fissures may try to arrest its downward slide by using war opportunistically. This tactic gives a state involved in a war license to declare marshal law that suspends the constitution. It can suppress dissenting voices that are insisting on political reform or regime change to solve problems, address injustices, reduce inequities, attack corruption, weed out incompetence, and so forth. Such measures have proved to be short-lived, however, and change comes about in time.

Resistance to War: Peace as a Trajectory for Social Change

He shall judge between the nations, and shall decide disputes for many peoples; and they shall beat their swords into plowshares, and their spears into pruning hooks; nation shall not lift up sword against nation, neither shall they learn war anymore.

—Isaiah 2:4

If we assume that life is worth living and that man has a right to survive, then we must find an alternative to war.

—Martin Luther King, Jr., "The Quest for Peace and Justice"[30]

War as an instrument of social change meets with resistance. This is not resistance to change, but resistance to the use of war to direct, deepen, and widen social change processes.

---

[30]The Reverend Martin Luther King, Jr. delivered this speech in accepting the Nobel Peace Prize in 1964. The quote from the book of Isaiah comes from the English Standard Version of the Bible.

Iris Summers' niece, Cecilia, announced sometime after her sixteenth birthday that she was a pacifist. Under no condition would she ever participate in war, nor would she do anything to help those who made war. "Would you be a nurse and help wounded soldiers?" Iris asked her. After some thought, Cecilia agreed that she would, but only if her patients promised not to return to combat. She wouldn't patch them up so they could go fight again. "What if somebody like Hitler attacked the United States? What if the Japanese attacked us again?" This was a hard question for Cecilia, but she was resolute. "No. The Japanese attacked our navy, not America. Maybe Germany would have attacked us, someday, but we could have beat them a different way, without war." Iris was taken aback, but she laughed and said she knew Cecilia would come to her senses when she got older.

There are many arguments for going to war, e.g., defending freedom, protecting hearth and home, and saving innocents from brutality. There are as many security arguments for making preparations for war, both as deterrence and to be able to pursue possible future wars successfully and minimize one's own losses. Those providing these arguments are not necessarily promoting war, and they would probably prefer another approach if feasible and likely to succeed. They tend, however, to oppose any weakening of a state's recourse to war. That does not make them opponents of peace. Quite the contrary, most would argue for peace and believe peace is possible only because of a willingness to use war as an instrument of defense and security.

There is another peace position, however, best understood as opposition to social change accomplished through war and the preparations for war. In its most extreme form, pacifism rejects war in all its forms and refuses to acknowledge even its positive consequences as morally acceptable. In a less absolutist version, opposition to war as an agent of social change provides the motivation for working to reduce or eliminate the reasons wars are fought. Many people embrace this position and work in various capacities to create social change that obviates the use of structured coercion and violence. Two broad categories of this kind of work are (1) efforts to change conditions that most often lead to war and (2) facilitating alternative means for settling disputes.

Addressing Material and Ideal Interests. Because the causes of war are so varied, changing the conditions that lead to war covers a great deal of territory, including conflicts over material and ideal interests. Material interests are the tangible things people need and many of the things they want. Arable land and earth resources, historical sites with conflicting claimants, access to markets, and a geography that offers greater security are some of the material interests over which recent wars have been fought.

Addressing material interests is the work of people who try to make more food available by improving agriculture and of practitioners of nonlethal technology transfer. Conservationists and environmentalists who urge lifestyle changes and the careful use of resources to reduce reliance on contested resources also fit into this category. Reproductive health practitioners and those working to mitigate population pressure can see the benefit of reducing competition for limited resources. Educators promote improved lifestyles and economic growth that better utilize human capabilities. Improved economic opportunities, even in competitive environments, offer possibilities for interdependencies and networks that cut across ethnic and sectarian lines, linking workers and businesses in ways that erode historical animosities and distrust. Helping to build small businesses, associations of workers, financial institutions, trade associations, and the like can be approached as peace work.

Ideal interests are more ambiguous but no less consequential. Being able to participate in one's culture, including practicing religion and speaking a customary language, is deeply valued by thousands of ethnic groups. Accepting and even promoting minority cultural practices, while possibly seen as potentially threatening to a regime, can calm an otherwise restive group that feels marginalized or discriminated against by a dominant culture group. Those who work to revive or recover lost art, music, history, crafts, literature, spiritual practices, and culinary tastes can be seen as doing peace work. These things address the ideal interests of ethnic or religious groups by providing evidence of their heritage and bolstering their image, thus reducing their sense of marginalization.[31]

Diplomacy, Peace Building, and Reconciliation. Diplomacy is the most visible means of resisting war by offering alternative ways of settling disputes. In its ideal form, it is a way of preventing social change through war by providing channels of communication and forums for discussion—even heated discussion and deep disagreements. Usually associated with state-to-state interaction, diplomacy is also carried on by nongovernmental organizations that work between groups to ease distrust, encourage contact and the sharing of perspectives and experiences, and engage in mutually beneficial projects.

Signing a peace treaty or suspending military operations is not the end of a war, especially when it has involved a civilian population as both

---

[31]Archeology also can be used as an instrument of ethnic assertion that feeds jingoism and widens fissures between groups when both are claimants to land and historical sites.

combatants and victims. The knowledge of crimes committed, the pain of victimization, deep fears, demands for vengeance, and the desire for compensation remain vivid. If unaddressed, these can sow the seeds for future conflict and even war. Truth and reconciliation and the "climb down" after war are a way to move beyond residual distrust and slake the appetite for vengeance following war.

The United Nations has no standing army. For decades, member countries have assembled armed forces to keep the peace in war's aftermath. Troops have been in Cyprus for decades, both before and following the war between Turkey and Greece in 1974. U.N. troops numbered in the thousands following the resolution of civil wars in Bosnia and Kosovo. Today, there are about 100,000 peacekeepers around the world: seven in African nations; one each in Cyprus, Kashmir, Kosovo, Lebanon, Syria, Afghanistan, and Pakistan; and Haiti—a peacekeeping mission since 1964. The international community has expanded the United Nations' peacekeeping efforts by sending in troops when fighting ceases to prevent a recurrence of violence and to end humanitarian crises caused by wars. Such interventions have become "a principal mission of the world's armed forces" (Sivard 1996: 30). Of late, U.N. forces have engaged in very selective "peacemaking" operations where troops have engaged armed groups in Congo's ongoing insurgencies that had their origins in the Rwandan genocide.

The U.N.'s third role is peace building, a relatively new assignment as the organization increasingly focuses on civil rather than international wars. People in the U.N. peace-building mission include economists, police trainers, human rights monitors, and advisors on good governance. They work with the state and civil society organizations to improve a war-torn country's capacity to sustain the peace and, especially, to ensure greater trust in, and effective operation of, the government.

In South Africa, Rwanda, Argentina, and elsewhere, truth and reconciliation processes have been set up to help people put war behind them but also to prevent future wars. They differ by the circumstances and preferences of the locale, but all of these efforts involve the same basic elements. A public format is arranged where perpetrators and victimizers can come forward. They are encouraged to do this by the promise or possibility that their crimes will not be legally prosecuted if their "truth" is deemed satisfactory and sincere contrition is expressed. Those hearing the confessions of guilt are both the public and the victims themselves who can speak to the accused, forcing them to face the suffering they have inflicted.

The truth and reconciliation process involves objectively investigating crimes committed during conflict without exonerating anyone. It led to the organization of tens of thousands of hearings and played a significant role in establishing the legitimacy of the new government in South Africa headed by Nelson Mandela. *Gacaca*, traditionally structured meetings for reconciliation in Rwanda, followed for several years the 1994 genocide there. In both cases, truth and reconciliation has met with mixed results (Graybill 2002; Staub, Pearlman, and Miller 2003). Truth and reconciliation has been applied on a smaller scale elsewhere, including South American countries in the aftermath of civil wars and government and paramilitary atrocities ("dirty wars").[32] The goal of healing wounds and moving away from a psychology of retribution seeks to create the possibility of a more peaceful society, able to accept the anguish of the past in order not to repeat it.

Topics for Discussion and Activities for Further Study

Topics for Discussion

1. Perhaps some of you in the class have a personal experience with war. Let those with the experience talk, and then let the others ask questions. What mental preparations are required to engage in a war zone? In a war zone, how does thinking about war change? What about reentry after leaving the war? How do you talk about war now?

2. After the 9/11 attacks, the United States declared a war on terrorism. What does this mean? Is it a war effort against anyone who uses terrorism as a tactic? Why isn't the use of terrorism considered a crime, like homicide, so that the response is to fight the crime rather than fight a war? What social changes have come about as a consequence of the "war" rather than the "crime spree"? Be sure you are talking about terrorism and not something else (see fn. 7).

3. Compare the positions of pacifism and militarism. Imagine and discuss situations that might call for a military response. What would a pacifist response be? Pacifism isn't necessarily nonaggressive. It doesn't necessarily mean nonresistance. Can you imagine engaging in resistance that uses military strategies, such as denying the enemy food and other things needed to carry on, but refuses to engage in violence, including taking a life?

4. In 2010, Defense Secretary Robert Gates talked at Duke University about the U.S. military. He pointed out that military personnel are increasingly from

---

[32]In El Salvador, the replacement of a right-wing regime with a more democratic government allowed TRC reconciliation but provided a blanket amnesty for all perpetrators.

the South, Mountain states, and small towns. He also raised questions about an all-voluntary armed forces, wondering if it was allowing the 99 percent of the population not in the military to "zone out" wars and the personal costs of fighting them. Do these trends seem problematic to you? Do you think there should always be a war tax passed before going to war? Should there be conscription? Why, or why not?

Activities for Further Study

1. If you live in a small or large city, it isn't too difficult to find people who are refugees from a war. You may also be able to do this in a smaller town. Have an extended conversation with a refugee. What was the war? What was their experience of the war? After having this conversation, do some research on the war. Does knowing someone who experienced a particular war change the way you understand the war?

2. Select a country that was a site of conflict in World War II, e.g., Italy, Ukraine, Russia, France, Spain, China, Japan, Philippines, Algeria, Greece, or Hungary. What changed during the war? What changes in the decades after the war can be attributed to the war?

3. There are several good journals devoted to research on conflict resolution and peace research. These include *Journal of Conflict Resolution, Peace and Change, Peace Review,* and *Journal of Peace Research.* Read the tables of contents of several issues of these or other peace-related journals and select an article that looks interesting to you. Read it, and then find a second article that is frequently cited in the first article you read. Write a report on the two articles and explain how the two are related to one another. Is one a refutation of the other, does it build on the other, or do they confirm each other?

4. Do some research on technology spin-offs from military research and development. Trace the path of a technology (remember, technology isn't always a device or instrument) from the military to a civilian application.

5. Research public opinion. Many polls are taken during modern wars that ask people about their opinion of a war, e.g., how they think it is going, whether they support it, or if a particular strategy should be used. Pick a war no earlier than the Vietnam War and look at public opinion over the course of the war. What trends or changes in public opinion do you find? To what are these changes attributed? Does this tell you anything about war and public opinion more generally?

6. Can you trace cultural styles to war? What elements of music, fashion, art, architecture, painting, or any other practiced art have been influenced by

war? Select a fairly specific period and place to study this, e.g., Serbia in the 1990s.

7. Go to the United States Institute of Peace website (http://www.usip.org). Look at the projects they have funded. How would you characterize their sponsored projects? In what respects are they following the ideas of Gordon Allport's contact hypothesis (in Chapter 3)? Are they peace-building or peace-keeping efforts?

7

Corporations in the Modern Era

The Commercial Transformation of Material Life and Culture

I hope we shall . . . crush in [its] birth the aristocracy of our monied corporations which dare already to challenge our government to a trial of strength and bid defiance to the laws of our country.

—Thomas Jefferson (letter to Tom Logan, 1816)

J ustice John Paul Stevens of the U.S. Supreme Court cited the third president of the United States in his strong dissent to the majority's 2010 decision allowing corporations unlimited spending on behalf of political candidates.[1] Quoting the court's earlier *McConnell* decision, Stevens wrote, "We have repeatedly sustained legislation aimed at 'the corrosive and distorting effects of immense aggregations of wealth that are accumulated with

---

[1]Jefferson's animus may seem curious in light of the history of British corporations that financed the settling of the first North American colonies and, as discussed in this chapter, are often credited with providing the model for representative government adopted by the framers of the U.S. Constitution (Tuitt 2006).

the help of the corporate form.'" The court's decision, Justice Stevens continued, "will undoubtedly cripple the ability of ordinary citizens, Congress and the States to adopt even limited measures to protect against corporate domination of the electoral process."

The essence of Justice Steven's dissent in the *Citizens United v. Federal Election Commission* is that corporations are legally devised entities that organize activities and are given special legal protections similar to those afforded individuals. Their rights and obligations are a matter for the state to decide. Justice Stevens emphasized how the First Amendment's guarantee of freedom of speech—the basis for the court's majority to rescind a hundred years of legislative limitations on election spending by corporations and unions—applies only to real individuals and groups of individuals. Corporations are "not natural persons, much less members of our political community . . . Although they make enormous contributions to our society, corporations are not actually members of it. They cannot vote or run for office . . . The financial resources, legal structure, and instrumental orientation of corporations raise legitimate concerns about their role in the electoral process" (Stevens 2010).

Corporations are recognized by the law as having economic rights otherwise reserved for individuals, e.g., to own property, make contracts, and be represented in civil and criminal court. Many people associate corporations with a kind of personality, and certainly corporate publicists spend lavishly to "brand" corporations by associating them with favorable characteristics: innovative, caring, trustworthy, loyal, helpful, friendly, thrifty, optimistic, brave, and so forth. Individuals' images are iconic—Betty Crocker for General Foods, Colonel Sanders for KFC, Ronald McDonald for McDonald's, Mr. Goodwrench for Goodyear—and help endow them with human qualities.[2] In some cases, real flesh-and-blood people *are* behind the corporate logo. Bill Gates is the face of Microsoft, Steve Jobs is Apple, and Oprah Winfrey is her media empire.

Corporations are often started by entrepreneurs who remain associated with the firm that bears their name or imprimatur. When ownership is made public through the sale of shares, or the firm is purchased by a large

---

[2]David Korten, a former executive and now a critic of large corporations, rejects the idea of a corporation as "almost a living person." It is an "illusion cultivated by corporate relations and given legal standing by court rulings . . . The corporation is not a person, and it does not live. It is a lifeless bundle of legally protected financial rights and relationships . . . It is money that flows in its veins, not blood. The corporation has neither soul nor conscience" (Korten 1999: 75).

corporation, the original owners may become corporate officers or are given special titles. Their role can evolve into meeting the public, giving speeches on behalf of the corporation, or lending their likenesses to company logos. Their power passes to boards of directors and shareholders who usually know little about the actual activities and operations of the firms, and to a managerial staff that does.

Corporations predate the nineteenth century, but their modern form and place in public, political, and economic affairs has changed and grown so enormously in the last century that it is impossible to understand contemporary social change without grasping the role corporations have played in the transformation of culture and social structure. Literally, they own most of the world and process most of its resources; they are the source of jobs and the vehicle for amassing wealth. They produce the goods and services that pervade our lives, disposing of the old and encouraging us to buy the new, better, faster, more efficient, trendier, safer, healthier, cleaner, fun things that make us happier, better looking, envied, sexier, and respected for our taste and ability to acquire them. William Roy, author of *Socializing Capital*, a study of the rise of corporations in the United States, observes that "few features of contemporary American society are more far-reaching or awesome than its large industrial corporations" (Roy 1997: xiii).

Understanding Corporations and Social Change

How corporations themselves—and especially large multinational firms—direct change is rarely explored in studies of social change. Rather, the things corporations produce in the way of technology have been studied in depth. Social and political theorists seek to decipher the role of corporations in the operations of the state. Labor research focuses on the changing nature of work as corporations have grown, created jobs, and organized the labor process. Organizational analysts chronicle the strengths and liabilities of corporate bureaucratic forms, and business studies cover the map of what works best to capture and hold markets, foster innovation, and motivate workers. All of these contribute to an understanding of important aspects of corporations and are discussed in this chapter in terms of their impact on social change. Corporations themselves as agents of social change, however, have worked under the radar for many citizens and scholars who seek to understand social change. Ironically, this may be due to their being so pervasive. They just seem to be the natural order of things in modern times.

Corporations as Evolutionary Systems

Much writing about corporations and social change is within what is described in Chapter 3 as the evolutionary systems perspective. It emphasizes the arrangement of parts, how they work together in processing resources, and how the whole adapts, grows, reproduces, and subdues its adversary.

Typical is Alfred Chandler, one of the most articulate and respected historians of American corporations, who shows in *The Railroads* (1965) how the corporate form developed in the late nineteenth century in the face of new challenges and opportunities. Its chief manifestation was the railroad corporation that built tracks spanning the North American continent, linking thousands of towns and cities. The success of a few corporations over others was a consequence of their bold innovations and the amassing of both financial resources and political power. In Chandler's account, they proved to be more fit than their competitors to do what the nineteenth century wanted and needed in order to progress.

The three basic elements of the evolutionary systems perspective—growth, complexity, and reproductive advantage—are prominent in Chandler's accounts. Much of the basic technology for transcontinental railroads had been developed by midcentury. The successful railroad corporations, led by determined executives, developed complex organizational forms that were both flexible and resilient. Industries such as iron production and railroad car manufacture that supported railroads' expansion also became corporate giants. A rapidly growing population eager to build farms and businesses across the vast continent provided the impetus. Those who owned the successful large corporations 125 years ago became fabulously wealthy, and it has remained so. Not just in the United States, Western Europe, and Japan, but throughout the world, "one's relationship to corporations is now the most important determinant of wealth" (Roy 1997: 13).

Corporations in the Conflict Perspective

In contrast, a conflict perspective, also discussed in Chapter 3, tells a different story. It emphasizes power, contradictions, and the triumph of particular economic interests in conflict with others. In this view, corporations are one of many possible ways to organize economic activity. Explaining how they came to dominate the global landscape recognizes the ambitions of individuals and the entrepreneurial and technically innovative efforts the "Captains of Industry"—the 1920s moniker coined by Thorstein

Veblen—exerted in building business enterprises. They very consciously promoted the corporate form for businesses, sought and gained legal protection for their efforts, and established themselves as powerful forces in the social landscape.

Understanding large corporations as agents of social change requires the examination of politics, financial power, and the mobilization of public opinion. Efficiency and economies of scale gave some, usually the largest, corporations an advantage over others, but this doesn't necessarily translate into an advantage based on a superior product or service. Dominating the story of corporations is the pursuit of wealth, not simply the desire to make a better mousetrap, satisfy peoples' needs, or foster a society that works in harmony with the things human beings cannot control.

In the conflict perspective, the essence of major corporations is their power. Because of their size and the scope of the money and resources they command, large incorporated businesses configure much of the physical and social landscape of daily life. Corporate power gives them the means to confront the challenges facing the firm itself, the people who rely on it, and the environment. Because of their wealth holdings, the sophistication and size of their workforces, their resistance to state regulation, the instrumental focus of their efforts to advance the agenda of the corporation, and their global reach, studying corporations and social change is a study in power: its strengths and limitations, ebbs and flows, its positive and potential uses, and abuses.

Greater than any other force effecting social change today, corporations have the ability to "determine the context within which decisions are made by affecting the consequences of one alternative over another" (Roy 1997: 13). This is the essence of corporate power. It is sometimes startling when individuals reject what corporations determine as the good life. It can be perilous to oppose corporate-driven social change , either as an individual or a social movement.

Nonetheless, more than a few independent artists accept the consequences of working and living outside the corporate sphere. Biodynamic and small organic farmers, entrepreneurs in small family businesses, antiglobalization activists, spiritual communities, and committed environmental preservationists count themselves out as well. Those who actively reject popular commercial entertainment and commercial advertising as too mind-numbing, offensive, or intrusive to deserve their time and attention fall into this camp. Their awareness of corporate power and resistance to corporate-driven social change may be a rear-guard action, or it may be a harbinger of the future.

Businesses, Firms, and Large Corporations

As many as five and a half million organizations in the United States today are incorporated. Only a handful of them, and a few thousand across the globe, engage in business on a massive scale. They are the agents of change addressed in this chapter. These are the large national, and more often multinational, corporations that employ most of the people, own most of the business property, exert most of the political influence, provide most of the personal wealth, and generate most of the commercial media that structures working life and leisure in affluent countries and, increasingly, in less affluent countries today.

Corporate advocates consider the creation of this "technology," the corporation, to be one of the most important innovations in the history of the industrial revolution and the expansion of material wealth of the past two hundred years. The advantages of corporations are many. For one thing, their method of organizational control, with a board of directors and a managerial staff, avoids what is seen by many as a too-personal management style of smaller family-owned firms. The elderly Henry Ford's effort to be involved in all aspects of his automobile business nearly led to its collapse before his son took over and imposed a modern corporate management plan. Large corporations are continually tinkering with the technology of the corporate form as they move inexorably around the world and ever deeper into our lives.

The Pervasive Corporate World. Before the collapse of the Soviet Union in 1991 and the transformation of China following the death of Mao Zedong in 1976, it was possible to say that a large portion of the world's resources and its industries and services were public endeavors. Since these events, however, much has changed. Emulating countries in the West, private ownership has taken a majority share in the economies of Eastern Europe, the Soviet Union's former republics, Russia included, and increasingly in China.[3]

The corporate form, developed in the United States and Europe, has been adapted to what were formerly state-owned and publicly operated

---

[3]Large corporations are not exclusively owned by individuals and other privately owned firms, as La Porta and his colleagues (1999) show. State ownership of corporate shares, sometimes known as sovereign funds, is substantial in Middle Eastern countries, as well as Austria, Portugal, Spain, Israel, New Zealand, Germany, Greece, Italy, Norway, and Singapore. In most countries, petroleum reserves and many other earth resources, as well as airlines, harbors, and railroads, are owned and operated entirely or in part by the state.

enterprises. In many cases, the former state managers have become the owners, while the workers under state socialism are now employees. Billboard, magazine, Internet, and television ads, along with glitzy stores and malls, exhibit a newfound consumerism in products and services provided by local and multinational corporations. Certainly the political changes in these countries are large, but the drive to "catch up to the West" in economic terms has fueled the changes most intimately experienced by ordinary people. The corporation has become pervasive in the everyday lives of Chinese, Kazaks, Russians, Hungarians, Croatians, Czechs, and Latvians.

The pervasiveness of corporations is easily seen in your immediate environment. The book you are holding is a corporate product, and the materials used to make it came from corporations that harvested the trees, produced the glues and dyes, and manufactured the computers on which it was written and then assembled by the publisher. Look around you. What can you see that is not the product of a corporation? Not much. Maybe you are eating a carrot from your garden. Maybe hanging on the wall is a dream catcher a friend made from scavenged twigs, feathers, string, and bits of debris. Maybe a small, hand-sculpted animal, bought in a far-away country, sits on your desk, or a bowl you made in art class, a shell from the beach, a crocheted potholder, or a crumbling doll made with garden weeds. Anything else?

Corporations "provide almost all our food, housing, transportation, entertainment, financial services, health care, communication, and national security items" in affluent countries (Tuitt 2006: xiii). Are you wearing anything that doesn't have a corporate logo or tag? A friend's knitted scarf? If your clothes were made abroad, they could well have originated in an unincorporated factory or clandestine sweatshop, but they were marketed by a corporation. How many corporate ads can you see in your immediate surroundings: informational messages, invitations to join in a sale, and brand names and logos posing as adornments? If your television or radio is on, or you are streaming on your computer, you are almost surely seeing or hearing ads and corporate messages. All of this seems so natural and hardly objectionable. It's just the way things are. Such is the power of corporations.

Retail corporations large and small account for 90 percent of all sales in the United States, sales that make up 70 percent of all economic activity. Of all employed Americans, 95 percent work for someone other than themselves. In the United States, firms with 500 employees or more (less than 1 percent of all firms) employ about half of all workers.[4]

Many, if not most, of the consumer choices people make during the day are structured by corporate actions, both in terms of the products available for purchase and the perceptions that advertising confers of what is valuable, desirable, and necessary. When the CEO at Goldman Sachs, Lloyd Blankfein, told a congressional hearing investigating the financial meltdown that precipitated the Great Recession, "We're doing God's work," it was widely ridiculed as hubris of the first magnitude. Religious references aside, the corporation is—along with the family and the nation state—the most important and influential social organization in the world today.

The Dominant Corporate Form. Even its critics recognize that the corporation is "an engine of economic growth." The largest corporations usually do not have the dynamism and enthusiasm for innovation found in smaller businesses, but the corporate form has proved to be "uniquely effective in rendering human effort productive" (Micklethwait and Wooldridge 2005: xx).

In simple terms, a corporation is an organization held together and operated in terms of a set of contracts. Management staff, not stockholders, sign the contracts and represent the corporation for legal purposes. When a corporate scandal or other act occurs that jeopardizes the public, such as the April 2010 explosion of the Deepwater Horizon drilling platform in the Gulf of Mexico, it wasn't BP's largest shareholders (owners) who testified before Congress. It was BP's executive officers. The corporation's legal status gives it specific protections and collective rights and requires it to operate according to legally prescribed guidelines. Business corporations accept obligations in order to pursue profit-making activities that give immunity to shareholders from accidents, illegal corporate practices, and bankruptcy. And, they work very hard to ensure their profit-making activities can proceed with as little interference by the state as possible.

---

[4]Mark Granovetter challenges the idea that most people work in large establishments and shows how the actual workplaces of most employees are smaller than might be thought, given the dominance of large corporations. "In the United States, in recent years, at least one in four private-sector workers find themselves in establishments with less than 20 employees, one in two in those with less than 100." In looking not at peoples' worksites but rather the size of the firm, Granovetter concedes, "The argument that workers are parts of much larger operations now than fifty or seventy-five years ago is probably correct" (Granovetter 1984: 327).

A corporation, as a legal individual, has the advantage of being able to own property and pool capital. It does this by offering itself to the public through the sale of stock. Buyers then become owners of a portion of the company. Shares can be sold with no consultation with the company or its officers, at a price set by others' willingness to purchase shares. Corporations in the United States and most other countries can own shares in other companies or own entire companies outright. They can borrow money and issue bonds in the form of debt that pay the purchaser a fixed return.

The prospect of profits lures investors to offer money that finances the operation and expansion of the corporation. More profitable corporations attract investors who speculate on future profits. Greater size and more capital give a competitive advantage, hence enhanced prospects for profitability. As a rational bureaucracy, corporations hire and fire employees as needed, barring agreements with unions that restrict their latitude, in order to remain profitable. They sell off parts of themselves and acquire new parts, spin off new corporations, and in myriad ways try to keep profits high, their taxes low, and investors satisfied.

Like actual individuals, a corporation has the right of privacy (e.g., proprietary knowledge critical to its being competitive) that keeps public eyes—as well as its employees'—out of its books, minutes of meetings, and laboratory secrets. Its actions are focused on generating profits, and it is legally required to do this as a responsibility to its shareholders. It is expected to abide by the law of the land, pay taxes, and conform to regulations, including scrutiny by governmental and regulatory bodies. Otherwise, it is free to do as it pleases and pursue profits wherever they may be found.

The greatest advantage of corporations is limited liability. Not only are risks of business failure shared, but responsibility for failure is limited to the corporation itself, not its investors. Bad corporate behavior does not taint or implicate investors. If sued in court, the corporation is represented as if it is an individual, though it cannot be sent to jail. Its debts and responsibility for lawful actions are confined to it alone. If it declares bankruptcy and must pay off its debts, these obligations extend only to its resources and not those of its shareholders. If its actions result in human injury or death, discrimination, or unfair business practices, it can be fined and must make amends, but its shareholders are not liable beyond what they have invested in the corporation.

Other Types of Corporations. There are several types of corporations, in addition to the business corporation just described. Closed corporations and "closely held" corporations are legally constituted to operate as corporations, but the stock is normally family owned and not publicly traded. Until 2006, the Cargill Corporation, a global giant in trading and shipping of grain, was the largest closely held corporation in the world. Today, Koch

Koch Industries

nonprof.

Industries, owner of the major lumber company Georgia Pacific and a worldwide investor in energy, has this distinction.

Nonprofit corporations also provide limited liability, are represented by a board of directors, and are recognized as legal individuals in order to acquire, hold, and sell property, to sue and be sued, and so forth. They issue no stock and have no profit distribution beyond operating the organization and, thus, are exempt from paying taxes. Nonprofits range from the American Red Cross to the Catholic Church, the Ford Foundation, which supports philanthropic and research activities, the Rand Corporation, which does research for the Department of Defense, charities, private hospitals, communities, some schools, and condominium homeowners' associations.

TORT VICTIMS AND THE ACTUAL PRICE YOU PAY

In *The Corporate Paradox*, Wouter Cortenraad examines the abuse of limited liability, what he calls "tort victims." He defines tort victims as those affected by "activities and behavior patterns on the part of the firm that are desirable from the shareholders' point of view but that adversely affect parties outside the firm" (Cortenraad 2000: 10).

Tort victims may be employees of a firm exposed to health hazards, injured, sexually harassed, unfairly terminated or reassigned, or who have their rights violated beyond what corporations are normally allowed to do for business purposes. Victims can be buyers or users of corporate products and services who are defrauded or harmed.

Penalties and restitution are limited to the assets of the corporation, not the personal wealth of those who own shares in the corporation. When committed by a large bureaucracy, responsibility for tort victimization (wrongdoing) is difficult to assign. Occasionally a company's officers, staff members, and chief executives may be penalized if it can be proved that they are personally culpable of not providing "honest services."[5]

---

[5] Jeffrey Skilling, past CEO of Enron Corporation, was tried for conspiracy, security fraud, insider trading, and making false statements to auditors. He went to prison in 2006 and is serving a 292-month sentence. In June 2010, the U.S. Supreme Court ruled in favor of Skilling whose lawyers challenged "honest services fraud" legislation, throwing into uncertain legal territory the question of responsibility of corporate officers. The court agreed that language in the legislation is too vague. Though it has been applied by the courts for four decades and most recently redrafted in 1988 to specify the meaning of honest services fraud, only bribery and kickbacks are clearly prohibited by the legislation, according to the Court. In the absence of more specific legislation, it is not illegal if corporate officers defraud investors, have a conflict of interest, and misrepresent their conduct in managing the business (i.e., causing "reasonably foreseeable economic harm").

The ability of corporations to externalize many of the costs of operation creates tort victims. These are much harder to identify for legal indemnity. Environmental damage and indirect harm to the health of persons adversely affected by a firm's activities are externalized costs. A factory's toxic discharge that gets into a stream, killing fish or making them dangerous to eat, is a cost but one that may be difficult to prove, especially if many other firms are doing the same thing and there is runoff from upstream farms and highways as well.

David Korten (2001) cites figures for annual medical costs from smoking cigarettes ($53.9 billion), unsafe vehicles ($135.8 billion), injuries and accidents in unsafe workplaces ($141.6 billion), and cancer caused by toxic exposures in the workplace ($279.7 billion). Most of these are never recovered by those who are hurt or become ill, or by their dependents. In that sense, the companies have been able to avoid the costs by shifting them outside their responsibilities, e.g., externalizing the costs, usually on to individuals and the state.

When purchased, an item will cost less if externalized costs have not been factored in. The purchaser is probably unaware of the externalized costs, inasmuch as these can be buried deep inside the production process. When consumer advocates or environmentalists raise the issue of paying for externalized costs, firms cite the additional expenses paying for them will incur and how this will bear on retail prices. It can be enough to dampen the public's concern for tort victims.

The Corporation's Varied History

People who now protest about the new evil of global commerce plainly have not read much about slavery and opium.

—John Mickelthwait and Adrian Wooldridge,
The Company (2005: xx)

After 1555, the Dutch, British, and other powerful trading corporations operated under state charter to raise money for voyages, carry on trade, and repel any resistance they encountered from other countries' trading companies and the local people at the sites of their excursions. To accomplish its task of establishing dominance around the world, the British East India Company, a privately owned corporation, at one time had more than a quarter million troops under its command. Put quite succinctly, the British crown, like the Dutch Crown, outsourced or "subcontracted imperialism to companies" (Micklethwait and Wooldridge 2005: 35).

Early Chartered Corporations. Trading corporations orchestrated colonialism and imperialism on behalf of states and for private gain. They carried out the slave trade, claimed most of the world's territory as belonging to their nations, decimated indigenous people, set up courts and imprisoned malefactors, created currencies and overrode traditional economic order, crushed or co-opted local political rule, and created their own governing administrations. Such was the power of the corporate form in partnership with the state. As today's antiglobalization activists are fond of pointing out, corporations continue to spearhead the global economy, though with somewhat greater oversight and less latitude to directly engage in war on their own behalf.

Corporations have a history that goes back centuries before imperial adventures. The problem of property—how to continue its legal status despite the deaths of those who operated it—was addressed by creating state-approved corporations in the Middle Ages. The corporate form solved the problem of perpetuating property rights outside the limitation of the human life cycle, a problem that didn't exist until private property had become a fact of life, and well before it seemed like the most natural thing in the world. A town, church, group of artisans, or university acquired property and kept it, withstanding death in a way humans cannot. Some of the earliest corporations continue today. Stora Enso, a Swedish mining corporation begun in 1347, still exists. The Corporation of London, not quite as old, continues to own the land of the city's financial district, three schools, and four markets.

For hundreds of years, corporations required the permission of the sovereign (the king or crown) to operate, in order to accomplish "a task considered critical for the public. They were given such privileges as monopoly rights, eminent domain, and the exemption on liability" (Roy 1997: 16). They could even be exempt from taxation. The Jamestown Settlement was founded by a corporation, the Virginia Company, chartered by England's King James I in 1607. In 1619, its democratic character—shared voting, the choice of co-officers, and approval of policies—became the model for the first government in that settlement. This move toward democracy angered the king, however, and he revoked the Virginia Company's charter.

Also established by a chartered corporation and equally problematic was the Plymouth Colony. These are the same Pilgrims celebrated at Thanksgiving for their 1620 founding of a colony at Plymouth, Massachusetts. The religious dissenters rejected democratic governance as well as private property, preferring what they believed to be a biblical form of communism and communal property: from each according to his ability, to each according to his contribution. In 1623, Governor William Bradford decried their practices and, to improve the colony's economic performance, established private property in place of communal property.

More favored than Jamestown and Plymouth, in 1629 the Massachusetts Bay Colony—actually a trading company—became a commonwealth of representative government. Accountable only to the crown and its stockholders, the corporations' pursuit of wealth guided much of the settling of the early North American continent by Europeans. They made it possible for newcomers to colonize the land by displacing, often with violence, the indigenous people (Tuitt 2006).

From Public Service to Personal Fortune. In 1791, private corporations were established in the United States, nearly a century before the first manufacturing corporations. They were intended to advance the public good, especially economic expansion. Corporate shares for investors were first bought and sold in 1798. Prior to the mid-1800s, corporations were much more likely to be formed in order to perform "tasks governments felt could not or should not be conducted privately." These tasks were "too risky, too expensive, too unprofitable, or too public." In short, corporations were chartered "to perform tasks that would not have gotten done if left to the efficient operations of markets" (Roy 1997: 41).

By 1800, there were 335 corporations in the United States, most financed with government bonds to build canals, toll roads, and bridges. The watershed came with the federal Companies Act of 1862, allowing corporations to form by filing the necessary documents with the courts in order to operate for general business purposes. No longer was "serving the public good" the prime role of corporations. The principal goal was wealth creation.

The corporation was transformed further with the building of the transcontinental railroad, an endeavor replicated in Europe. As in the United States, massive state support was budgeted for railroad construction in France, Germany, and England. They, too, provided limited liability for business failure and accidents and to raise the capital needed to get railroads built (Herrera 2006: 58, 108). Railroads not only pioneered modern corporate organization and governance; they created a model for shareholder finance, the creation of wealth, and the collaboration of the state and private investment that became integral to industrial corporations of the twentieth century.

By the 1880s, there were declining railroad profits, and a roller-coaster economy had wiped out fortunes made in both speculation and sound enterprise. The railroads linked cities and towns across the continent, but once built there was less demand for iron and steel and the products of manufacturing firms that railroad construction had supported. In order to increase prices in the face of declining demand, corporations created trusts— collusion among private businesses to set prices, divide markets among trustees, and in other ways control competition favoring those belonging to the trust. They were declared illegal with the 1890 Sherman Antitrust Act. The pursuit of personal wealth moved on.

Richard Edwards' *Contested Terrain* (1979) explains how the unfettered competition, spurred by antitrust legislation, led thousands of companies to ruin in the last years of the nineteenth century. This problem was solved by innovations in the corporate form. Corporate operations encouraged mergers, developed vertical integration,[6] initiated a legal patents regimen, and learned how to use their power to solicit advantageous legislation favoring the largest of them.

Tobacco firms characterized the new large corporations' vertical integration. They produced or purchased directly the raw materials they needed, processed these to manufacture finished products, and then distributed the goods to retailers for sale. The emerging chemical corporations did the same. "Alkali producers secure[d] their own extractive plants to supply them with raw materials . . . [T]hree of the first five ammonia-soda plants were financed by glass interests, and the very first electrolytic plant was bought by a paper maker (Noble 1977: 14–15). Ohmann explains this shortening of the chain between production and final sales as a means for "corporations to achieve autonomy—control from inside . . . partly because they stabilized and controlled distribution" (Ohmann 1996, 74).

Critical to corporate success is what Roy (1997: 74) calls *marketing capitalism*, the integration of production and sales. Large corporations developed their own distribution systems and did their own advertising. Montgomery Ward, Sears, Woolworth, and other retail corporations were summarily created to sell the products of large corporations. Over the years, advertising agencies relieved the corporations of the task of figuring out how to sell their products to the public, themselves becoming a major corporate sector in the economy.

As corporations grew, they were able to acquire manufacturing technologies larger and more expensive than what smaller companies could

---

[6]Vertical integration reduces the number of business transactions between production and final sales. Large corporate manufacturers bought forests, mines, and whatever else was needed to become their own source for production. In time, vertical integration extended to distribution and marketing.

afford. In order to fully take advantage of economies of scale and benefit from lower per-product costs, large corporations had to increase market share. In time, corporations wanted not only to reduce competition but, as much as possible, actually control the demand for the products they made and the services they offered. Advertising was and remains one of the most important means for doing this. Corporations "sought not so much to supply the market as to organize it" (Williams 1980: 191).[7]

In William Roy's phrase, private capital was transformed into "social capital," collectively administered by an executive board, and the corporate form took off. "Industrial capital merged with investment capital and sparked the corporate revolution." He calls this a "liminal period for the American economy," i.e., a period of uncertainty and indeterminacy, when the rules were changing and no one knew what wealth or penury the changes would bring (Roy 1997: 198, 257). The day of the giant corporation was dawning. "Capital in publicly traded manufacturing companies . . . jumped from $33 million in 1890 to $260 million the following year . . . and hit over seven billion dollars in 1903" (Roy 1997: 4–5). Between 1890 and 1905, ten large corporations were joined by what had been family-owned businesses and partnership industries, becoming corporate giants.

Though corporate growth temporarily slowed down after 1905, the new corporations of the emerging automobile industry—including Studebaker, General Motors, Standard Oil, and BF Goodrich—increasingly found support from a sophisticated set of corporations in electrical manufacturing and industrial chemistry. The corporate form, organized with an intention of societal improvement and economic expansion, was now adapted for a much more direct economic function. Corporations were the key to personal fortune.

Monopoly Capitalism

There was and is a great deal of competition in the entrepreneurial sector of twentieth- and twenty-first century economies. Small manufacturers, local retailers, cafes and restaurants, repair shops, building contractors, and

---

[7]The direction in which modern economic activity goes is obviously neither natural nor inevitable. In a capitalist economy that followed the trajectory outlined here, a path-dependency analysis helps to show why consumer goods and an emphasis on sales to growing and new markets has guided the mobilization and use of economic resources. Less profitable areas of social life—such as the needs of the poor, public services, and environmental protection—have been much less the focus of private economic activity.

installers are mostly family owned, even when taking on the corporate form. They are highly vulnerable to the larger economic environment, often borrowing money on a regular basis, sometimes expanding and hiring when opportunities occur and cutting back during downturns, and with a high rate of bankruptcy for new businesses.

Much of an affluent nation's economy, however, is dominated by large corporations that operate in a far less competitive environment. Their power over capital, labor, markets, and legislation confers advantages in ways unavailable for smaller businesses. Most important is their ability to dominate a market with the goods and services they provide. "Technically, such a situation is called an oligopoly (dominance by a few firms), but the economic power these few firms exercise is virtually identical to that of a true monopoly . . . On the basis of market domination, monopoly firms are able to restrict competition and charge higher prices for their goods" (Hodson and Sullivan 2008: 359). This happens despite decades of legislation designed to enhance competition and reduce monopolistic practices.[8]

Technology plays a role in this dominance, given the cost and complexity of many economic activities. Economies of scale, i.e., the ability to produce a good or service at a lower cost by having the capacity to produce more of these than a competitor, is often possible only when a massive amount of capital is available, when huge research and development efforts can be undertaken, and when the high cost of production and marketing can be absorbed until sales receipts begin repaying the investment.

Size is not always an asset for corporations, but it has many advantages. Among them are economic diversification and the capacity to move investments from one activity to another, to acquire competitors' businesses through mergers, to price goods below cost ("predatory pricing") in order to gain market share and diminish a competitor's earnings, to engage in effective public relations and ad campaigns, and to influence regulations and legislation through access to political power. These activities all favor larger, wealthier corporations.

---

[8]The Interstate Commerce Act of 1887 recognized the state's obligation to regulate businesses involved in interstate commerce. The 1890 Sherman Antitrust Act led to the breakup of Standard Oil into thirty companies, and the 1914 Clayton Act and Federal Trade Commission Act prohibited monopolistic practices, establishing the Federal Trade Commission to enforce antitrust laws. In 1945, the Corporate Control Act extended government oversight of U.S. public corporations, and in 2002 Congress passed the Sarbines-Oxley Act, in response to the recession following the dot-com bubble, providing government oversight of the auditing and accounting industries.

Finally, the largest corporations have become so integral to the economic well-being of a nation that the state and the public, including labor unions and workers, are highly dependent on their profitability. Tax revenues and government services, jobs, retirement benefits, and the economic health of communities are tied to large corporations. There is a collective loathing to let large corporations fail when the state can keep them afloat with supportive legislation, loan guarantees, tax breaks, regulatory concessions, and tariff controls. With their new power to give unlimited sums for and against political candidates—the *Citizens United* decision Justice Stevens bemoaned—their power is enhanced further.

Who Runs the Corporation? A debate about who actually controls the corporation has been going on at least since the Companies Act of 1862 was passed. When family-owned and partnership companies dominated the economy, it seemed clear who was in charge. But when Quaker Oats became part of the American Cereal Company, Carnegie Steel became United States Steel, the Deering's and McCormick's farm implement enterprises became International Harvester, and Shredded Wheat merged into the National Biscuit Company, things seemed to change.

David Nobel describes the bold entrepreneurs of the electrical industry.

> In the period between 1880 and 1920 the first and second generations of men who created and ran the modern electrical industry formed a vanguard of science-based industrial development in the United States. These were the people who first successfully combined the discoveries of physical science with the mechanical know-how of the workshop to produce the much-heralded electrical revolution in power generation, lighting, transportation, and communication; who forged the great companies which manufactured that revolution and the countless electric utilities, electric railways, and telephone companies that carried it across the nation. (Noble 1977: 6)

The fields of industry were enormously competitive, however, and falling prices threatened growth and investment. Through mergers, acquisitions, and the buying of hundreds of patents—and thus keeping them from competitors—the largest corporations shifted from "entrepreneurial capitalism to financial capitalism to managerial capitalism . . . The nineteenth-century individualistic entrepreneurs, the traditional captains of industry who had jealously guided every facet of their companies' activities, largely lost their domination over large-scale U.S. business" (Laird 1998: 203–204).

Managers with skill sets different from the companies' innovative and inventive founders were required for firms larger in size and increasingly dependent on new scientific possibilities. With greater technical and organizational complexity, a new managerial class developed, educated in science-based engineering. William Wickenden's 1930 study found that between 1884 and 1924, "two-thirds of the [engineering] graduates had become managers within fifteen years after college" (cited in Noble 1977: 41). The chief proponents of a progressive reformism saw in the engineers-as-managers a source of scientific socialism, combining industrial might with universal social welfare and the development of human potential (Veblen 1967; see also Berle and Means 1932; Chandler 1977). The new executives and managers, however, remained answerable to shareholders who focused on profits, stock prices, and dividends, rather than the operations of the company as a source of innovation, providing enriching employment, and remaining integral to a healthy community.

The modern corporate form also appeared to separate ownership from control by making an executive staff answerable to a diverse board of directors. Bankers began investing heavily in industries transitioning from family to corporate ownership, the most prominent being U.S. Steel. The (Alexander Graham) Bell system that spawned the American Telephone and Telegraph Company and General Electric that originated as (Thomas) Edison Electric Light Company were heavily financed by the wealthiest bank in America, that of J. P. Morgan. The bank, like others, put its people on the boards of directors (Noble, 1977: 9, 12).

The suspicion was that financial interests would become paramount, eclipsing the firms' focus on providing innovative products of high quality.[9] In hindsight, anxiety over financial establishments controlling corporations assumed a difference of interests among corporate elites. In reality, financiers share with executives, top managers, and major shareholders an outlook and set of interests about the operation of large corporations and their role in society and the world. All of them work hard to ensure the dominant ideology remains in practice.

Financial considerations once again seem to guide the operations of large corporations today, especially firms whose boards of directors and finance offices are packed with executives and experts who recognize the firm less in terms of what it makes or the service it provides and more as a balance sheet and source of profits. The growth of finance and banking as a portion of the U.S. economy, and the strategy of improving the value of company stock by selling off less productive assets and replacing long-term employees

---

[9]William Roy writes about a period of financial control at the turn of the twentieth century, but one that was short-lived: "The power of financiers relative to that of other economic actors was transitional, a temporary stage of finance capital between family capitalism and managerial capitalism" (Roy 1997: 250).

with contract, temporary, and contingent workers, has again raised the question, Who controls the corporation?

How independent are managers of major corporations today? The "managerial revolution" that gave rise to highly visible corporate executives who seemed to speak for the company and found little opposition to their authority from shareholders, including banks, separated ownership from control in ways long felt to be threatening to capitalism. Managers are, in the final analysis, employees, and their power can be interpreted as worker self-management, a hallmark of democratic socialism. In the real world, management is more often pitted against the workforce that competes for profits, better working conditions, and a greater voice in company affairs. Paying executives with company shares makes them owners as well and has been an opportunity for their amassing great personal wealth in recent decades.[10]

The Ways Large Corporations Direct Social Change

Rather than orderly change within an existing institutional structure in which many tactical adaptations accumulated into substantial change, [corporate-driven] change took the form of restructuring the American industrial order.

—William Roy (1997: 257)

In social science terminology, the economy is a social institution, and specific corporations are organizations; the corporation is an organizational form. Similarly, the family is an institution. Nuclear families, extended neolocal families, and so forth are different forms of social organization. But there is a major difference between corporations, as powerful drivers of social change, and families. The family is the repository of tradition, the first agent of socialization, and a primary investor of culture. It is tasked with the most important responsibility of the human species: reproduction of its own kind, not just biologically but culturally. Families change, and older family forms, e.g., extended family households, patriarchy, generations living in the same community, give way to new forms. The family, however, is not a force for

---

[10]La Porta and his colleagues (1999) also find that, worldwide, a great deal of shareholder control exists by virtue of one family (often the company's founders) or a small group of people holding a controlling interest in the corporation. Indeed, worldwide major shareholding families often assume managerial roles. The United States is a significant exception, in part because of its having better "shareholder protection" legislation.

social change. Just the opposite: it is a force of permanence, predictability, replication of its own kind, and stability.

In contrast, the large corporation is very much an instrument of change. Despite the ways corporations may resist change to themselves and the supporting structures that make their operations possible, as will be seen later in the chapter, their current form developed in the rapidly changing environment of industrialization. They both took advantage of and fostered a robust era of technical inventiveness and scientific achievement, applying these to new products and services that found or created a buying public.

In search of new markets and greater market share, large corporations have devised organizational operations, product development strategies, advertising and public relations, labor management regimens, and ways to influence the political system unimagined 150 years ago. Most obviously, the everyday life of anyone living in an affluent or aspiring nation is constructed by and adorned with corporate products. These products, not the natural world, are the environment most people live within. Their messages, music, and images are the most real thing, linking people into communities of consumption, channeling both public and private desires (Ewen and Ewen 1992).

Large corporations are a force guiding social change in at least five ways. First, they are intimately involved in reconfiguring and finding uses for existing technology and developing new technologies that facilitate and guide what we do as users and consumers. Second, for more than a century corporations have organized and contested labor, deciding how work will be done and the terms governing working life. Third, their size and economic power rest on the control of capital and the purposes to which capital is applied. Fourth, major corporations dominate the aesthetic landscape with images of themselves, their products and services, and the sensibilities that encourage everyone to participate in the world of material consumption to the best of his or her ability, and beyond. Finally, they are integral to the political process and play an outsized role in advancing legislation that benefits themselves and opposing legislation and regulation that could diminish or impair their operations as privately owned, profit-seeking organizations.

Technology and the Corporate Dynamic

Technology does not occur in a vacuum, nor is it neutral to those who possess and use it. Discoveries, inventions, and technological applications acquire their significance in the ways they are used and the social contexts of their uses. One context confers social distinction on the inventor, another envelopes him in a cloud of social ridicule. Ideas of what is sacred, immoral, socially inappropriate, or threatening have scuttled new practices and made their adherents infamous. Ideas about what is important, beautiful, pleasurable, and empowering have the opposite effect. In today's world, corporations play a major role in deciding which technologies are developed and adopted, and which are not. Corporate interests have been the most important advocates for new technologies, a creative force for research and development, and insatiable advocates for innovation. Much of the contour of modern societies owes its form to corporate embracement of technological change.[11]

Corporations and Profitable Technology. Seeking efficiency, being able to produce a good or service at lower cost, more quickly, or of better quality, can mean more profits. When technological development is costly, large corporations have the resources to invest in it. Joel Mokyr, in *The Levers of Riches* (1990), argues that material "progress" is a primary reason the state facilitated the corporate form of business ownership (itself a technology) in the nineteenth century. It has been a recurrent argument for lenient enforcement of antitrust laws throughout the past century. A corporation that can convincingly show its need for large market share in order to have the resources to invest in new technologies makes a good case. The largest pharmaceutical companies, for example, are allowed many exceptions to normal competitive practices by the countries in which they operate, given the research cost of developing new drugs and the expensive field trials required before new drugs can be marketed, to say nothing of research that meets a dead end.

In the social context of a highly commercialized economy, technologies are most likely to be evaluated by corporations on the basis of expense and profitability. Technology research and development, however, is a contested terrain. A technology of a weaker competitor that threatens the profits of powerful economic interests can face barriers of restrictive laws and negative public relations campaigns. Patents and copyrights for rival technologies can be acquired and squirreled away in order for them to be forgotten.

James Duke, namesake of Duke Tobacco and the founder of the American Tobacco Company trust in the late nineteenth century, purchased the Bonsack

---

[11]Sometimes pioneering corporations use cutting-edge technology to expand their market. Taking a bold lead, however, may find them embroiled in cost overruns or out of sync with existing systems of distribution and application, what Danny Miller (1990) calls the "Icarus paradox" in reference to the mythical Greek inventor who mastered flying but lost his son who flew too close to the sun.

machine to make cigarettes but also purchased the patent rights to the potentially competitive Allison machine, keeping it out of the hands of his competitors. Senator Frank Church of Idaho conducted hearings in the 1970s on patents and copyrights for alternative energy technologies, what are now called green technologies. *Petroleum Industry Involvement in Alternative Sources of Energy*, prepared for his Subcommittee on Energy Research and Development, documented how energy companies and others who were heavily invested in carbon-based energy purchased sustainable energy technology patents, ostensibly to develop them, but many were left undeveloped and languished for decades (Crane et al., 1977).

It is an economic truism that greater profits accrue to the replacement of human with mechanical production. Machines can be more precise, work longer hours, require no retirement benefits, and get a tax write-off as they age. Integrated machine tools, computer numerical control machines, and robotics have reduced labor costs, just as Duke's Bonsack machine allowed three workers to do the work of forty or fifty people hand-rolling cigarettes (Roy 1997: 227). As is discussed shortly, however, shop-floor technologies have resulted from choices and corporate decisions about how to organize labor, not only for greater efficiency but for greater control over the labor process.

Technologies as Consumer Goods. One of the most fascinating aspects of corporations and technology is the way in which things previously unknown have become objects of necessity. The search for new markets is not only the search for new consumers. Current consumers are available for new products. Corporate advertising helps them learn what they don't know. Who thought they needed a reaper? Cyrus McCormick had to show farmers a better way to do what they had been doing for centuries. Who needed a sewing machine? Only when the wonders of owning a Singer machine were demonstrated to women could a family have a conversation about what it really needed, which turned out to be a sewing machine (Sivulka 1998).

Today's technologies that facilitate social networking go well beyond improvement in the basic telephone. Smartphone users can access a vast store of information in any locale with a few taps of the thumb. They make it possible to communicate experiences almost instantly. Being able to stay informed about the comings and goings of nearly anyone has become a necessity that no one thought important a few years ago. Like scores of other technologies, smartphones are changing the way people live, interact, communicate, and think about the world. Their source is corporations vying for consumers by producing new, unimagined technologies of consumption.[12]

Control and Investment of Capital

Like human beings and natural resources, capital is a valued resource. Poor countries are desperate to obtain investment capital. Small businesses can respond to new opportunities only by obtaining capital. Families go hat in hand to get the capital needed to buy a home or car, pay for college education, or get through a health crisis. Major corporations possess and control the use of much of a nation's available capital, allowing or requiring them to decide what products and services will be produced, how they will be marketed, what resources will be used to operate firms, what buildings will be built, how the landscape will be designed, what jobs will be available, and how much people will be paid.

Growth of Capital. A single modern corporation's holdings can exceed the size of many countries' economies. Globally, corporations are fifty-one of the one hundred largest economies in the world; national economies are the other forty-nine. This means that nearly 150 nations have economies smaller than the fifty-one largest corporations. Except for ten nations, the two hundred largest corporations have greater holdings than all other countries in the world combined.[13] Their annual sales are more than a quarter of the world's gross domestic product (GDP). In 1999, the two hundred largest corporations in the world specializing in services, such as banks, financial companies, and telecommunications, accounted for nearly half of all sales in services

---

[12]A fascinating example is the man's wristwatch. With the proliferation of cell phones and smartphones, a finely made wristwatch has lost much of its utilitarian value. For those who want to wear a watch, digital watches are superbly accurate and inexpensive. The economic threat to mechanical watchmakers was obvious, and corporations responded creatively. Using movie stars, golf pros, and brilliant intellectuals in glossy ads, they transformed the mechanical analog display watch into an expensive object of adornment and status for men.

[13]In response to its advertising restrictions on cigarettes, Philip Morris International threatened to sue Uruguay. As Duff Wilson (2010) pointed out, that can be intimidating. Philip Morris' annual sales are twice the size of Uruguay's GDP.

worldwide (D'Arista 2007). "The six largest financial corporations in the United States account for 55 percent of all banking costs" (Madrick 2011: 70).

In the 1870s and 1880s, railroads were engines of growth and the greatest source of capital accumulation the world had ever known. The steam engine itself was so expensive to build and operate that, despite being recognized as a practical invention in 1804, it was not used widely until decades later (Herrera 2006: 54). Thousands of investors, including states like Pennsylvania and dozens of communities, lost money on railroads that failed, but those who became the dominant builders and operators were well rewarded. In England, investing in railroads became so popular it was given the name "railway mania."

Rather than being a historical anomaly or blip, railroads became the model for accumulating capital and applying it to wealth creation that corporations emulated for the next century. In 1870, when the U.S. GDP was less than $7 billion, twenty-five railroads had capital holdings of more than a billion dollars. By 1889, the railroad industry was worth $8.56 billion, far eclipsing the $5.7 billion total value of the nation's manufacturing industries. "One of the preconditions for the corporate revolution at the end of the century was the centralization of wealth into a form accessible to publicly traded corporations. The railroad system mobilized and centralized the expanding quantities of wealth" (Roy 1997: 97, 104).

William Roy's illustration is telling of this economic power and its significance. "Half of every dollar a farmer or merchant paid to ship a bushel of wheat or a barrel of nails was expropriated into corporate capital . . . Instead of millions of farmers and merchants making the decisions that determined how the wealth would be appropriated, a few hundred railroad executives and financiers decided" (Roy 1997: 105). That power has only increased. In their use of vast sums of capital, corporations play an enormous role in directing and blocking social change.

Corporate Concentration and Capital Accumulation. Concentration and wealth holdings go hand in hand. Data collected for the Bureau of the Census' Statistics of U.S. Corporations document that 891 large firms had sales/ receipts of $2.5 billion or more in 2002 (a tiny fraction of the 5.6 million U.S. firms). These 891 companies employed more than twenty-six million people, or 23 percent of the U.S. workforce, and had sales of $9 trillion, 42 percent of all sales of U.S. firms.

Mergers of several firms have been one of the chief means for this enormous growth in receipts for a handful of corporations. This is especially the case in the financial industry. As Jane D'Arista's research shows, the top ten financial institutions in the United States all were created by mergers with other financial institutions. Their operations manage money in such a way that profits accrue to investors, while company officers and employees share the rest.[14] The more money they manage, the greater the profits. "In 1984 the top 10 banks account for 26 percent of the total assets of the sector, with 50 percent held by 64 banks and the remaining 50 percent spread out among the remaining 11,387 smaller institutions" (D'Arista 2007: 4). An average of 440 bank mergers per year reduced the number of banks by four thousand in 2005, and by then five banks controlled the lion's share of all capital available for investment.[15]

Transformation of the Labor Process

A basic sociological principle is that we *are* what we *do*. That simple idea has many meanings. It can mean that, from the perspective of others, our identity is based on our actions: rob a bank and you are a robber. Grow a garden and you are a gardener. Win a sporting event and you are a winner. Care for your sick mother and you are a saint. Another meaning of the simple idea is that we become the person we are because we engage in activities that mold our worldview, forge our self-concept, engage us with a select group of others, and place us in the social hierarchy. Because so much of our life is spent working, the work we do is a major determinant of who we are. That is why, beyond the rewards of pay and promotion, it's good to do work that helps you become the person you want to be.

One of the most important ways large corporate organizations have effected social change is by organizing work and the labor process, profoundly

---

[14]The twenty-five best-paid hedge fund managers took home on average $1 billion each in annual compensation for 2009, led by David Tepper's $4 billion from Appaloosa Management. Other corporate executive pay "winners" in 2009 included Oracle's Lawrence Ellison, who received $84,501,759, and Viacom's Philippe Dauman and Thomas Dooley, who received $84.5 million and $64.7 million, respectively, in pay, stock awards, and other compensation. Occidental Petroleum's Ray Irani got $31,401,359, Robert Iger at Disney took home $25,578,471, and Robert Stevens at Lockheed Martin received $20,473,451 in the depths of the Great Recession. The CEO of an average-to-large corporation today takes home more money by noon on January 1 than the average employee will earn in an entire year.

[15]"By mid-2008, five banks had become the dominant institutions in the market in terms of total assets and as holders of ninety-seven percent of the total amount of notational derivatives such as interest and exchange rate swaps, collateral debt obligations (CDOs) and collateral default swaps (CDSs)" (D'Arista 2007: 4).

and intimately influencing peoples' lives, including the sense of who they are, their social worth, their friendships, health, and life chances. Because the labor process has changed so much over the past century, it is only reasonable to think that people and the lives they live, too, have changed a great deal.

The ability to control the labor process, vested in the power of the corporation, has significant structural consequences as well. For example, the U.S. educational systems' curriculum, as well as its methods of evaluating and sorting young people, serves the corporate world that hires individuals on the basis of certified qualifications. Schooling prepares employees to accept regular evaluations of performance that determine promotion or dismissal. Classroom discipline and a tolerance for routine, timed tasks in schools can be readily applied in the workplace (Bowles and Gintis 1976).

What is taught in universities, which departments grow and receive additional resources and faculty positions, and what research agendas receive institutional support and encouragement increasingly reflect a fluctuating job market and the need for corporations to have well-trained employees. This is especially true in for-profit universities, community colleges, schools of engineering, agriculture, business, and law, and it is increasingly influencing the liberal arts and sciences. In order to prepare students for jobs, institutions of education respond to the decisions of corporations that determine demand in the job market, the kind of work employees will do, and the way it will be done.

From Agriculture to Industry to Services. Iris Summers grew up on a farm. Her husband worked in industry. Her training was in nursing—a service—and nearly all of her grandchildren provide services, e.g., teaching, retail sales, health care, insurance. Her family is a microcosm of the changing labor market of affluent countries. Two out of five workers were engaged in agriculture a hundred years ago. Today, about two out of a hundred do the work of growing food. The agricultural labor market shifted first to manufacturing and industry and then to services, with a bifurcation of services into personal and retail services and professional and highly technical services. Currently, more than three-quarters of all workers in the United States provide a service. Information rather than machine tools is the material with which they are engaged—creating it, using it, and providing it to others. This has fundamentally altered the way work is performed.

In addition to the migration of people off the farm, many immigrants had agricultural skills when they came to the United States a century ago. They and their children entered a new world of work. African Americans left the South by the thousands during and soon after World War I, many finding jobs in industry. In the mid-1920s, 90 percent of Detroit's African American men employed in industry were working for Ford Motor Company (Edwards 1979: 127). This is social change on a massive scale.

In the nineteenth century, domestic work often meant farm work. As farms were consolidated and families moved to towns and cities, domestic work was (compared to the skills needed on the farm) deskilled. The answer to the decennial census question about the occupation for nonfarm women working in the home was "no occupation," despite their caring for children and elderly parents, shopping and preparing food, doing laundry, and cleaning for their families. In the public mind, and increasingly in her own, the "unoccupied" or unemployed housewife was a consumer.

Women have always worked. Between 1905 and 1930, a quarter or more of all women in the United States were in the paid labor force, disproportional numbers of whom were unmarried white women, poor women, single mothers, and women of color (Domenico and Jones 2006: 2). Farm girls like Iris Summers and her sisters trained to be nurses, secretaries, bookkeepers, telephone operators, and teachers, joining the urban labor force of factory workers and office staff. The character of their work, along with that of their brothers and husbands, changed as well, as corporations developed new ways to use their labor power.

Technology and the Struggle for Control. Richard Edwards' (1979) study of the transformation of the labor process explains how forms of corporate control evolved in the making of a modern workforce. Early factories were often not much more than the gathering of dozens of artisans who previously had been in separate workshops. They came with their experience, knowledge, apprentices, and tools. Owners were largely powerless to control the speed with which they worked, and there was little management of the production process. Only a coercive foreman and piecework pay could wrest control in the company's favor, and never satisfactorily.

In contrast, water-driven power looms created a labor process where machine tenders, often children with minimal skill, worked according to the pace of the machine. Adult newcomers to factory labor balked at such conditions, as well as the low pay for a sixty-to-seventy-hour workweek. Corporate authority over the mechanized labor process was not assumed to be a natural phenomenon. From 1880 to 1920, the United States saw a massive resistance of workers across industries and regions of the country. Labor stoppages and worker strikes erupted in response to long hours and low wages. Objectionable, too, was the regimentation of work and the discipline that enforced the work regimen.

In the scholarship of labor historian David Montgomery (1979), three periods of labor activism were especially tumultuous: 1901–1904, 1916–1920, and 1934–1941. The strikes, lockouts, sit-ins, and boycotts were usually not initiated by an organized union but broke out when groups of workers refused to continue working under intolerable conditions. They were likely to be successful when other workers joined in sympathetic support based on their own work experiences. As Edwards describes,

> After the 1901 steel strike, leadership in aggressive unionism passed from the craft to the industrial unions. With the formation of the Socialist Party in 1901 and the Industrial Workers of the World in 1905 . . . [and] particularly after the IWW's Lawrence [Massachusetts, textile mill] victory and Eugene Deb's impressive tally in the Presidential election—both in 1912—the future shape of industrial control appeared to be much in question. (Edwards 1979: 51)

Corporate Liberalism. In response to labor militancy, hundreds of corporations tried to develop a safer, cleaner, and more habitable working environment, what urban reformers and critics of corporations were demanding both inside and outside the factories. Firms adopted policies—James Weinstein (1968) dubs this the "corporate ideal in the liberal state"—that would not only improve the health and safety of the workforce but generate worker loyalty and dissuade workers from forming and joining unions. "They promoted social-welfare legislation in order to reduce the burdens and antagonism of working people . . . thereby hoping to substitute orderly and predictable negotiations for industrial warfare" (Noble 1977: 61).

Corporations provided "recreational services, clinics and health care, pensions, stock-sharing and other savings plans, housing, and educational and other benefits and services" (Edwards 1979: 91). One 1926 survey of 1,500 firms, not all of them corporations, found that 80 percent had some form of company welfare program (benefits) costing, on average, about 2 percent of workers' income (Edwards 1979: 95). Justified as cost-cutting measures and because they had not been effective in quelling worker dissatisfaction, in time most of these practices were diluted or shifted to the responsibility of the state. In the 1930s, they took the form of Workman's Compensation and Social Security.

Machine Control. More effective for corporations were the changes involving the application of machines that shifted worker hostility away from disciplinary foremen. Machine technology reconstructed the workplace to make the routinization of jobs appear both rational and impersonal. Technology played a major role in what Harry Braverman (1974) calls the "habituation of labor," i.e., training people to work with machines that set the pace, determined the routine, and in effect became the repositories of the knowledge and skills previously possessed by artisans. Becoming habituated to the routinization of jobs also involved gaining workers' acceptance of corporate authority.

The effort to closely manage work led corporations to employ time-and-effort studies, the most famous going under the title of Taylorism, named after Frederick Winslow Taylor. Taylor's nemesis was "soldiering," i.e., the practice of working with less than maximum effort or giving less than "a fair day's work." Taylor's plans for the work process reflected his bias against the working class and their apparent unwillingness to work as hard as he thought they should. Taylorism was a largely impractical scheme to dissect the motions of work and reduce tasks to a few repetitive actions, with many individuals doing simple bits of standardized activity previously performed as a complex skill set by one person. Taylor's efficiency calibrations extended to speed and endurance studies in order to determine what was possible, rather than accepting what workers actually did and the practiced skills to do them. Deskilled labor was more calculable and controllable.

Few companies ever tried to fully adopt Taylorism, but they did learn the importance of management having the *knowledge* previously monopolized by artisans. Companies seized upon the economic benefit of hiring unskilled workers, training them in a day or two and paying them the bare minimum under threat of being easily replaced. Finally, Taylorism recommended a workplace entirely outside the workers' control. Braverman (1974: 114) describes this as the complete "separation of conception from execution" with owners and managers making the plans and devising the activities and workers carrying them out.[16]

Continuous-flow production, commonly thought of as the assembly line, didn't need Taylor's studies. The idea emerged decades earlier with the standardization of parts for the assembly of firearms, the disassembly of animal carcasses in meatpacking plants, and in response to the popular demand for bicycles. Continuous-flow production was developed most extensively in automobile production after the turn of the century. Henry Ford's massive automobile works were built to organize labor into stationary tasks while the product moved through the factory on carts or conveyor belts. It proved to be very profitable. Even into the 1920s, nearly every Model A was sold before it was made (Gartman 1979). Labor turnover was

---

[16]Based on his fieldwork and analysis, Michael Buroway takes issue with some of Braverman's ideas. Much ethnographic research (e.g., Hodson 2001) supports Buroway's conclusions that workers do not remain passive and totally acquiescent in situations of technical control. Their latitude is "within narrow limits" and often takes the form of hidden resistance and sabotage that amounts to no change in the labor process itself (Buroway 1979: 94).

high, exceeding 100 percent annually, despite Ford's paying workers the astronomical figure of five dollars a day. Nonetheless, technical control grudgingly but inexorable came to be accepted as the way things are done.

Bureaucratic Control. Workers who objected could work in other sectors of the economy, most obviously in small businesses that retained more familial relations and in which the labor process was less regimented. They could also avoid what Edwards describes as the final form of labor control, bureaucratic control. This organizational technology puts the labor process within a rational scheme of rules and procedures, provides an elaborate classification of jobs, and offers the means for individual, rather than group, advancement in pay and responsibility.

By retaining some elements of corporate welfare benefits, increasingly relying on sophisticated machines, and embedding work within a bureaucratic maze, labor was transformed into something that was done for someone else, in the way he wanted it done. By the time industrial unions entered the U.S. economy as major players in the 1930s and 1940s, there was little argument about this. Unions wanted a say in work rules, the setting of pay, and the generosity of benefits, but the legitimacy of corporations controlling the labor process was rarely in dispute.

Michael Buroway's aptly titled *Manufacturing Consent* (1979) captures the habituation to the labor processes that large corporations design and carry out. Most people came to see themselves as employees, not workers. They belong to the middle class, not the working class.[17] They are more comfortable emphasizing individual achievement rather than group solidarity. Conflict in the workplace is overwhelmingly a matter of grievances over management's breaking or ignoring rules, not management's making the rules in the first place.

Most people do not work on an assembly line today, but technical control is everywhere, enveloped in a seemingly rational bureaucratic maze. It is accepted by most people as the consequence of undisputed technical rationality. As the economy has moved more and more toward the provision of services and the ranks of professions have grown, the routines and

---

[17]Surveys have for decades found that, when asked to which social class they belong, most people identify themselves as being middle class, even some who are exceedingly wealthy and many whose income and type of work would appear to place them in the working class. "Middle class" self-identifies people, rich and poor, with a particular lifestyle, set of values, and admired but reasonable aspirations. It has become more a cultural shorthand for widely respected personal virtues, character traits, and attitudes.

bottlenecks of bureaucracy might be castigated, but bureaucratic control, too, has become a fact of life.

From Workers to Consumers. Work is not interesting for a great many people. It is neither challenging nor fulfilling, and it may be done better by machines in the not-too-distant future. Still, most people are happy to have a job, any job that pays the bills. This frame of mind is what British sociologists (Goldthorpe et al. 1969) first called an "instrumental orientation to work."[18] A job lets them do what they need and want to do when they are away from their job.

Whether instrumentally or not, most people work very hard. And they work long hours. Juliet Schor's statistical studies of *The Overworked American* (1991) and Arlie Hochschild's more ethnographic *Time Bind* (1997) portray lives tied to work, not for its intrinsic value but for what it provides in the way of material goods and services. By almost any measure, full-time employed persons in the United States are working more than ever before and more than anyone else in the world.[19]

Hochschild's studies also reveal the irony of the modern intersection of work and family. Work itself can become the "family" that has been lost or diminished in two-earner households where commuting long distances, microwavable meals, and the household's information and entertainment technologies allow everyone to escape into personal domains. The increasing number of hours family members spend working and commuting, added to by the hours spent using hand-held and desktop computers at home, have greatly contributed to changes in home and family life.

---

[18]The concept of an instrumental orientation may be a naïve or nostalgic image of an earlier time when work was meaningful and empowering and not merely a means to enjoying life away from work. Or it may be visionary, akin to Karl Marx's idea of transcending alienated labor in a socialist future when machines do the drudgery and people are free to work creatively, spontaneously, and as masters of their own efforts. Richard Sennett's *Corrosion of Character* (1998) paints a compelling portrait of work today as insecurity, unreasonable expectations, and coping. Not everyone agrees, preferring to emphasize the way employees create a satisfying social atmosphere and insert personal meaning into their work, no matter how routine or insecure. "There is little evidence that people are less concerned about, interested in, or committed to work . . . [T]he notion that work is a purely instrumental activity and that people work simply in order to consume is difficult to sustain" according to Paul Edwards and Judy Wajcman (2005: 41).

[19]Robinson and Godbey (1997) analyzed multiple cross-sectional studies done every ten years between 1965 and 1985. The time diaries data led them to conclude that people in 1985 were working *fewer* hours than in the past. This finding is not borne out by other research and data, including that of the U.S. Bureau of Labor Statistics.

Many people would like to work more than they do in order to earn the money they need to live. Some of them are the underemployed and the working poor whose wages cannot keep them out of the ranks of poverty. Others are not poor but are barely able to make ends meet. They have little or no savings, huge credit card debt, and may be carrying a mortgage that makes them homeowners in name only. At the other end of the spectrum, many people in affluent societies work beyond necessity, putting in long hours in order to acquire things that define the good life. Parents lavish their children with goods as a consolation for the time they cannot spend with them (Pugh 2009). People gauge their worldly success by vacation locales, wardrobes, homes, automobiles, season tickets, and any number of expensive objects and experiences. As Thorstein Veblen (1899) described more than a hundred years ago, tapping into people's pursuit of social repute through material consumption is one of the corporation's greatest triumphs.

Advertising and the Corporate Creation of Culture

Think for a moment about something you would like to have. It may be physical fitness, a close friendship, better communication with your parents, or more sleep. Can you buy what you want? Is there a product that can help you get it? Not everything, perhaps, but a remarkable number of the things you want have a corporate commercial solution. That is, they can be acquired, learned, experienced, or facilitated by purchasing a thing or a service. If it has commercial value, it is probably made available by a large corporation. If you don't know it is available, you may not be paying attention. Attention to what? Corporate advertising!

Advertising as Information and Persuasion. In its simplest form, advertising is information. Commercial information communicates the availability of products and services for purchase. From its earliest inception, advertising has sought to make people aware of things they might not know are available. Better yet, it helps people know what they need to have. Maybe they didn't perceive the need or didn't know there was a way to satisfy the need. As soon as the need is recognized and the option of satisfaction presents itself, there is a want. It may be weak or strong. The purpose of advertising is to enhance the desire to fulfill the want. Persuasion characterizes all advertising, even the most subtle or factually accurate.[20] The Rolling Stones

---

[20]World War I is sometimes seen as a watershed in advertising. Raymond Williams (1980) and Pamela Laird (1998) attribute the success of building public opinion in favor of the U.S. entrance into the European war as yielding valuable new persuasive techniques for commercial advertising.

are wrong. You *can* get satisfaction. It is affordable, it is necessary, you deserve it, it will have additional benefits, and time is short, so buy now!

Persuasion to buy products and services is a staple of affluent societies dominated by the corporate-driven culture of consumption. Like corporate power more generally, it has become so pervasive that it may barely be noticed (Galbraith 1956). Its challenge is to "cut through the clutter" to gain the buyer's attention. Advertising is clever, employing the finest writers, digital animators, videographers, dancers, and technicians. It is often very funny. The best advertising strikes at the core of feelings about love, family, responsibility, excellence, security, beauty, and the other things people cherish. It speaks to them and for them.

Believe it or not, body odor (BO) is a commercial invention. Its entrance into the nation's lexicon dates from the 1920s, following the successful campaign to solve the newly discovered problem of bad breath. What a Chicago ad agency dubbed "halitosis" was solved by using a new product, Listerine, every morning. Responding to "the new urban attention to cleanliness," Lever Brothers promised not only to clean and disinfect, but to "protect the regular user of Lifeboy soap from the disgrace of smelling like . . . a body" (Sivulka 1998: 158). Later to come were deodorants, followed by antiperspirants, i.e., metal-laced scented aerosols to clog your pores.

One of advertising's most persuasive techniques—instilling fear and anxiety and then "stilling" it, as Sivulka (1998) terms it—became a staple of selling. This has proved to be especially effective in selling to women who can never be thin enough, young enough, or sexy enough, even when dutifully applying themselves to commercial products in pursuit of the "beauty myth" (Faludi 1991).

Advertising for Economic Gain. The benefits of advertising for large corporations are indisputable. Corporations spend hundreds of billions of dollars worldwide to ensure their products are recognized and selected. In the competitive sphere, brand loyalty keeps buyers from going elsewhere. Targeting new buyers, especially preteens and adolescents, is key to maintaining and expanding market share.

When cigarette companies argue that they advertise to keep smokers from switching brands, not to lure young people into acquiring an addiction to nicotine that will keep them buying the product for a lifetime, they are being only partially disingenuous. As George Duke knew long ago when he pioneered the first massive effort to market the cigarette as a nicotine delivery device new to most people, advertising works. Somewhat ironically, Duke learned this when he engineered the American Tobacco Company trust by incorporating many of his competitors into a single company. The effort to stifle competition by creating a massive monopolistic corporation should reduce the need to advertise, but Duke thought otherwise.

As Pamela Laird (1998: 243–245) explains, advertisers in the 1890s were initially hostile to trusts because they replaced dozens or hundreds of businesses with a few behemoths. They stifled competition. There were hundreds of competing bicycle companies, and they advertised intensively during the years of 1890–96 when bicycle sales increased from 100,000 to four million. On the other hand, Standard Oil, a trust engineered by John D. Rockefeller that controlled 90 percent of the U.S. petroleum business, advertised little. Advertisers set about to convince the monopolistic corporations that people would buy more of their products if they advertised.

Duke knew people didn't need cigarettes, so he needed to convince them to smoke. The most popular tobacco product at the time was the plug or chew. A chunk of processed tobacco was bitten off or a bunch of shredded leaves were lodged in the mouth, requiring occasional spitting to expel unwanted brown saliva. Duke presented cigarettes to the public as cleaner and of higher quality than a plug or chew. They were even higher quality than roll-your-own cigarettes, as evidenced by the manufacturing they required and their packaging. When Camels targeted women in the 1920s, the message was enlarged to include images of sophistication and independence, decades before a similar tact was taken to sell "liberated women" Virginia Slims cigarettes. The economic impact was a successful expansion of the tobacco-consuming market to the other half of the population that could increasingly afford to buy cigarettes.

Newspapers were the first to recognize the economic importance of advertising and gradually changed their format to accommodate larger images and messages than could be placed in a column of type. With the benefit of increasing public literacy and growing affluence, by 1900 magazines became the most popular way to advertise. The 3,500 magazines, previously dependent on subscriptions and newsstand sales, began to develop a reliance on advertising revenue. This is the mainstay of magazine publishing today and possibly Internet publishing in the future.

Andy Warhol and others who went on to great fame began their careers in advertising, but many thousands more have found it impossible to engage in their art "for art's sake" or as a viable artistic endeavor. In affluent nations today, advertising is the largest employer of persons trained and working in the arts. Graphic artists and designers, musicians and composers, actors, dancers, writers, filmmakers, and others in the creative fields spend their lives in the service of utilitarian commercial interests. Their talent has generated a very powerful force on contemporary culture. Raymond Williams (1980: 184) is not far off the mark when he says that "advertising is, in a sense, the official art of modern capitalist society."

Advertising's Creation of Culture. "Advertising is transformative . . . of culture, and over the course of the last century it has, in its own way, transformed the way in which Americans live" (Cross 2002: xiv). Mary Cross could extend her observation internationally to include countries that only a few years ago were devoutly anticapitalist. Not only in the United States and other affluent nations, advertising increasingly displays a life to which even the world's poorest individuals aspire. Everywhere, short of the most remote regions, advertisements are ubiquitous. Billboards festoon highways, logos proudly adorn clothing, and company banners are the scenery of sporting events. What to buy is announced in store windows and on buildings, with product placement insinuating brands into movies, online games, dramatic plays, fast-food meals, and musical performances. Advertisements flood the airways and are pushing hard to dominate the World Wide Web. A typical thirty-minute excursion in a supermarket exposes a shopper to about thirty thousand items that carry a commercial message. As Juliet Schor's *Time to Buy* (2004) shows in detail, children are inundated with commercial pleas even before they are school age, and this never ends.

Not confined to the present, advertising often looks back, nostalgically suggesting a way to return to better days. And it looks forward, to a future that is both unknown and full of possibilities—good and bad. Financial planning and insurance present a forked road, with one path leading to security and contentment and the other filled with despair and ruination. Commercial ads for health and weight loss products contrast past or present misery with a new you, if only you will buy in. Tranquility, if not the outcome of a life well lived, fortunately can be purchased.

Corporate advertising is part of the culture because its messages and images are communicated, understood, repeated, and acted upon. They deeply color the cultural landscape. It is very difficult to systematically measure the degree to which advertising has created modern culture, but it is indisputable that a large portion of everyday material culture, language, and dominant attitudes about relationships, status, well-being, and happiness involve the ideas and images purveyed through corporate advertising.

A half century ago there was much discussion of mass society and mass culture. Everyone seemed to be moving toward the same things, speaking the same way, eating the same food, wearing the same clothes (Olson 1963). Regional and ethnic differences were being lost, and the intellectual and aesthetic levels of what was considered dominant culture were declining. Slang, poor taste, cheap goods, popular music, bad manners, casual dress, and celebrity worship were crowding out everything of real value. The critics charged large corporations with bringing a kind of consumer democracy to the provinces and the heartland, born on the wings of advertising. The worry about mass society (e.g., Kornhauser 1959) gave way to the counter-culture of the 1960s, and little has been heard about it since, but no one has retracted their earlier view that advertising tries to sell all of us the same things. Essentially, it feeds the same hopes and anxieties we all experience. That is what expanding market share is all about.

Consumer Dreams. If the labor process left little to be desired for many workers, there was always the consolation prize: consumption. Corporations are no less powerful because those who take up the message *want* to be consumers. The integrated corporation reaches out *beyond* final sales of commodities to a metamorphosis of aspirations and imaginations (Ewen and Ewen 1992; Ohmann 1996: 91).

The idea of self-consciously embracing the identity of a consumer goes back a century, with the birth of the consumer movement of the Progressive Era that focused on deceptive advertising and defective products. It was part of the social and political movement that sought to have the state assume more responsibility for public health, workplace safety, banking practices, and standards of urban housing (Sivulka 1998: 117–119). It accepted the structural relationships of the modern era in which most people work for others, with their wages paying for what they need and, increasingly, what they want as consumers. Most importantly, it accepted the corporate domination of the economy. Its radicalism was in espousing a corrective to corporate abuse, excess, and neglect. Akin to the public interest movement with which its social movement organizations often align, the consumer movement grew throughout the twentieth century, and it continues to lobby for legislation and regulation enforcement on behalf of all of us consumers.[21]

The most effective advertising today has to contend with an increasingly sophisticated audience. At its most effective, it doesn't appear to be advertising

---

[21]"It is as consumers that the majority of people are seen. We are the market" (Williams 1980: 187). Certainly, this is the image held by advertisers and professionals who work in advertising. James McNeal, author of *On Becoming a Consumer*, explains in approving academic language how children "perform in the consumer role constantly from the day they are born . . . Most kids are reared in the company of markets and are expected not only to accept this environment but to be glad of it" (McNeal 2007: 257, 274).

at all. Among the best at this is the Nike Corporation, deconstructed by Robert Goldman and Stephen Papson in *Nike Culture*. Nike and similarly successful corporations say little about their products, but they say a great deal about hopes and dreams. "Advertisements are structured to boost the value of a product brand name (a commodity) by attaching it to images that possess social and cultural value (sign value) . . . Disconnected signifiers (images) and signifieds (meanings) are fed, broken up, then recombined to create new equations of meaning" (Goldman and Papson 1998: 24). The flag, provocative dancing, a lover's embrace, laughing children, amber waves of grain, beautiful youth, hip celebrities, extraordinary athletic feats—these need only to be associated with a company brand in a kaleidoscope of images and sounds. Products hardly need displaying.

Ads tell a story, and the story becomes a popular myth. Like myths for all times, "it tells us who we are . . . It provides vignettes of social life, ways of dealing with relationships, a great deal of advice about work, maturation, marriage . . . In the absence of traditional authority, advertising has become a kind of social guide . . . in conditions of incessant change" (Berman 1981: 12, 13).

Political Power and Agenda Setting

The fifth and fundamental way corporations influence the direction of social change is by exerting power in the political sphere. The state's power is formidable, especially given its authority over coercion, taxation, and regulation. The historical challenge of democratic governments has been to wrest control of the state from a small group of elites, at the top of which are aristocrats, dictators, or both. Once democracy is undertaken, modern nations cultivate multiparty electoral politics, inquiring independent media, and civil society organizations. The democratic state responds to the constellation of power and the pushes and pulls of social forces and economic interests. Corporations are among the most powerful organizations in determining what the state is, does, and does not do.

A small change in the corporate tax rate can mean the difference between having millions or billions of dollars to reinvest and distribute to stockholders and executives or these same moneys going to public infrastructure, social programs, and the military. Financial regulations and the willingness of the state to "securitize" or back investments and wealth holdings may be esoteric to the average person, but they are enormously important to those who own large stakes in major corporations. Keeping national monetary policy and global markets favorable to corporations, especially ensuring their access to petroleum reserves, ores and minerals, and forest products and other commodities, can require the state's diplomatic effort or military force. The state is a site of "contentious interaction," and corporations are very successful contenders.

Tracking Power and the Power Elite. In no society will everyone participate in politics, and in many mature democracies at least half the population doesn't bother to vote regularly, the political act that requires the least effort, understanding, and involvement. Among the voting half, the amount of influence varies widely. Beyond voting, some people contribute money and time to candidates and causes, read political news regularly, attend political events, write letters and e-mails to representatives and newspapers, or do any combination of these things. Others who contribute large amounts of money are invited to dinners, symposia, and other gatherings where they see, meet, or talk with political representatives. Still others are consulted by members of the political bureaucracy about issues, legislation, and regulations, while a few write documents that analyze and encourage policies and legislation favorable to the organizations for which they work. In some cases, powerful organizations and major corporations are able to write documents that are introduced as bills to be passed.[22]

Knowing who these powerful organizations and people are, what they are seeking, how they do their work, and how much influence their actions wield has occupied political scientists, sociologists, and journalists for decades. A key part of understanding how corporations drive social change involves tracing the power of corporations to influence politics and the actions of the state to advance or limit what corporations do.

Corporate Elites and Social Class Interests. Because great wealth comes through corporate ownership, the small group of a nation's ultrarich has much to benefit from their political influence. These people are often seen as an upper class who share a similar level of great affluence, live in exclusive

---

[22]For example, the American Legislative Exchange Council (ALEC) staff crafted Arizona's controversial 2010 legislation aimed at illegal immigrants, the Support Law Enforcement and Safe Neighborhoods Act. Along with elected legislators, ALEC participants include dozens of corporations and nongovernmental groups, e.g., the National Rifle Association, that pay tens of thousands of dollars to belong to ALEC and participate in its conferences. Through ALEC they are able to help write legislation that is then taken by legislators back to their assemblies. The nation's largest private prison corporation, Corrections Corporation of America, a longtime ALEC member, was among those participating in the ALEC meetings where the Arizona bill was crafted at ALEC headquarters outside Washington, D.C., and later lobbied for its passage (Sullivan 2010).

communities, vacation in many of the same resorts, are educated at and send their children to elite schools, help one another socially network for opportunities, and occupy positions on the same corporate boards of directors. G. William Domhoff (1967) provides compelling evidence that they constitute what C. Wright Mills earlier designated a "power elite." Michael Useem (1984) calls them an "inner circle" that is only a segment of the upper class but is not in conflict with others of their class. They coordinate major business associations, not in pursuit of small-business interests, but on behalf of issues important to the largest, increasingly globalized corporate firms.

They share the same interest in influencing the state and are vastly overrepresented in government, not so much as elected officials but as appointed directors, senior staff members, and advisors where their influence garners less public scrutiny (Domhoff 1990: 20). They move in and out of governmental posts, running corporations, Wall Street financial firms, and partnering in prestigious law firms when not occupied in government.

Individually, their particular business concerns may diverge from those of others in their social class. In terms of overall economic interests, however, they support one another and work on behalf of corporate capitalism. Their opponents are populists who champion smaller businesses, environmentalists who challenge their operations, labor unions that vie for a greater share of corporate profits and oppose the movement of jobs overseas, and any other advocates who seek to limit their discretion and ability to pursue corporate activity. Those who see a power elite at work think they know who runs the corporation. They argue that executives and senior managers, as much as owners, share the same ethos of corporate capitalism. Like the major owners, they benefit handsomely when the value of a corporation's stock grows.

Avenues to Political Influence. How, then, do corporations and their interests find a willing ear and helping hand in the halls of government? Domhoff (1978) offers a list of four ways this is done: acting as a special interest; involvement in the policy formation process; participating in candidate selection and elections; and sustaining the pro-corporate ideology. These capture most of the tactics by which corporations influence the political process, directing social change in ways that advance their agendas.

Special Interests. Corporations, unions, public interest groups, foreign governments, and others use what is broadly known as lobbying to achieve their goals. They seek audiences with political representatives; offer legislation and help to write legislation; hold receptions, parties, and fundraisers for candidates; and host travel junkets and golfing vacations where they spend time with legislators and their staffs. They buy print ad space and broadcast messages that encourage people to "write or call your congressman."

Instigating what was described in Chapter 6 as "Astroturf" (in contrast to grassroots) social movements, they often work with minimal visibility (Vogel 1996; Walker 2010) to gain access to and sway legislators.

When actual figures are tabulated, it is clear that, with some competition from labor unions, corporate interests overwhelm the opposition in terms of money spent, number of lobbyists, the range and sophistication of their efforts, and effectiveness. For example, the oil and natural gas industries, recipients of a complex and vast set of tax breaks, spent $340 million for lobbying from 2008 to July 2010. In 2009, the pharmaceutical industry spent $188 million ($67 million for television ads), and the U.S. Chamber of Commerce spent $144 million—mostly from large corporate donations—to lobby Congress (CRP 2010; Lipton, McIntire, and Van Natta, Jr., 2010). This included $86 million from the health insurance industry to defeat health care legislation. Wall Street financial firms spent $2.7 billion from 1999 to 2008 on lobbying (Madrick 2011: 70).

When the national health care bill was being prepared in 2009 and 2010, pharmaceutical companies, hospital corporations, medical equipment firms, and of course insurance corporations lobbied heavily for their interests. The final bill was highly favorable to both insurance companies—that now will be insuring millions of new policyholders—and pharmaceutical companies that negotiated an agreement prior to the final bill being written. The 2010 financial regulatory legislation, passed in the wake of the financial meltdown that triggered the Great Recession, was intended to prevent another financial crisis in the future. Half a billion dollars was spent by financial corporations—banks and investment firms—to pay for a battalion of lobbyists to work the halls, cloakrooms, and offices of Capitol Hill. They successfully beat back the most onerous regulations that would tax stock market activities, make financial transactions more transparent, and restrict the speculative practices of banks and financial corporations.

Hundreds of high-ranking men and women in the armed services find jobs as lobbyists when they leave the military, working for industries that rely heavily on military spending. They lobby for weapons systems and other budgetary matters that benefit their new employers.[23] When the extension of federal oversight by the Federal Communication Commission of broadband Internet communication came before Congress, the major

---

[23]Bryan Bender (2010) studied the careers of 750 four- and five-star generals and admirals who retired over the past two decades and found an increasing proportion working as consultants or defense industry executives over this period. Four out of five senior officers who retired between 2004 and 2008 were working in some capacity for defense corporations, many lobbying the Pentagon on behalf of their corporate employers.

corporations that would be affected began a major lobbying campaign. "The Sunlight Foundation, which tracks industry lobbying, reported that cable and phone companies had 276 former government officials lobbying for them in the first quarter, including 18 former members of Congress and 48 former staffers of current members of Congress on committees with jurisdiction over the Internet" (NYT 2010).

Policy Formation. Given the myriad problems in the world and issues facing communities, how is attention directed at some problems, and not others, in order to effect a change? Policy formation is contested terrain to determine what things are worthy of the state's attention and how these can best be addressed. Corporations do this in many ways, not the least of which is through the activities of their policy centers. Through research papers, websites, newspaper and magazine articles, and policy recommendations drawn up by people in their employ, corporations seek to frame problems as well as solutions that reflect their perspective and promote their interests.

"Think tanks" and organizations of like-minded corporations—e.g., the Business Roundtable, Heritage Foundation, and American Enterprise Institute—are funded by firms and individuals with the explicit purpose of formulating and advocating for policies favorable to large corporations. These organizations employ people with the academic degrees and experience to make credible their views in the offices of legislators and government agencies and the court of public opinion. Sheathed in an aura of expert opinion and objective analysis, their impact on policy formation is less obvious because it "avoids special interest pleading" and the "helter-skelter special-interest process" (Domhoff 1978: 61).

Campaigns and Elections. Given the enormous amount of television time candidates and issue advertising occupy before elections, these contests may seem to be very public events. Only a small portion of the public, however, is aware of and involved in the full scope of electoral politics. The campaign and election process varies widely from state to state, with different procedures for political parties to select their candidates and different means for other candidates to appear on the ballot. Regardless of the process, candidate selection, campaign financing, and prominent endorsements are important everywhere. Corporations are involved in the entire process.

In many countries, campaigns are of short duration and largely or entirely paid for with public funds. Not so in the United States. This provides an opportunity for individuals to contribute, as millions of people do in every election. Most money, however, comes not from small donors but rich individuals and large organizations. Most candidates spend an inordinate amount of time raising funds, most often from corporate donors, unions, and wealthy elites. The expectation of such donations is that candidates will be sympathetic to their interests. In planning for the next election, politicians stay in touch with the most generous campaign donors and vice versa. For example, in anticipation of his becoming the House majority leader, John Boehner of Ohio received more money in 2010 from Wall Street financial corporations than the 434 other members of the U.S. House of Representatives, including those representing New York City and Wall Street.

Corporations are, along with labor unions and employee associations, the largest contributors to candidates and campaigns. The U.S. Chamber of Commerce spent $64 million in the 2010 midterm elections, while the Federation of State, County, and Municipal Employees spent $11 million and the Service Employees International Union spent $10 million (Luo and Palmer 2010). The U.S. Supreme Court's *Citizens United* decision allows corporations, wealthy individuals, and unions to spend unlimited sums on elections with no paper trail of who gave how much to whom. Anonymous donors gave $135 million this way, helping make the 2010 elections by far the most expensive midterm elections in history. More than $4 billion from all sources went to candidates, political parties, and organizations that orchestrate publicity to elect and defeat candidates and ballot measures.

Corporate Ideology Formation. This is the fourth means by which corporations exert political influence. In the United States, notions such as "individualism, free enterprise, competition, equal opportunity, and a minimum reliance upon the government" are staples of the dominant ideology and are identified with everyone's interests, but they are particularly vital to the narrower interests of large corporations (Domhoff 1978: 170).

An ideology is most effective when it is nearly invisible. Ideas taken for granted are the most powerful in providing a foundation for particular policies. Whether these ideas are learned in the standardized, one-dimensional version found in school textbooks, becoming the right answer to a quiz, or the one thing everyone can agree on at the outset of a heated discussion, the dominant ideology of "fundamental truths" is rarely a topic of discussion or debate. In a capitalist society, and increasingly in a world where large corporations are such an integral and major part of political, economic, and cultural life, their way of seeing the world becomes the dominant ideology everyone recognizes as second nature.

Today's talk radio hosts and cable TV pundits, magazine and newspaper editors and writers, and online bloggers provide an endless stream of platitudes and bumper-sticker-size political messages, invectives, and attacks on ideas, urging support, opposition, praise, and ridicule. Almost no one, however, questions the corporate form. For those on the left of the political spectrum, "corporate" is a negative term and comes in for criticism, being linked to a variety of social, economic, and environmental ills. Moderates and those on the political right rarely use the term, favoring instead the image of corporations as merely successful businesses with which we can all identify. They are the creators of jobs and community well-being. When they leave a community for a foreign locale where labor is cheaper, taxes are lower, and environmental regulations are more lax, it is less of their own volition than a necessity based on sound business principles or the failure of a state or community to be sufficiently business friendly.[24]

Large Corporations and Resistance to Social Change

As with technology, social movements, and other drivers of social change, there are cases of opposition to social change on the part of large corporations themselves. There is also resistance to the social change process brought about by the activities of corporations.

Corporations Working Against Change

Most large corporations can be counted on to support the status quo. After all, they *are* much of the status quo. Journalists and scholars have for years tried to explain how corporations block or diminish policies aimed at regulating their power and control of resources. Public interest groups have recommended diminishing their influence over such things as tax policy and trade legislation, imposing more regulations, and increasing government and citizen oversight. Corporations, acting alone or through trade groups, often successfully oppose these measures and any change in their operations. In this way they are resistant to social change.

Corporate opposition to legislation addressing climate change (carbon taxes, cap-and-trade, mileage standards, and reducing emissions from burning fossil fuels), smoking and secondhand smoke, ozone depletion, and acid

---

[24]Many excellent studies have chronicled the demise of communities that experience the loss of a major corporate employer. One of the best is Dimitra Doukas' historically fascinating *Worked Over* (2003), the story of the Remington Company, for more than a century located in Ilion, New York, that was bought by, and became a subsidiary of, the DuPont Corporation.

rain are well documented (e.g., Oreskes and Conway 2010). Adoption of a single-payer health care system that would effectively extend Medicare to everyone in the United States, and less sweeping changes in health care access and delivery, saw hundreds of corporations line up in opposition. This opposition is expressed in all the ways examined earlier that corporations influence the political system.

In some cases, their interests in government action include deregulation or nonregulation of economic activities, i.e., having the state do nothing. Evidence and analysis are marshaled to demonstrate the folly of any restrictions or regulations. Better than beating back unwanted proposals about how the state should deal with a problem, it is sometimes preferable to large corporations that the problem never receives an airing. Any suggestion of a problem is dismissed outright as a nonissue or outside the bounds of the state. This is what Peter Bachrach and Morton Baratz (1970) call "nondecision making," the ability to determine what is *not* considered in need of, appropriate for, or worthy of the state's attention.[25]

Organizational Entropy

The economics of capitalism impel many companies to grow in order to survive. Controlling competition may necessitate mergers with and buyouts of one's competitors. New and expensive technologies, research and development, and market testing are more affordable for a large firm. The irony of size is that it is usually inversely related to flexibility and tolerance for new, bold, and certainly idiosyncratic initiatives, in part because greater size requires expanded bureaucratization in order to coordinate the organization's communication and flow of resources. It's harder for new ideas to get a hearing.

Farsighted corporations look upon social trends as economic opportunities. Richard Ohmann's description of early corporate advertising remains true today. "Manufacturers advertised on a large scale when they saw a chance to situate their products with a way of life that was becoming the norm (Ohmann 1996: 91). Rarely do large corporations deliberately engage their resources as agents of social change. The corporate form is a kind of technology that accumulates power and, if used well, provides a long life for a corporation. They are, as bureaucratic organizations whose prime purpose

---

[25] "A nondecision is a decision that results in the suppression or thwarting of a latent or manifest challenge to values or interests, [thus] preventing demands for change in the established order from entering the political process" (Bachrach and Baratz 1970: 44).

is to make money, conservative and careful not to venture too far from what has been working for them and others.

When originators of successful businesses merge with or are bought out by larger corporations, a consequence can be organizational entropy. The people, ideas, and organizational culture of the acquired firm are marginalized. Security and stability are valued over uncertainty and change, as the story of the Xerox Corporation and the home computer illustrates.

CORPORATE CULTURE VERSUS INNOVATION

Douglas Smith and Robert Alexander's *Fumbling the Future* (1988) documents a famous case of playing it safe and bureaucratic obstruction in the Xerox Corporation's rejection of the opportunity to develop the desktop computer. The story they tell starts with the Xerox Corporation having the foresight and good fortune to invest heavily in a new research facility located in what became Silicon Valley, California, far from Xerox's corporate headquarters in Connecticut. The Palo Alto Research Center (PARC) hired the best of the early computer scientists to experiment with ways to apply the recently invented microchip to process and store information. Xerox was waging a small bet on the future.

Computers at the time were massive machines filling large rooms that operated with information received on punched cards. This would soon change dramatically, and Xerox expected to be part of it. With ample resources at their disposal and a very casual, egalitarian management style, PARC researchers in California experimented in a very open and intellectually vibrant atmosphere. They engaged their computer experiments to do their day-to-day work, modifying them as they used them to develop designs and solve technical hardware and software problems. In a few years, PARC created the world's first desktop computer. By this time, however, Xerox wasn't particularly intrigued.

Xerox's corporate culture was—typical of bureaucracies—hierarchical. There were well-defined lines of communication and a pecking order that encouraged employees to defer to the executives who had made Xerox the number-one manufacturer of copying machines in the world. Its success made its owners and executives very wealthy, perhaps causing them to be less interested in developing a new product that, unlike everything else at Xerox, was not a copier.

According to Smith and Alexander, the laid-back style of work at PARC offended some coat-and-tie Xerox executives. PARC's director was not the best person to convince them about the new ideas and innovative technologies. As well, Xerox's other divisions were completely unfamiliar with computer technology and had little interest in establishing new marketing and manufacturing capabilities. They, too, were doing very well with copiers. Why change?

It didn't take long for the young Steve Jobs to see, when visiting PARC, the potential for a brand-new product, one that would become a staple in American households, businesses, and schools within twenty years. He hired several of PARC's staff to develop and launch his Macintosh computer from the new Apple Corporation. The rest is a well-known story of entrepreneurship, innovation, and social change. Steve Jobs has continued to lead his corporation and invest in technological innovations, maintaining an atmosphere of creativity and organizational flexibility that eludes most major corporations. If history is any guide, when Jobs finally departs the Apple Corporation, it too will succumb to bureaucratic entropy.

Resistance to Corporate-Driven Change

The Southern Baptist Convention and other Christian evangelical churches, normally aligned with conservative political sentiments, are increasingly rejecting the corporate message of the limitless pursuit of material goods, recognizing and criticizing the environmental costs this incurs (NPR 2010). They are not alone.

Personal Resistance. In *Walden*, Henry David Thoreau's 1854 account of his two years at Walden Pond (with many forays back and forth to his mother's home), the message is clear: live a simple life. Looking around at all the things corporations create and market, you might ask yourself, "Who needs all this?" Most Americans tell researchers they are in too much of a hurry. They are too busy to have a casual conversation, read a book, or sit back and think. The pace of life, the stress of multitasking, and the feeling that there is never enough time seems to be accelerating. Certainly, the promise a century ago of expanding leisure that accompanied early hopes for new industrial and household technologies has not borne fruit. The only people who have time are too often the unemployed and underemployed.

Individuals, social movements, and subcultures that yearn for a reorientation of personal and societal priorities have long targeted corporations as their nemesis. An extreme form of resistance goes under the title of communistic societies, often religiously inspired before and after the Pilgrims landed at Plymouth Rock. From the nineteenth century Oneida community to twentieth-century counterculture, communal groups have rejected the pursuit of material goods, faith in technological solutions, and the social, political, and economic relations that support and accompany private rather than social property (see Fogerty 1972; Nordhoff 1966).

There are many less radical forms of personal resistance. Most people accept commercial advertising as a fact of life, but others take pains to ignore or even avoid it as much as possible. Households turn off their televisions or use software to skip the ads when recording programs. Pop-up ads are blocked on the Internet, and no-call lists prevent advertisers from soliciting by phone during specified hours. A few magazines, especially literary and arts magazines, depend almost entirely on subscriptions and newsstand sales rather than advertising revenue. City councils restrict or ban billboard advertising. Public radio and television stations exhort people to donate to them in order to avoid commercial advertisements. In these and many other ways, individuals are directly and indirectly resisting the power of large corporations to define the good life and prescribe the route to getting there.

The corporate world itself is stressful, demanding, and often unforgiving. A large corporation is what Lewis Coser (1967) calls a "greedy organization" that takes as much as it can, and everything it gives carries a high price. Less radical resistance than communitarian "dropping out" can be found in thousands of businesses that deliberately remain small, operate in a highly competitive environment, and hire, train, and remain loyal to employees who work alongside the firm's owners and their kin. Not only small businesses but workers themselves reject careers in large corporations and the changes in their own work and family lives demanded by large corporations and other greedy organizations.

It is a recurrent, uplifting story that a successful corporate executive, after years of toil, departs the firm and starts a new career. She may have decided to direct her talents to problems of health care, education, or the environment. She will commit herself to cleaning up her city, state, or nation's political system. Or she could be venturing into a cutting-edge and highly interesting new business venture that will remain small and intimate, free of the stifling bureaucracy and worship of the bottom line. This, too, is a form of resistance, though few such adventurers see themselves as anticorporate. It's just no longer what they want to do. But in creating startups or going into social entrepreneurial ventures, many talented people are creating social change in a new way.

Social Movement Resistance. Sarah Soule begins *Contention and Corporate Social Responsibility* by recounting the Boston Tea Party, pointing out that it was an anticorporate social movement. "At the most basic level, the tea-dumping activists in 1773 were frustrated with the East India Company's ability to exert influence over the government and they were angry, more generally, at the unchecked growth of corporate power—power that was coupled with political influence"(Soule 2009: 2).

Anticorporate movements want to make corporations act with more social responsibility, i.e., to behave "in a way that benefits the greater society" (Soule 2009: 19). A common refrain in anticorporate social movements bemoans the size and power of corporations and the lack of public oversight. In many cases, the culprit singled out by anticorporate social movements is enabling legislation, a failure of the state to regulate, or state approval of a corporate practice, in which case the state itself is targeted. Soule describes three categories of anticorporate protest that oppose the direction of social change driven by large corporations (Soule 2009: 55–64).

The first category, antinegligence social movements, takes aim at corporations that "cut corners, leading to accidents, loss of life, destruction of the environment, and, in general, the harming of health and welfare of citizens" (Soule 2009: 62). Mountaintop-removal mining in Appalachia is seen by environmental activists as a shortcut, in lieu of more responsible but more costly ways to mine coal. The second category, antiproduct movements, is the most common. Like the Nestlé baby formula boycott and 1960s antiwar protesters' focus on napalm and Agent Orange made by Dow Chemical, these movements draw attention to a product made by a corporation. The campaigns reject particular products, not only for their immediate impacts but as jeopardizing the future in ways they collectively oppose.

Anticorporate policy movements, the third category, mobilize public opposition to corporate practices that adversely affect a group of people either within or outside the corporation. Social movements against policies that maintain the gender "glass ceiling," aimed at the Sears and Wal-Mart corporations in recent years, have implications for women's opportunities throughout the economy. The social movement against corporations that are aggressively marketing unhealthy foods to children is of this type, as well as having a larger goal of changing children's eating habits and ending the "obesity epidemic" in the United States.

Corporations are criticized for being too large, too influential with politicians and newspaper editors, too immune from prosecution, too big to let fail, too generous in paying their CEOs, and too impervious to the public interest. Though many unpleasant things may be said about the corporation, the corporate form per se is rarely at issue. What is at issue is their seeming acts of aberrant behavior. They are moving events in the wrong direction, events that can have negative implications for more widespread social change.

Antiglobalization and Corporations. A fourth type of anticorporate social movement, probably the most "transgressive," focuses on the corporate form itself, especially the very large, multinational corporation.[26] Much of the antiglobalization movement voices opposition to the cumulative effect of global corporate domination. Movement organizations object to the spread of a commercial culture that suffocates local art and traditions. Corporate control of capital, ownership of patents and copyrights, links to unpopular governments, unfair labor practices, and corporate actions that threaten human rights, health, and the environment are targeted (Burbach and Burbach 2000). The corporate form promotes industrial agriculture rather than healthy local food, creates sweatshops, ignores human rights violations, generates pollution, and despoils the environment.

Three decades ago people in the United States stopped being surprised to learn that various countries participate in the assembly of an automobile marketed by major American automakers as their own. "Made in Japan" had lost its pejorative connotation, and people were getting used to clothing tags with the names of countries they couldn't find on a map. Only gradually, and with conflicted feelings, did they recognize that jobs were "going south" as corporations pursued cheaper labor costs. The tradeoff, through no democratic discussion or vote, was the elimination of jobs long performed by American workers, in return for goods they could more readily afford. The processes of wage stagnation and the national imbalance of trade, shifting billions of dollars to China, Taiwan, and other emerging economies, had begun. This continued, grew, and became a focal point for antiglobalization resistance by the 1990s.

Even the positive changes corporations can claim to be making are insufficient to justify their many negative consequences for antiglobalization

---

[26]The G-20 Conference in Toronto in July 2010 was the "most expensive day in Canadian history" (Democracy Now 2010), costing $1–2 billion, mostly for security: 19,000 security personnel (five times the number at the Pittsburg G-20 Conference in 2009). A four-mile security wall was erected around the convention center where the conference was held. The G-7 Summit prior to the G-20 was in a remote Canadian town, as has been the practice since the "Battle of Seattle" World Trade Organization meeting in 1999.

activists. They should be replaced with some version of economic democracy, local control, reduced consumption, and environmental stewardship (Buttel and Gould 2006). The sum of antiglobalization issues leads some people to the conclusion that corporations themselves are the problem.

The Environmental Crisis
and Corporations of the Future

> *The ultimate purpose of business is not, or should not be, simply to make money. Nor is it merely a system of making and selling things. The promise of business is to increase the general well-being . . . Businesspeople must either dedicate themselves to transforming commerce to a restorative undertaking, or march society to the undertaker.*
>
> —Paul Hawken, *The Ecology of Commerce* (1993)

Today, a principal opposition to the way corporations operate comes from what is loosely called the environmental movement. From April to August 2010 nearly a quarter of a billion gallons of crude oil and gas spewed into the sea a mile under the surface of the Gulf of Mexico when BP's oil and gas well exploded. The worst environmental disaster in U.S. history once again made clear that large corporations are not in the business of making sure the environment is protected. Many people running corporations, however, no longer agree with Milton Friedman (1962: 133), one of the twentieth century's most important conservative economists. "There is one and only one social responsibility of business—to use its resources and engage in activities designed to increase profits so long as it stays within the rules of the game."

Corporations have led the way in the trajectory of an environmentally costly and totally unsustainable way of life. One does not need to join or even sympathize with antiglobalization protestors to recognize the tremendous toll on the environment—from global climate change to catastrophic oil spills, from the Bhopal disaster to the fouling of Nigeria's Gulf of Guinea—that comes with the way affluent nations acquire, process or manufacture, consume, and dispose of earth resources and goods and the role large corporations play in encouraging and facilitating this activity. As the world's natural environment changes, at least in the short term for the worse, efforts to alter the current direction in which global corporations are effecting change will surely intensify. On occasion, these efforts come from corporations themselves.

Older corporations like Johnson and Johnson, and many newer firms often headed by young entrepreneurs, take seriously corporate destruction of and threats to the environment. They are finding ways to be profitable and also socially responsible (Vogel 2005). Corporate social responsibility can be nothing but a public relations ploy. If taken seriously, however, it can lead to more sustainable practices while benefiting a corporation in tangible ways.[27] It recognizes that corporations are the most powerful organizations in the world. If solutions to major problems are to be found, corporations must point the way.

Some successful businesspeople have reached the conclusion that the ways goods currently are being produced are too unsustainable to continue indefinitely (e.g., Hawken 1993, Korten 1999). Well-known activists who have charted innovative, more sustainable business models, from microfinance to small-scale appropriate-technology manufacturing, argue that economic pursuits are the key to improving peoples' lives in the poorest parts of the world. Because the problems are often linked to economic activity, so must be the solutions.

Social entrepreneurship is sometimes described as the creative application of business sense to problems that have little likelihood of profit but offer improvements to seemingly intractable issues like global poverty, illiteracy, and hunger (Bornstein 2005). The term also describes similar nonprofit efforts that combine with profitable activities, such as the Newman's Own Foundation that makes a range of food products and gives all profits to charities. When an organization of women in a remote area of India produces objects of traditional art—figurines, baskets, jewelry—for a tourist market, profits can go to the women's households but also fund a health clinic, a school, or a safe house for abused women. A coffee import company selling raw beans to roasters can plow back some of its profits into the health and well-being of coffee producers in poor countries. It can work with coffee farmers on improved, sustainable farming methods. Can this be done by large corporations? Is the corporate form a possibility in social entrepreneurship only when the organization is of modest size? This has

---

[27]Practically all major multinational corporations today have embraced some aspect of corporate social responsibility (CSR) or corporate citizenship, ambiguous concepts variously defined as voluntary efforts to "do more to address a wide variety of social problems than they would have done in the course of their normal pursuit of profits" (Vogel 2005: 4). The World Economic Forum identifies not only those directly associated with the firm but "communities and future generations" as part of the relationships that obligate corporations as significant forces in societies (Zerk 2006).

been a challenge, inasmuch as size, managerial control, and the requirement to maximize shareholder profits that seem inevitably a part of large corporations conflict with social entrepreneurship (Light 2008).

Paul Hawken and others insist on a radical alteration of thinking and practice, replacing the "delusion that business is an open, linear system: that through resource extraction and technology . . . economic growth can be extended indefinitely into the future" (Hawken 1993: 32–33). Rather, a much stronger commitment to the reality of natural limits and honesty about what has been done, possibly irrevocably, to despoil the environment require thinking and practices that embrace cycles of growth and decay, building and dismantling, using and restoring. As a very successful corporate entrepreneur, Hawken is a realist about the power of corporations. "Business is the problem and it must be part of the solution . . . Ironically, business contains our blessing. It must, because no other institution in the modern world is powerful enough to foster the necessary changes" (Hawken 1993: 17).

Topics for Discussion and Activities for Further Study

Topics for Discussion

1. The labor process and labor market have changed a great deal in the last hundred years and will continue to change in the twenty-first century. Discuss the changes you think are most significant. Given the trends discussed in this chapter, what should a young person do to have a satisfying, secure livelihood? What grievances could a social movement address that would help to ensure a decent working life in the future?

2. How have lifestyles changed with the proliferation of a consumer technology? You can start with the obvious: hand-held computers with myriad applications. Discuss other consumer technologies. What do they allow people to do that they couldn't do before? Have they created a new sense of want (i.e., urgency to satisfy a need)? What lifestyle changes do you perceive?

3. The film *The Corporation* covers several of the topics in this chapter. If your school has a copy, watch it and discuss it. The film is divided into several segments. You might want to watch only a few and have a discussion about these before proceeding on.

4. Some people argue that no matter how sustainable corporate operations become, the very idea of corporate-driven material consumption will always threaten the environment. No matter how energy efficient affluent societies

become, they will inexorably be on a dead-end path. If they are right, what is the alternative? If they are wrong, discuss the ways they are wrong.

5. Talk about ads. Do they matter? Many people say no, that sophisticated audiences make choices based on what's best for them and in line with their values. Is advertising able to alter this? Does it?

Activities for Further Study

1. In 2010, the Cargill Corporation led a campaign against reducing salt in food. Given the health risks of eating too much salt, this campaign seemed narrow-minded. Specifically, the salt industry (Cargill processes and transports salt) objected to any food additive guidelines on package labels and other regulations. Look at this or another instance of a corporation campaigning for its interests, in competition with health, safety, environmental, or financial regulation, or any other public interest advocates who have sought (and perhaps succeeded, as with tobacco) to regulate corporate behavior. Note: there may be corporations lining up on the side of the advocates.

2. Much has been said in this chapter about corporate influence on the political process: the revolving door of corporate executives going into and leaving government; funding campaigns and causes with large amounts of money; creating "think tanks" to carry out research that is widely publicized to promote a corporate interest or counter an unwanted issue. Books like Naomi Oreskes and Erik Conway's *Merchants of Doubt* explore this with skill and critical zeal. Do some research on corporate political influence regarding an issue that interests you and try to answer the question, Is this how democratic governance should work?

3. Many corporate officers and entrepreneurs are as critical of corporations as their most vocal opponents. Some are part of very successful corporations, e.g., Patagonia, Smith and Hawkins, Newman's, or Thanksgiving Coffee. What are the corporate critics doing that is different from corporate business as usual? Why are they doing this?

4. Opensecrets.org is the website of the Center for Responsive Politics (CRP). It is a fascinating compilation of fundraising and spending for lobbying, campaigns, and other activities. Do an original project with these data. Look into an issue that garnered contributions from major corporations. Use a good newspaper electronic index to find some newspaper articles on this issue, and see how the reported figures (if any) compare to those collected by the CRP.

5. In pairs, take opposite positions on the question of information technologies and family life. What is sometimes called "web-centric homelife" is decried by scholars like MIT's Sherry Turkel, author of *Alone Together* (2011). Not everyone agrees. Do some reading pursuing opposite viewpoints. Read each other's analysis and respond, either in writing or a class debate.

8

States and Social Change

The Uses of Public Resources for the Common Good

Roosevelt shepherded fifteen major laws through Congress, prodded along by two fireside chats and thirty press conferences. He created an alphabet soup of new agencies—the AAA, the CCC, the FERA, and the NRA—to administer the laws and bring relief to farmers, industry, and the unemployed. "He also took America off the gold standard, created the Tennessee Valley Authority, established two major public works programs, and . . . provided regulation for stock issues. The first advances were made toward a minimum wage, a ban on child labor, and legal support for union organizing . . . This period, known ever after as 'The Hundred Days,' made profound changes in government's attitude toward the citizens and created the ideological conflict that animated American politics through the twentieth century to the present day." The Roosevelt revolution created modern America.

—Russell Baker (2009), quoting
Adam Cohen's *Nothing to Fear*

> *We only played against black teams. I guess the thing that I remember most is that we had to stay at people's houses rather than in a hotel. If there was a black restaurant, we had to eat there. Otherwise, accommodations weren't that good . . . That system, I think, just kind'a passed down. I guess for myself, I would have to say I followed what I saw in front of me. You know, I had parents who told me I could do anything, but I could watch them and also see that they couldn't.*

> —Jack Alexander, speaking of his youth
> during the Jim Crow era[1]

A strong state, whether democratic or authoritarian, whether in the United States or elsewhere, can wield enormous power and dramatically alter the course of peoples' lives. Because the modern state has both real and unrealized power, it is continually pressed to address the issues of the day, whether large or small, to make things right or to put things in order. It can seek to reduce suffering, work for social justice, or further enrich the most favored and powerful. It is no wonder that those intent to change their world so often focus attention on controlling the state or seizing its power.

Strong States and Social Change

With the growth of the national state, the scope of governments that carry out state functions has grown enormously. Modern states monopolize the means of coercion. Making war is a prerogative reserved for nations. The state collects taxes, regulates finance, and sets monetary policy. It builds and operates schools, waste treatment plants, highways, airports, bridges, parks, prisons, and what is generally called public infrastructure. Local, state, and federal government regulations affect private land use, the development or conservation of publicly owned land and earth resources, commerce and corporations, the design of buildings and homes, the content of food products, modes of travel, patient care, and pharmaceuticals.

Government services deliver mail, respond to disasters, provide some level of support to those most in need, engineer dam-building and land reclamation projects, patrol the streets and investigate crime, rescue the lost, and bury the poor. States support research, administer pension programs,

---

[1]Jack Alexander grew up in Topeka, Kansas, a few years before *Brown v. Topeka Board of Education* (1954). This quote is from the broadcast aired on KTWU Channel 11, Topeka, Kansas, in 2004 (Alexander 2004).

ensure health care to many or all of the citizenry, and authorize courts to adjudicate contested civil suits as well as criminal actions by fining, imprisoning, and in some cases executing those found guilty. States sign agreements that obligate them to act on international accords, join international court proceedings, and engage in military operations.

Much has been said in previous chapters about the state and its involvement in social change, regardless of the primary force driving change. The modern state encourages technological development. States sponsorship helps in the development of supporting technologies in education, communication, health, transportation, and other areas. In most if not all of these, the state has financial, research, and regulatory involvement. Corporations are the creation of the state, a legal construct. State incentives for corporate investment attract a wealthy elite who play an outsized role in directing the state's activities. War is a state-organized mobilization of resources to counter other nations' and nonstate actors' war-making powers. As seen in Chapter 6, the act of involving the citizenry in war contributes to loyalty and patriotism that can build and sustain the legitimacy of the state as an agent of social change. All of these things make the state and its governmental bodies the focus of social movement grievances.

This chapter focuses on the direct actions of states to effect social change by using public resources to advance collective goods, i.e., improve the well-being of the nation. Three of the case studies describe historical periods in the United States where a strong state made fundamental changes in people's lives. The first improved public health, the second increased public resources, and the third expanded civil rights. Examining social change in China since the Mao Zedong era offers a contrast to social change in a multiparty democratic capitalist system. In the past thirty-five years, China has experienced enormous economic growth by a tightly controlled private sector and sovereign use of vast accumulated capital under the direction of a single-party political system.[2] It has much to teach about national and global social change in the years ahead.

---

[2] A strong authoritarian state like China, especially one that organizes itself to pursue modernization and economic expansion, has the luxury of an opaque and compliant one-party political system. Civil society, the political voices and forces found in the space between corporate power, private life, and governmental operations in a democracy, is miniscule and easily quelled by an authoritarian state. An authoritarian state has significant autonomy to act as it wishes, with few checks and balances, by controlling and deciding unilaterally on the use of the resources it needs to operate, including coercion.

The Role of Strong States in Modern Times

With the exception of anarchists and nihilists, almost no one across the ideological spectrum—liberal, conservative, libertarian, and green—disagrees that the state should provide protection and physical security for its citizens. Even advocates of minimalist government accept the rule of law, the role of courts to adjudicate disputes and bring wrongdoers to justice, and the need for day-to-day protection. They would not tear down jails and prisons. Few would abolish the military, though its size and how it is used are major issues of contention.

A priority of strong states is to undergird and support economic activity to a lesser or greater degree. In the promotion of the common welfare, strong states are obliged to advance social improvement. Democracies and authoritarian regimes alike base their legitimacy, or at least reduce the chances of rebellion, by engaging in popular actions of widespread benefit.

A nation unable to respond to disasters, without an information highway or modern transportation system, without flood control and irrigation, without schools and research facilities, with no provisions for the poor, indigent, those with special needs, and the elderly, is a nation unable to provide what most of the world's people expect. While there are some profitable opportunities in satisfying aspects of each of these, most responsibility falls to the state.[3] Strong states are able to do this; weak states are not.

The State and Social Change:
The United States in the Twentieth Century

Much has been said in earlier chapters about the use of state power, especially when it is the object of corporate influence and the avenue to address grievances of social movements. Laws and their enforcement restrict and impel patterns of behavior. Political rhetoric alters public opinion. The operation of the state—including war—makes it a significant economic force.

In the accounts that follow, the state was a powerful driver of social change by solving problems of the common good. Social movements,

---

[3]When there has been an opportunity for private profit in the United States, what the state had begun often shifts to the private sector. This has been much less the case in most European countries, where the role of the state in orchestrating universal health care, regulating economic activity, and reducing social and material inequality is more robust.

technology, corporations, and the conduct of wars—the other drivers of social change—are closely tied to the social change activities of the state. Despite the effort to disentangle these five, the following three accounts make apparent the combination of forces that moved the United States toward what it became during the lifetime of Iris Summers. They recount how the state has used several means to effect change, invariably interpreted as improvements and progress that followed the popular mandate for a strong state: as supporter of research and the creator and enforcer of regulations, by designing and funding public works, and through judicial decisions and legislation.

Public Health: Reducing Sickness and Death

When Iris Summers was born, chances that her mother would die in childbirth or from complications of childbirth were staggering. Six to nine women died for every one thousand births. A woman giving birth to six children had a one-in-twenty chance of dying that way, leaving her children without a mother. Similarly, for every one hundred births, ten to fifteen infants died. Another 10 percent of all babies died before the age of one. For poor, nonwhite, and immigrant women and infants the rates were much higher, and they remained so for nonwhite women throughout much of the century. By the end of the twentieth century in the United States, however, fewer than one in one thousand infants died at or soon after birth, and fewer than one mother in ten thousand died in childbirth (Kotelchuck 2007: 105, 124).

Similar dramatic changes occurred for motor vehicle accidents. In 1923, 22,600 people in the United States lost their lives in auto fatalities, when far fewer Americans drove far fewer miles than today (Dellinger et al. 2007: 344). In 2008, there were 37,261 motor vehicle fatalities (five thousand were motorcycle accidents), with more than 20 percent attributable to alcohol abuse. What was nearly a massacre in the 1920s became a heartbreaking but less common tragedy due to better-built cars, safer roads, driver education, federal motor vehicle safety standards, safe driving laws, and good law enforcement.

During Iris Summers' childhood, the cause of most deaths in the United States was from infectious diseases like tuberculosis and diphtheria and diarrheal illnesses like cholera. Typhoid fever and other food-borne diseases killed and sickened many. Today these causes of death are rare. People in the United States are living, on average, nearly thirty years longer now than early in the twentieth century and are more likely to die of cancer or cardiovascular diseases, e.g., stroke or heart attack. Unfortunately, these causes of death are often attributed to smoking cigarettes (which few people did in 1900), a combination of overeating and having too little physical activity (not major problems for most people a hundred years ago), and environmental pollution that is a constant reminder of the wages of material abundance.

Occupational death was far from uncommon. In 1907, 2,534 miners were killed in coal mines in the United States, 869 of them in eleven accidents alone. Mining remains the nation's most dangerous occupation. In 2009, thirty-four miners were killed on the job, including the Upper Big Branch mine disaster in West Virginia that took the lives of twenty-nine miners and gripped the nation's attention. Occupational fatality rates overall declined tenfold in the twentieth century (Robbins and Landrigan 2007: 219, 211).

State power has been applied to improving the health and safety of the nation, sometimes grudgingly and nearly always with contention. In collaboration with and spurred on by many voluntary and nonprofit organizations, the state took on this responsibility. Fewer deaths by automobile, disease, and unsafe working conditions are to a great extent directly attributable to governmental action at the local, state, and federal levels.

Most early public health initiatives began at the local level, where cities took the lead in public sanitation, collecting vital statistics, curbing infant mortality, and promoting motor vehicle safety. These efforts led to statewide regulations and the creation of state bureaus and agencies and, finally, to federal involvement. Public health is one of the most important ways the state has driven social change in peoples' everyday lives.

Local Public Health in the Progressive Era. In 1900, forty states had health agencies, usually at the county level, as did many cities. Two cities particularly stand out as initiators of public health: New York and Chicago. At the turn of the century, only 6 percent of people in the United States had access to filtered water, and the sewage of only 25 percent of households was treated. Overcrowded slums and no regulation for waste disposal bred infectious diseases, to which young children were most susceptible (Melosi 2005). Progressives sought out the source of widespread problems and often found them in government corruption, but more often they were brought on by the rapid changes that accompanied an economic transformation from agriculture to industry, urban growth, and the influx of immigrants. Poverty, poor housing, and especially polluted water, piles of garbage, and poor sanitation became publicly recognized problems in need of collective solutions.

A hotbed for Progressive Era politics[4] and progressive citizens applying social science to problems in the 1880s, New York City initiated a major effort to improve child health by creating the Bureau of Child Hygiene. The city undertook measures to clear the streets of garbage, improve sanitation and water, relieve overcrowding and create better housing, and establish access by the poor to medical services. The germ theory of infectious disease was new and unknown to many people in 1880, but a general idea of communicable disease and the connection of filth and illness were widely shared.[5]

The Progressive's emphasis on cleanliness led the city to enlist a corps of social workers, nurses, and others who made home visits and held neighborhood meetings to discuss and demonstrate the importance of using soap and clearing trash. Street cleaners were outfitted in white uniforms, and "the public began identifying the workers with doctors, nurses and others in the health professions" (Melosi 2005: 54). It was accepted that the city government had a responsibility to help people be safe from infectious diseases.

New York and other cities began massive projects to collect garbage, process sewage, and provide clean water in place of easily infected local and private wells. Food, especially milk consumed by infants and young children, was identified as a frequent culprit for infection and death, leading to widespread adoption of pasteurization and, later, to fortified processed food such as enriched flour and iodized salt. Infant and children's mortality dropped significantly in a very short time. The next step was to improve maternal health through cleanliness, better access to healthy food, and medical supervision of pregnancies and childbirth. Public health departments undertook this effort, again with much success (Kotelchuck 2007).

---

[4]The Progressive Era dates from the early 1890s to about 1920. It was a national movement that fielded candidates for local, state, and national offices. Its voice was often the "muckraking journalism" that involved investigations of urban slums and political malfeasance. Progressives pursued more open, accountable government, innovations in education, and a more "scientific" approach to solving social problems, taxation, and government spending. Their efforts contributed to the passage of the sixteenth through the nineteenth amendments to the U.S. Constitution.

[5]Water-borne infections were only vaguely understood and not widely accepted until late in the nineteenth century. People believed many diseases were caused by "miasma" or foul-smelling air, and the high rate of tuberculosis in polluted areas with putrid air seemed to confirm their fears. When New York had a major outbreak of cholera in 1848–49, (smelly) pigs were banished from the city (Wilford 2008).

Mulberry Street, New York City, circa 1900.

Source: Library of Congress, Prints and Photographs Division, Detroit Publishing Company Collection

Chicago also was a leader, not only in Progressive Era muckraking for better government, but also for infant and maternal health and the regulation of food processing—especially of meat and milk (Wolf 2007). It led the way in obtaining the vital statistics needed to understand health problems and solve them epidemiologically: by tracing outbreaks of illness to their source and studying the interactions and causal relations among biological, behavioral, and environmental factors.

The federal and state governments did this in some areas such as workplace diseases and injuries (Sellers 2007), but private groups of Progressives, aided by investigative journalists, academics, and social activists, led the way in creating the organizations that collected, analyzed, and disseminated information needed to gain public attention and political support for public health issues. Chicago was also the first city, in 1933 at Northwestern University, to have modern traffic policing research, information that led to a systematic approach to traffic laws, motor vehicle safety, and enforcement of driver behavior, including driver licensing and testing.

Federal Public Health Initiatives. Washington's involvement in public health came later than local and state actions. The politically conservative but progressive Teddy Roosevelt, himself a sickly child, called for federal action on infant health in 1909, to no avail. In 1912, the U.S. Public Health Service was launched but confined itself mostly to quarantine efforts to halt infections pandemics. The first school of public health, The Johns Hopkins University, was at a private institution and funded by private donations, including the Rockefeller Foundation. Federal funds for local efforts began in 1917; the Social Security Act of 1935 expanded the Public Health Service's mandate; and in 1946 the federal government created the Center for Disease Control (CDC). Today the CDC is perhaps the nation's most important organization for documenting health and supporting actions to combat illness.

The creation of Medicare and Medicaid in 1965, part of the federal War on Poverty, and the establishment in 1970 of the Environmental Protection Agency (EPA) and the Occupational Safety and Health Administration (OSHA) mark important steps taken at the federal level to expand public health, enhance personal safety, and extend access to medical services. Many federal public health activities today are undertaken in partnership with state public health and environmental protection agencies (Ward and Warren 2007: x).

The role of science and technology, as well as the impact of social movements on public health, contributed to enhancements in job safety, stopping the spread of disease, better reproductive health, and longer lives. Though a miniscule portion of all health care expenditures in the United States (1 to 2 percent of the more than $1 trillion spent annually), the accomplishments of public health have been considerable. University research facilities—the sites of most basic research and the employer of most of the nation's Nobel Prize winners for medicine—depend almost entirely on federal funding to support biomedical research. In contrast, clinical medicine in the United States is mostly private. It depends heavily, however, on public research for new drugs, surgical techniques, diagnostic equipment and other technological breakthroughs, and ways of delivering health care more effectively and to those most in need.[6]

War and Public Health. Somewhat ironically, the state's engagement in wars has been a significant contributing factor to the rise of public health and the social changes it has wrought in the way of shifting social demographics, increasing longevity, reducing disease, the collection and analysis of vital statistics, and the expansion of public health infrastructure. The Veterans Administration is the largest public health agency in the United States and is available to millions of persons who served in the military. During and following World War II, the federal government's Emergency Maternal and Infant Care Program for wives and children of lower-grade service personnel made health care available to millions through state health departments. Briefly suspended in 1949, its popularity led to the military dependents' Medical Care Act in 1956 that was incorporated nine years later into a broader Civilian Health and Medical Program for the Uniformed Services (Kotelchuck 2007: 115).

Conscription and the requisite draft physicals in the First World War alerted the nation and public health agencies to nutritional deficiencies endemic in the lower social classes. Their health problems were seen as a threat to national security. Veterans hospital research on circulatory and especially heart disease in the 1960s was instrumental in directing the focus to diet as well as early diagnosis. The public research contributed to an understanding of the health risks of toxins and carcinogens such as Agent Orange to which soldiers and civilians were exposed during the Vietnam War.

Two levers of the state—taxation and regulation—are most important in effecting changes in health and safety. Nothing better illustrates this than the

---

[6]The relationship between improved health for the population at large and the activity of public health agencies, aided by legislation and funding, is unequivocal. When efforts have lagged, indicators of health and well-being have remained flat or declined. This is especially true for infant health and women's reproductive health. For example, as the focus shifted to preventive care and successful pregnancies under the Healthy Start initiatives of 1991, infant mortality rates improved, but other features of maternal and infant health remained stable after a Republican Congress opposed expansion of public health efforts, ultimately defeating the Clinton administration's Health Security Act. Though access to health care expanded with the passage of the Affordable Health Care for America Act in 2010, the United States remains the only affluent nation in the world that does not guarantee health care for everyone. Comparative national health statistics reflect this.

efforts to reduce smoking as a health hazard and steps taken to improve auto safety.

Smoking and Public Health. Iris Summers started smoking soon after the first research on smoking, in 1939, linked lung cancer to inhalation of tobacco smoke. She was hooked. She loved to smoke. She knew it was unhealthy, but "so is driving a car," she would tell her scolding grandchildren. By 1964, more than seven thousand research articles confirmed the link (Eriksen et al. 2007: 424). Antismoking advocate Allan Brandt (2007) calls the 1964 Surgeon General's Report a watershed that confirmed what researchers and doctors had long known. Smoking is a significant health risk, especially for lung, laryngeal, and esophageal cancers and chronic bronchitis.[7]

By the mid-1960s, new sophisticated statistical techniques provided convincing findings unchallenged by contrary evidence. The analysis of massive random-sample survey research data sets, panel studies of doctors, the use of control groups in experimental designs, and longitudinal panel studies that followed smokers and nonsmokers as the former developed life-threatening illnesses sooner and in much greater numbers than the latter were carried out. They came to an unequivocal conclusion: smoking causes serious health problems and is the nation's greatest preventable cause of death.[8]

More than eight million people in the United States are sick or disabled due to tobacco use, and nearly half a million die annually from this preventable disease (Wilson 2007). In the United States, this has an enormous cost in terms of personal and public health expenses ($96 billion), loss of work ($97 billion), taxes lost ($71 billion), support for surviving dependents ($2.6 billion), property damage, and shortened return on public education.[9]

---

[7]Research since 1964 has linked smoking to an even wider range of health issues and illnesses. The 2004 Surgeon General's Report concludes that "smoking harms nearly every organ of the body" (quoted in Brandt 2007: 452).

[8]In 1930, a point when smokers had not had time to develop significant numbers of lung cancer cases, annually there were 4.9 lung cancer deaths for every 100,000 men in the United States. By 1990, this number had increased almost twentyfold, to 75 per 100,000 men per year (Brandt 2007).

[9]Private health costs are at least $28 billion. Annual public costs from smoking include Medicare, $27 billion; Medicaid, $31 billion; and VA (veterans) health care, $10 billion (Lindbloom 2010).

The most significant inhibitor of cigarette smoking is cost. Thus, the most effective action taken by the state is taxation, which raises the cost of cigarettes. An increase in the price of a pack of cigarettes by 10 percent results in a 4 percent reduction in smoking. Antismoking campaigns also have been able to influence public opinion and legislation, including eliminating price supports to farmers and corporations growing tobacco.

Nearly thirteen billion dollars for advertising in the United States in 2009 by tobacco corporations and the addictive power of nicotine, as well as the allure of rebellion and glamour associated with smoking, have kept cigarette smoking in vogue. Smoking remains a popular addiction,[10] despite conclusive research that smoking is a major health hazard, warnings on cigarette packages, restrictions on advertising and the distribution of free samples, smoking bans in public places, and years of public service campaigns, In the three-way struggle among corporations, individuals who smoke, and the antismoking movement, the outcome is mixed. Since 1964, the overall rate of smoking in the United States has declined by half. Still, more than 20 percent of adults smoke regularly in the United States, and nearly 20 percent of high school students are smokers.

Automobiles and Public Health. Traffic safety is a story of two narratives. On the one hand is the "nut behind the wheel" paradigm, to quote the public interest crusader Ralph Nader. The driver is responsible for accidents due to bad behavior, including reckless driving, poor driving skills, and alcohol impairment. The other "cars kill people" paradigm sees dangerous, badly designed and maintained vehicles and unsafe roads as the problem. In the history of public health efforts to reduce death and injury on the road, both approaches have found favor, with mixed results (Albert 2007).

Nearly three million people have died in car crashes in the past century. The silver lining is that, for miles traveled, there was a 93 percent reduction in automobile deaths between 1923 and 2000 (Dellinger et al. 2007: 344). As a public health issue, the state played a major role, though not at the outset. Early in the twentieth century, auto corporations, through the National Association of Automobile Manufacturers, lobbied Congress for auto safety standards, but a conservative Congress declined. This was considered a matter for local jurisdictions, despite the

---

[10]Worldwide, many nations have expended less effort to stop smoking, coveting the tax revenues the practice generates. This is not the case in a few places, e.g., Singapore, where laws are much more restrictive and penalties more severe. Recent trends in advertising and packaging around the world suggest a growing concern about smoking and the public costs it incurs (Wilson 2010).

growing proliferation of automobiles and their growing importance in peoples' lives.

In 1924, Commerce Secretary (later president) Herbert Hoover convened a conference to set uniform traffic laws. It had little influence, and local jurisdictions continued to write their own regulations, including those for road construction, creating a maze of different laws. There was little or no product liability; *caveat emptor* (let the buyer beware) prevailed. The promise of safer driving was a persuasive argument used by the petroleum, auto, tire, and related industries in early lobbying to expand the nation's highway system to include interstate highways.

Insurance companies and the automobile industry made efforts to create safer cars. In most cases of liability claims, however, the industries argued that traffic accidents were the consequence of poor driving. Public health initiatives should focus on better driving, speed limits, rigorous enforcement of traffic laws, driver education, and restrictions on select classes of drivers. In the 1950s, Ford Motor Company briefly ventured away from the nut-behind-the-wheel paradigm when it sought, without success, to improve sales by emphasizing that safety makes for a better car. Instead, emphasizing speed, style, and personal freedom sold more cars, and the idea was shelved.

The public interest movement, led by Ralph Nader, pushed for greater public accountability of automobile corporations (Vogel 1978). Only in 1966 did the federal government establish the Highway Safety Act and the Motor Vehicle Safety Act, enforced after 1970 by the new National Highway Safety Bureau. Almost immediately highway accidents declined. Not until the 1990s did auto safety again emerge as a way to sell cars, this time in competition with foreign automakers like the Swedish carmaker Volvo, a company that promoted its benefits of safety above all others.

As the builder of highways, the state has been able to establish requirements to increase the safety of roads. The federal government also sets requirements such as the use of seatbelts and speed limits in appropriating highway funding to states. Using the Interstate Commerce Clause, the federal government has been able to require warning and backup lights and other automobile features that make driving saver, even for nuts behind the wheel.

The Public Watering of the West

The canals that provided commercial transportation early in the nineteenth century in the eastern United States were considered a public good, hence the provision of a corporate charter to their builders. Like schools, roads, sanitation systems, and national parks, they were, for Teddy

Roosevelt and other progressives, public works acting as instruments of progress for the nation. The stories of engineering the rivers in the eastern United States—for example, the Tennessee Valley Authority (TVA) and other massive electricity-generation engineering feats (e.g., Nye 1990)—and the creation of levees to control the Mississippi (McPhee 1989) chronicle personal ambition and the collective power of the state that changed the lives of millions of people. Similarly, the extensive system of dams constructed on nearly every major river in the West was for the public good. They were built to control flooding, provide for irrigated agriculture, create recreation sites, and generate electricity. They were part of one of the most concerted efforts by the state to effect social transformation through public works.

Marc Reisner's *Cadillac Desert* paints a dramatic scenario of the impact of Hoover Dam and Lake Mead on the Colorado River, a vital source of water and electricity for much of California and the Southwest.

> If the Colorado River suddenly stopped flowing, you would have four years of carryover capacity in the reservoirs before you had to evacuate most of southern California and Arizona and a good portion of Colorado, New Mexico, Utah, and Wyoming. (Reisner 1986: 125)

In the United States, the colonization of the West was described as manifest destiny and had national security implications. Most of all, though, it was all about progress. It captured the public's imagination, offered new economic livelihoods for would-be farmers and ranchers, and provided opportunities for a few to amass fortunes. It evoked national pride and patriotism.

Colonizing the West. A national ideology supporting colonization guided thinking about the role of the state and social change. When Thomas Jefferson arranged for the purchase from France of all the land drained by the Mississippi River, he was pursuing a colonization agenda that in many respects mirrored what Spain, Portugal, Britain, the Netherlands, Belgium, and other European states were doing globally.

In "reclaiming the wilderness," the state was securing territory and resources for growth by diminishing, displacing, and isolating indigenous peoples and resettling the nation's citizens in newly acquired territory. That is a pretty basic definition of colonization. The state's intention and obligation was to facilitate the settling of land, often at great cost and with considerable violence. Military support—the war against the Indians and shepherding of ethnically European settlers—is a well-known exercise of the state's coercive power. It was a prelude to, and made necessary, the public watering of the West.

From the outset, colonizing the West required the engineering of water, euphemistically described as reclamation. Land and water were to be put under human control and made available to those who would use them productively. This was not a replication of what had already happened in the East. Rather, "reclamation was a social experiment that set forth water resource development as the basis for a new civilization in the West" (Robinson 1979: 32).

Decades before Woody Guthrie sang about California as a Garden of Eden, agriculture in the state was well developed and heavily dependent on irrigation. It had, however, reached diminishing irrigation opportunities by 1890. Growth and a continuation of its population boom required water from out of state (Pisani 1984). As happened throughout the West, initially irrigation was financed by private investment that could afford to dam streams and excavate drainage canals. Larger projects had difficulty attracting family and corporate money, and each state's separate and limited efforts usually came to naught.

In order to enlist the western states in reclamation, in 1894 the U.S. Congress passed the Carey Act that "granted up to 1 million acres [of federal public land] to each of the 10 arid states if they caused the land to be irrigated, settled, and cultivated." It emphasized the creation of and benefits to family farms and sought to prevent corporations from expropriating the benefits. Like earlier measures by the federal government to aid small farmers, however, the Carey Act was a "dismal failure . . . The states were unwilling or incapable of administering the complex program" (Robinson 1979: 16, 9). More direct action from Washington would be needed.

Power and Politics, from Roosevelt to Roosevelt. When Teddy Roosevelt assumed the presidency after William McKinley's assassination in 1901, reclamation had the advocate it needed to push a strong bill through Congress. That measure was the Reclamation Act of 1902, establishing the Reclamation Service and the fund to pay for projects too large for individuals and states to undertake. Eastern and midwestern state legislators in Congress argued it was unfair to spend money in the West and not equally in their region, that there was a surfeit of food in the country and more agriculture wasn't needed. They argued that federal reclamation could only be implemented by trampling on "states' rights." Teddy Roosevelt, a conservationist in the mold of hating to "waste water," would have none of this, and the Reclamation Service was launched.

This legislation invoked the spirit of John Wesley Powell, the man who led the first expedition to map the Colorado River and its main tributaries.[11] Powell had proposed using the water of the major rivers to "civilize" the West by settling the yeoman farmer there. Irrigated land fell under the 1862 Homestead Act that required families to make improvements on their land in return for the right to make their homes on what had been land owned by the entire nation. Watering the West was framed by its advocates as a continuation of the pact between the homesteaders and the nation.

California, Utah, Colorado, and other states believed the answer to their need for irrigation water lay in damming the Colorado River. Plans for channeling much of its water to California via the All-American Canal were discussed soon after the Reclamation Service (later renamed the Bureau of Reclamation) was formed. Water was highly coveted in the West, and there was considerable opposition to California's receiving water from a river that never passed through the state. Fortunately for California growers, those who promoted the idea of reclamation and making the West bloom proved stronger than the opposition. It would take some time, though.

In its first twenty years, more than two dozen projects were authorized and begun in a pattern that resembles much of the "pork" of Congress today. Every senator and member of Congress coveted a project for his or her district, and a good deal of horse trading was necessary to secure approval. Political expediency was as much the criteria as sound hydrologic engineering, economics, and the needs of farmers and ranchers. The reclamation fund, initially provided for by the sale of federal lands and to be replenished with the government's sale of irrigation water from the projects, soon ran into financial problems. The fund then borrowed significant sums from the Treasury. Loan repayment periods for irrigators were extended from ten to twenty, thirty, and finally fifty years, adding to the fund's insolvency. By 1922, less than 10 percent of the loans had been repaid, and 60 percent of the irrigators were in default (Reisner 1986: 120–123).

The enthusiasm with which frontier conservatives in the West embraced and fought for the spoils was ironic. Reisner calls the watering of the West an "experiment in pseudosocialism." The frontier spirit and sanctity of individual initiative, letting people's poor choices fall on their own shoulders, and a congenital distrust of government have long been staples of politics in the American West, from the Sagebrush Rebellion to the fight over wolf reintroduction in Yellowstone National Park. As is often observed, however,

---

[11]Powell's *The Exploration of the Colorado River and Its Canyons* remains one of the great and most gripping epics of human courage and endurance.

"What the government does for them is progress, while what it does for others is socialism" (Lauer 1991: 206).

Three years after the stock market crash that ushered in the Great Depression, President Franklin Roosevelt, distant relative of Teddy, was determined to spur economic recovery. He sought to do this through government financing and public works projects in order to reverse the growing joblessness, business failures, farm foreclosures, and emerging sense among a portion of the population that capitalism was a failed economic system. Roosevelt's National Industrial Recovery Act poured tens of millions of dollars into construction projects to create jobs and promote economic recovery. The New Deal efforts included the Works Progress Administration (WPA) that housed the Federal Writers' Project and the Federal Emergency Relief Administration (FERA), the Civilian Conservation Corps (CCC), and the Resettlement Administration and other agencies that became the Farm Security Administration in 1935. As discussed in Chapter 2, it also employed artists to chronicle the depths of the Great Depression and anthropologists to collect the stories of former slaves.

Herbert Hoover, defeated for a second term by Franklin Roosevelt in 1932, had used his considerable administrative skills as secretary of commerce in 1922 to secure the Colorado River Compact. This agreement among states of the Colorado River Basin, reworked many times and strongly opposed by Arizona, set out the proportion of stored water each state would receive and later divided the electricity generated by the Hoover Dam.[12] Construction began in 1931, and the project became a cornerstone of Roosevelt's massive river-basins legislation that authorized dozens of dams providing electricity, flood control, and irrigation water—to say nothing of federal construction contracts and jobs—over the next two decades.

Monuments of Power. When completed and dedicated in 1935, the Hoover Dam, the "most ambitious government-sponsored civil engineering task ever undertaken in the United States" (Stevens 1988: 20), was the largest hydroelectric plant in the world. With massive spillways and diversion tunnels, the dam itself is 726 feet high, equivalent to a sixty-story building. The dam is 660 feet thick at the bottom and tapers to forty-five feet at the top

---

[12]It was called Boulder Dam until its dedication in 1930 and christened Hoover Dam by interior secretary and Californian Ray Wilber. This set off "a partisan battle that would last seventeen years, during which the dam's name would be changed back to Boulder Dam before it was permanently made Hoover Dam in 1947" (Stevens 1988: 34).

where it is nearly a quarter of a mile wide. Its façade was designed by an English architect, Gordon Kaufmann, to emphasize the dam as "a monument to twentieth-century technology, a symbol of man's triumph over nature and his ability to shape and control his environment . . . The theme of power" (Stevens 1988: 30).

At one point in its construction, more than five thousand men (Chinese workers were barred, and only thirty African Americans were hired) worked on the project. The much-diminished Industrial Workers of the World, the Wobblies, tried without success to unionize its workforce. More than a hundred men died on the worksite. In total, 4.4 million cubic yards of concrete were poured in a marathon of building that continued almost day and night, at 220 cubic yards per hour, for two years (Reisner 1986: 135). It required the excavation of 3.7 million cubic yards of stone, and 45 million pounds of structural steel and pipes were laid. No wonder it remains a major tourist attraction today.

Boulder Dam, 1941 (officially named Hoover Dam in 1947). Photographer, Ansel Adams.

Source: National Archives and Records Administration, College Park, Maryland. http://www.archives.gov/exhibit_hall/picturing_the_century/port_adams_img110 .html.

Grand Coulee Dam. Upon completion in 1936, Hoover Dam produced more electricity than any hydroelectric facility in the world, though this was soon eclipsed by another federal public works projects. Grand Coulee Dam on the Columbia River was an even larger project, with a spillway wider than the Hoover Dam. The structure is nearly a mile across, 550 feet high, and contains enough concrete to build a highway from Seattle to Miami. For many years it was the largest hydroelectric plant in the world and has remained one of the largest since its expansions in 1966 and 1974.

During World War II, the United States bombed the Third Reich's power plants in an effort to stop Nazi Germany's development of a nuclear weapon. Safely outside the range of German and Japanese aircraft and artillery, Grand Coulee Dam and its power plant are credited with providing the massive amount of electricity needed for the Manhattan Project, the U.S. atomic bomb project that created the weapons used to destroy Hiroshima and Nagasaki, Japan. Equally important, the smelting and manufacturing of aluminum—a tremendously energy-demanding process—for the construction of bombers and other planes for the U.S. war effort was possible because of Grand Coulee's electricity generating capacity. Thus, the state as a facilitator of social change extended beyond public works to the conduct of a war that did so much to set the course of world events up to the present.[13]

Years later, the contamination of land at the Hanford Plant where much of the radioactive material for nuclear weapons was produced, and seepage of toxic waste into the aquifers and rivers, made this area an early candidate as a Superfund site. Once again, war and preparations for war are linked to the environmental crisis (Hooks and Smith 2005), as cleaning up the Hanford Plant continues to absorb hundreds of millions of dollars annually.

---

[13]Geoffrey Herrera's *Technology and International Transformation* examines how the conduct of war, state-supported scientific efforts, corporate collaboration, and the Nazi's expulsion of Jewish atomic physicists from universities and research institutes were critical to the United States' rather than Germany's developing and using the first atomic bomb. In analyzing the Manhattan Project, he concludes, "If the Industrial Revolution pushed the state into the management of industry, the atomic bomb extended that reach into science . . . The development of new technologies under the impact of the science-industry-state collaboration has become routinized and state-directed" (Herrera 2006: 183, 190).

The Judicial Road to Civil Rights

> *Under our constitutional system, especially with an activist-minded Supreme Court, the judiciary may be the most important instrument for social, economic, and political change.*[14]

> —Lewis F. Powell, Supreme Court Justice,
> 1972–1987

Among the most significant ways the state directs social change in the United States is by passing and enacting laws. Laws are interpreted and enforced, and occasionally tossed out, by the judicial system. Courts may decide cases on the basis of legal precedent and, more significantly, conformity to the nation's Constitution. Social change is continuous, so there is often some catching up to be done by legislatures and the courts (Breyer 2010) to bring into conformity peoples' practices and legal guidelines. The state is also a force for directing social change that appears held back by organizational entropy, cultural mores, and unjust institutional practices, as well as anachronistic interpretations of the Constitution (Dworkin 2008). As Justice Powell recognized a year before joining the Court, judicial decisions can be a powerful force by which the state effects social change.

An apt illustration is marriage, a status conferred for some by religious bodies but by the state for everyone. The state determines who can marry and claim the legal entitlements of marriage—tax breaks, shared responsibility for children, medical decisions for one another—to say nothing of social respectability. In increasing numbers, lower courts are upholding same-sex marriage laws and disallowing laws denying marital rights to same-sex couples, usually under the equal protection clause of the Fourteenth Amendment to the Constitution. Put most simply, "equal protection" means that the laws apply to everyone equally, including laws regulating marriage (no one can marry one's brother, mother, or six-year-old neighbor), without exceptions for sexual orientation. In this case, like so many, the courts are both catching up to contemporary practices and setting the path for marriage in the future.

Jim Crow and the State. The modern civil rights movement in the United States began well before Martin Luther King, Jr., became its de facto leader and major spokesperson in 1955 and well before Irene Morgan,

---

[14]This is from the 1971 "Powell Memo" to a friend who was then the president of the U.S. Chamber of Commerce. Available at http://reclaimdemocracy.org/corporate_accountability/powell_memo_lewis.html. Accessed May 4, 2011.

Sarah Keyes, Rosa Parks, and others refused to give up their seats on public buses to Whites and move to the back of the bus. It probably began in the 1930s, but its seeds were planted much earlier in the founding of the National Association for the Advancement of Colored People (NAACP), the experiences of black soldiers who went to Europe in the First World War, the gains of African Americans who supported and began to benefit from trade unions, and the movement of hundreds of thousands of people, most of them poor farmers from the South, to cities in the Midwest, North, and West. They worked in industry both during and in the decades after the war and found a more open, though still discriminatory, social environment that broke the sense of inevitability about the "color line" separating people and denying many of them equal rights.

From the end of Reconstruction a decade after the Civil War to the first decades of the last century, a process of legal attrition incrementally denied equal rights to black persons of African decent,[15] both formerly free and slave. Laws and judicial decisions nullified the intentions of the Thirteenth, Fourteenth, and Fifteenth Amendment, known as the Civil War amendments (ratified between 1865 and 1870) to the U.S. Constitution that ended slavery and were intended to ensure legal equality, including equality to participate in the political system.

The violence and intimidation of the Ku Klux Klan, local citizens' councils, and law enforcement personnel directed against Blacks; thousands of lynchings; chronic obstruction of justice; enforcement of white-supremacist social norms and customs; and denial of equal education, employment opportunities, loans, and the right to organize to pursue collective grievances steadily created a starkly stratified society. Jim Crow was systemic racism. Its most visible manifestations were the prejudicial beliefs, attitudes, and expressions most people associate with racism and the violence used to maintain white supremacy. The North and West were certainly not immune from Jim Crow, but the most vicious racism was in the South and border states and practiced against Blacks to an escalating degree in the late-nineteenth and early twentieth century.

---

[15]This extended to many other nationalities and ethnic groups as well; Chinese, Japanese, Jews, Irish, Italians, Mexicans, and many others were denied equal access to the political and economic systems and suffered the same kind of violence and intimidation, as well as the legal barriers discussed here. Often working together, and sharing the legal gains of one another, the end of Jim Crow is only the most visible and well-chronicled of the stories of each of these groups, to say nothing of struggles for equal rights for women, gays and lesbians, and persons with special needs.

Violence against African Americans required Blacks and liberal Whites to form the NAACP in 1909. Throughout its first decades, the sociologist W. E. B. Du Bois was a central figure in the organization that, in addition to its antilynching campaign, focused on suffrage, fair treatment in the courts, educational access, and employment opportunities. While the NAACP remained committed to all of these, its legal challenges in ending segregation and political exclusion proved its most effective tactic of contention.

Of the many grievances that might have been raised and directed to the Court, the right to vote was chosen first to test the waters of social change by challenging the nation on its own terms to live up to its vaulted words. In 1913, the general counsel of the recently formed NAACP argued before the U.S. Supreme Court in *Guinn v. United States* that laws in southern states were intentionally and effectively denying Blacks the right to vote. The Court agreed in 1915 that the so-called grandfather clause in several southern state constitutions was a "repugnant violation" of the Fifteenth Amendment.[16]

In the meantime, Jim Crow continued to be widely practiced, including in the halls of government. President Woodrow Wilson, a Democrat who identified himself as a progressive, endorsed segregation of restrooms and dining facilities in several government buildings. Segregated accommodations were extended under the next two Republican presidents, Warren G. Harding and Calvin Coolidge. In the 1920s the Supreme Court heard and rejected challenges to miscegenation laws barring marriage (often interpreted as any form of affection) between persons of color and Whites. The depth of white supremacy beliefs, discriminatory practices, and the denial of equal opportunity was greatest in the South, but throughout the country racism was a dominant ideology and fact of life. Somewhat ironically, however, the conservatism of the courts provided the openings for a successful legal challenge to Jim Crow.

---

[16]*Guinn v. United States* resulted in the suspension of state laws that required people to take a literacy test before being certified as eligible to vote if their grandfathers had not voted prior to the abolition of slavery, served in the military, or been born abroad. Obviously, these requirements—the "grandfather clause"—were designed to require a literacy test of Blacks, to the exclusion of illiterate Whites and newly arrived (white) immigrants. While not disallowing all forms of obstruction to voting, the Court ruled in *Guinn* that the grandfather clause violated the Fifteenth Amendment and offended the due process clause of the Fourteenth Amendment requiring that persons not be denied their rights without first having judicial due process to represent themselves, be found guilty, and so be rightfully denied the right to a particular status or action.

From 1880 to the 1930s, the Supreme Court, like much of the established order and elite consensus of the nation, took a laissez-faire view of economic relationships. Contracts made between consenting parties were binding, regardless of the inequality between the parties, e.g., corporations and the individual worker "contracted" to be paid subminimum wages or to work sixty or more hours a week. Despite the Progressive Movement's arguments and a handful of laws to the contrary, economic relationships were deemed outside the state's jurisdiction. That contracts might breach social customs was not for the Court to decide. If legally executed, they are binding, according to the court.

In what today might be considered a libertarian rather than a liberal decision, the NAACP's second legal victory, *Buchanan v. Warley*, was won in 1917. At the time, there were many laws barring Blacks from living in designated whites-only neighborhoods. When William Warley, a white homeowner, tried to reverse the sale of his property to William Buchanan, a black man, Buchanan sued. The state court of Kentucky decided against Buchanan, but on appeal, the Supreme Court reversed the state court. It forbade state courts from enforcing racially restrictive housing laws—therefore nullifying them—that contravened contracts between individuals. Racially restrictive neighborhood covenants were later written to obviate this ruling,[17] but the message of the courts was clear. A greater good, achieved through the free enterprise system, superseded the discriminatory practices of Jim Crow.

The Doctrine of Separate but Equal. The legal context of these and other civil rights cases was dominated by the 1896 decision of the U.S. Supreme Court in *Plessy v. Ferguson*. Jim Crow had established segregation as a white-supremacist norm of social behavior for some time. The Court's decision made this legal. Homer Plessy—a man who identified himself as a Mulatto, not a Black—was told by the conductor to leave his accustomed seat on a Louisiana train, based on a newly passed law, and move to another car designated for "coloreds." He refused and was put off the train. Plessy contended, in good economic terms, that the law degraded his personal status (and economic value as a salesman) without due process,

---

[17]In deciding *Corrigan v. Buckley* (1924), the Court upheld racially restrictive covenants, on the laissez-faire interpretation that privileges individual contracts and property rights. Not until 1948 were racially restrictive covenants determined to be "unenforceable" by courts, in *Shelley v. Kraemer*. Such covenants, often barring Jews and persons of various nationalities from buying property in neighborhoods where covenants applied, continued to be honored for many years after 1948.

violating the Fourteenth Amendment. The Court rejected his contention, establishing the legal authority nationwide that separate facilities were not necessarily unequal. "The majority reiterated . . . that segregation was an abridgement only of social equality, with which the Constitution was unconcerned" (Tsesis 2008: 128).

During Jim Crow, towns and cities in many parts of the country segregated not only public transportation but every possible public entity.

> Blacks were separated in parks and theaters, prohibited from using most white hotels and restaurants . . . Residential segregation ordinances were used to separate white and black neighborhoods. Statutory segregation covered fishing holes, boating spots, racetracks, pool halls, and circuses. In some states hospitals were segregated. Oklahoma went so far as to require telephone companies to install separate telephone booths for whites and blacks." (Tsesis 2008: 129–130)

Separate-but-equal may sound innocuous, but in the context of white supremacy, the emphasis was always on *separate*, not *equal*. Churches might seem equal though segregated, but schooling was anything but equal. In the late 1940s, Yazoo City, Mississippi, spent $3 of public money on every black student and $245 on every white student (Payne 1995: 42). Three percent of Kentucky's higher-education budget went to black schools; 97 percent went to white schools. In 1947, Florida spent barely a third ($390) on each black undergraduate's college education of what they spent for each white undergraduate ($1,220).

Building Legal Challenges to Jim Crow. The NAACP authorized a study of unequal education in 1931, *The Marigold Report*, that documented wide gaps in spending between racially segregated schools. In 1934, one of the NAACP's most important legal counsels and dean of Howard University's law school, Charles Hamilton Houston, took several of his young protégés, including Thurgood Marshall, the man who twenty years later would argue the NAACP's most important legal decision, *Brown v. Topeka Board of Education*, on a trip in the South. They documented with data and photographs the vast inequality of schools. *The Marigold Report* reflected the liberal belief that setting forth the facts was the most important step in fighting injustice. When people knew the truth, they would do the right thing.

Charles Houston's legal strategy was to chip away at segregation by first attacking it in higher education. Graduate education and particularly law school admissions yielded progress in the 1930s and '40s. Houston's first victory was in *Murray v. Pearson* in 1935. The decision required

Maryland—because the state had no law school for anyone who was not white—to admit a black applicant, Donald Murray. Three years later, in *Missouri ex rel. Gaines v. Canada*, Lloyd Gaines won his suit for admission to law school at the University of Missouri. The school agreed that it had refused Gaines admission solely because he was black, but Missouri had no separate black law school. It therefore proposed to send him to another state until it could build a black law school. The Court ruled in Gaines' favor.

In 1940, Thurgood Marshall, leading the NAACP legal team, successfully argued his first case before the U.S. Supreme Court. He gained the dismissal of a murder conviction based on confessions achieved through torture. Beatings of suspects to gain confessions were common, especially when the accused was black. In *Chambers v. Florida*, the Court took the unusual step of countermanding lower courts' verdicts on behalf of the plaintiffs. The case received widespread publicity and exposed many Whites to previously unrecognized physical abuses of Blacks that they would not tolerate happening to themselves.

Also in 1940, A. Philip Randolph's threatened march on Washington to demand fair employment practices for defense contractors, discussed in Chapter 6, caused President Roosevelt to create the Fair Employment Practices Committee. The committee was replaced in the Eisenhower years with the Fair Employment Board and a decade later by the Equal Opportunity Employment Commission. The mandate of these agencies was specifically to investigate and remedy cases of racial bias in employment, one of the major issues of the modern civil rights movement.

During these years, the Supreme Court handed down other decisions that whittled away at the political barriers of Jim Crow. The South since Reconstruction was dominated by the states' Democratic Parties, staunchly opposed to anything that would reduce white supremacy. It was a region of a single party, and to be barred from primaries that selected party candidates was tantamount to being barred from running for office, i.e., political participation. In addition, voting in the general election often required voting in the primary, an impossibility for most Blacks. By deciding *Smith v. Allwright* in favor of Donnie Smith's 1944 suit against an election official in Harris County, Texas, the Court rejected the idea that a political party primary was equivalent to a club meeting that could freely choose who to admit and who to bar from participation. This and related suits continued to lay the foundation for the voting rights legislation to come.

A Rights-Centered Liberalism. Most early twentieth century cases involving the denial of rights had been argued in terms of due process and its effect on interstate commerce. After 1937, the Court adopted what Kevin McMahon (2004) calls a "rights-centered liberalism" that interpreted the Constitution as a document of individual rights, including rights denied under Jim Crow.[18] In order to foster a healthy democratic society, the Court said, these rights should be considered paramount. Increasingly, segregation was recognized for what it really was (in its empirical manifestation) as exclusion from participating in ways promised by democracy. Racial segregation was a well-documented personal and collective humiliation and degradation that required violence and intimidation to maintain it. Both were anathema to the rule of law.

President Harry S. Truman, who succeeded to the presidency when Roosevelt died at the start of his fourth term, supported the expansion of the rights of Blacks and was less concerned than Roosevelt about the political ramifications of his actions.[19] For Truman and many Americans, the Second World War was fought for freedom's sake. It opposed the fascist ideology that led to the systematic murder of more than seven million people, mostly Jews but also hundreds of thousands of Roma (gypsies), homosexuals, the disabled and mentally ill, and others whom the Nazis deemed racially, mentally, and physically inferior.

In 1946, Truman formed the Civil Rights Section (later Division) in the Department of Justice. This was significant, not because it initially investigated and tried many cases. It didn't. But it provided a means by which the federal government could intervene in issues previously considered off limits and in matters of local jurisdiction. Henceforth, a spotlight would be focused on

---

[18]In his first years in office, Roosevelt strenuously criticized the conservatism of the majority of Supreme Court justices who opposed his New Deal legislation. After threatening to add justices to the nine-member Supreme Court, Roosevelt's popularity was challenged. But in time his New Deal efforts and the extended length of the Great Depression won out. After Roosevelt was able to make some new appointments, the Court changed the way it decided cases involving discrimination, increasingly recognizing the rights of Blacks and other minorities (McMahon 2004).

[19] Roosevelt recognized the threat to the Democratic Party's stability if the conservative southern Democrats were to bolt because of federal support for the dismantling of Jim Crow. His failure to sign antilynching legislation and to integrate the military during World War II, along with other missed opportunities, are often cited as evidence both of his lack of sensitivity to the plight of African Americans and his concern for the political impact of addressing their needs. McMahon (2004) argues that Roosevelt's broader agenda, however, paved the way for the erosion of Jim Crow.

egregious cases, especially violence against Blacks in the waning days of Jim Crow.

The Terror of Lynching. The most hideous form of Jim Crow violence was lynching. The particulars are ugly and gut-wrenching in their brutality and the mob violence that often accompanied them.[20] In the southern states after Reconstruction (1877), approximately a hundred men and women were lynched every year prior to 1900. In one state alone, Mississippi, twenty-five people were lynched in 1889 and twenty-four in 1891. Lynching declined somewhat, to an average of seventy annually in the southern states between 1900 and 1920 (Rapper 1969), but lynchings became more brutal, more often involving mutilations and "highly inventive forms of torture" (McMillen 1989: 232).

After the turn of the century, two periods, 1904–1908 and 1918–1923, saw spikes in the number of lynchings. In the first period, someone was lynched every twenty-five days, on average, in Mississippi. Across the South, sixty-one Blacks were lynched in 1922 and twenty-three the next year, a decline attributed in part to anticipation of a federal antilynching law that never materialized (Payne 1995: 19). In 1935, eighteen Blacks were lynched in the United States, but fewer than nine people died by lynching in each of the next twenty years. The horrendous crime of lynching was increasingly confined to a few states.

The higher price of cotton (there is an inverse relationship between cotton prices and the annual number of people lynched) and the threat of federal intervention have been credited for fluctuations in lynching, but the overall decline is probably attributed to the dwindling need to control the labor of Blacks, accomplished at its most extreme by the threat or commission of lynching (Payne 1995: 17). Cloth made of synthetics was finding its way into the market, and cotton farmers turned to other crops requiring less labor. Tractors, flame cultivators, cotton harvesters, and other machinery reduced the need for farmhands. At the same time, young Blacks were growing increasingly hostile to Jim Crow and were more likely to avoid contact with Whites. Rarely overt, their more subtle ways of "pushing back" created a growing hesitation by Whites to attack them with impunity.

---

[20]Arthur Rapper's classic study in 1933, *The Tragedy of Lynching*, compiled not only narrative accounts but statistics that shocked many people and brought lynching to a new height of public awareness. Neil McMillan's *Dark Journey* (1989) provides an excellent chapter on lynching, "Judge Lynch's Court."

POLITICAL GENERATIONS IN THE MODERN CIVIL RIGHTS MOVEMENT

"In countless ways their self-respect and pride receives daily blows . . . They are hurt in ways their grandparents could not have imagined." This observation by Hortense Powdermaker (1968: 333), a social anthropologist who did research from 1932 to 1934 in a place she called Cottonville (Indianola, Mississippi), poignantly captures the generational changes of Blacks during Jim Crow. Her work was among the first to chronicle the evolving attitudes and changing self-image of those dominated and demeaned by segregation and discrimination in locales where white supremacy reigned. It was in Indianola that the first Citizens Council was formed, one month after *Brown v. Board of Education*, to reject and oppose the intentions of racial equality.

Powdermaker didn't try to disentangle age, period, and cohort effects in analyzing the differences in generations of those she studied. Her work and that of others, however, in contrasting three generations, makes apparent the impact of World War I, the decreasing isolation of the rural South and Delta region, and the exposure to the growing contrast between the values the United States purported to espouse and the reality of black life in places like Indianola.

The oldest generation, elderly even when Powdermaker did her work and captured in the audio recordings of the WPA's Federal Writers' Project discussed in Chapter 2, were born in and soon after the abolition of slavery. Neil McMillan speaks of this generation in terms of tenancy, a version of the antebellum plantation system that held black and poor white farmers in economic bondage. The difficulties of purchasing or even renting land left poor farmers with little recourse but to be tenant farmers, working for and splitting the annual harvest with the landowner. This generation expressed no nostalgia for the old days; "Just the same as his slaves" is how one older person described it (McMillan 1989: 124).

By and large, the oldest generation was accommodating to white supremacy. They spoke to Powdermaker of the good white people on whom they could depend for help. Lacking confidence and often illiterate, their opportunities had been so constricted there was little reason to chafe at what could not be changed.

For adults born soon after the turn of the century, the middle generation, "servility was widely regarded as a loathsome but necessary act" (McMillan 1989: 27). Somewhat better educated than their parents, the people of Indianola knew, from those who had migrated north and the veterans who had seen a different world during the First World War, that Jim Crow was indeed loathsome practices and beliefs.

The middle generation recognized that fear among Whites was usually at the core of their violence against Blacks, and this fear helped frame daily restrictions and exclusions. Powdermaker found this generation of middle-aged adults tactful, diplomatic, and often silent in the presence of Whites. They had assimilated the creed of equal opportunity and deeply resented its being denied them. While their children's less tactful expressions of the same resentments caused them to worry for their safety, "many parents found it unacceptable" to pretend subservience to Whites (McMillan 1989: 27).

The youngest generation Powdermaker studied, those born after World War I who would later serve in the Second World War, participated in an even greater migration to the North and West. They took the first direct actions that became the modern civil rights movement. These young people often kept away from Whites rather than act as expected. In confidence and among one another, they readily expressed their opposition to Jim Crow and looked forward to a time when it would be a thing of the past. Their parents "could hide what their children can't" (Powdermaker 1968: 333).

Whites' attitudes continued to remain stagnant, and they failed to perceive the impending changes, as Powdermaker's survey of Whites' attitudes attests. Not so with Blacks. Either things would change, a cataclysmic conflict would ensue, or perhaps both.

Scholarship of the modern civil rights movement, such as Taylor Branch's *Parting the Waters* (1988) and Kenneth Andrews' *Freedom is a Constant Struggle* (2004), describe the emergence of a fourth generation, born in the late 1930s and 1940s, that might be considered the second half of the Depression and War Babies generation. These are the young people who took charge when direct action was needed following *Brown*. They joined the movement in order to confront white resistance and the reactionary conservatism of southern racists as well as the passivity of the Eisenhower administration and go-slow attitude of most Whites in the 1950s.

They formed the Student Non-Violent Coordinating Committee (SNCC) and the Congress for Racial Equality (CORE) when the NAACP

(Continued)

(Continued)

and the Southern Christian Leadership Council (SCLC) counseled restraint and the need to work "through the system." This generation and more aggressive members of their elders' generation—people like Bayard Rustin—pressed their challenge of Jim Crow practices. They initiated sit-ins and marches, freedom rides, and not only voter registration drives but a presence at the polls to ensure that Blacks could vote without intimidation. Black pride, and what looked to many Whites like radical militancy, expressed a refusal to accommodate, be silent, wait, and moderate their insistence on social justice.

Racism in the Political Landscape. Hortense Powdermaker's *After Freedom* documents how virulent racism in the South—including approval of lynching as a useful device of white domination—continued in the 1930s. The isolation of the South was ending, however, and the conservatism of the South's Democratic legislators and courts was increasingly a national embarrassment on the world stage, something President Truman took very seriously. With the formation of the United Nations and its Declaration of Human Rights, the statutory discrimination against African Americans was increasingly objectionable. U.S. Cold War propaganda, contrasting the freedoms of the United States with dictatorships and communist-party rule in the USSR, was readily countered by Soviet propaganda about racism in the United States.[21] As discussed in Chapter 5, black migration was also creating political opportunities for Blacks as electoral politics outside the South increasingly had to take their vote seriously (McAdam 1999a).

In 1948, President Truman addressed Congress to urge passage of civil rights legislation: making lynching a federal crime, protecting voting rights, passing laws against discrimination in commerce and employment, and creating a permanent civil rights commission. Conservative legislators balked, and the legislation was not enacted. Undeterred, Truman used the executive power over agencies and the Pentagon to achieve many of his aims, including the end of segregation in the armed forces.

The power of southern Democrats was not the only reason Truman's proposed legislation failed. In the 1930s and '40s, white public opinion

---

[21]Nearly one hundred years earlier, in 1854, Abraham Lincoln expressed the same sentiment, calling slavery a "monstrous injustice . . . [that] deprives our republican example of its just influence in the world—enables the enemies of free institutions, with plausibility, to taunt us as hypocrites" (quoted in Foner 2010).

nationally supported segregation, especially the forms that would prevent intimacy between young Blacks and Whites. A 1948 Gallup poll found that 82 percent of those asked opposed Truman's civil rights program (Tsesis 2008: 240). His agenda caused southern legislators to bolt the Democratic Party and form the Dixiecrats prior to the 1948 presidential election. Their candidate, Strom Thurman, ran for president. Despite this, Truman won and devoted his 1949 State of the Union speech to his civil rights program.

The End of Separate but Equal. The earlier decisions in *Murray* and *Gaines* seemed to mark the end of racial segregation in higher education, but in actuality didn't. Ada Lois Sipuel was admitted to the University of Oklahoma School of Law but then assigned to Langston University in Oklahoma. She objected and filed suit. Upon appeal, the Supreme Court agreed, in the 1948 decision *Sipuel v. Board of Regents of the University of Oklahoma*, that she was denied an educational opportunity. She began law school with white classmates at the University of Oklahoma. The university, in retaliation, assigned her to a seat with a "colored" sign on it and would not allow her to eat with the other students in the school cafeteria, a ludicrous and insulting measure often violated when her white classmates joined her.

Like Ada Sipuel, George McLaurin won his case in district court to attend graduate school at Oklahoma State University. McLaurin was seated, however, outside the classroom and in designated areas of the library and dining hall, a ridiculous affront.[22] He appealed to the Supreme Court, and in *McLaurin v. Oklahoma State Board of Regents* (1950) the Court rejected this treatment. The decision was handed down at the same time that a companion case, *Sweatt v. Painter*, was decided.

Herman Sweatt was refused admittance to the University of Texas Law School because Texas law barred Blacks from attending the school. The Texas court delayed a final decision (to admit him to the University of Texas) and gave the state time to create a separate law school in Houston, now Texas Southern University. Sweatt appealed to the Supreme Court on the grounds that the schools were not comparable, and he won.

Brown v. Board of Education. The days of segregation were numbered, but a definitive, more encompassing decision was still needed to dismantle segregation in all walks of life. That came in 1954, in the most famous civil

---

[22]Much of the civil rights movement resonates in photographs and documentary films. *The Road to Brown* chronicles many of the cases and personalities described in this account of the movement.

rights case of the twentieth century. At the time, seventeen states required segregation of educational facilities in public schools, all of them in the Deep South and border states, including Missouri. Eleven states, most of them in the Great Plains and the West, had no legislation, and sixteen forbade segregated schools. In four states, including Kansas, segregated schools were required at the lower grade levels in larger towns.

The NAACP, with future U.S. Supreme Court justice Thurgood Marshall arguing the case, represented nearly two hundred plaintiffs in four states—Kansas, Virginia, South Carolina, and Delaware—and the District of Columbia. Of the five suits, the Court selected the Kansas case in issuing a decision, *Brown v. Topeka Board of Education*, that applied to all the cases. Oliver Brown, Linda Brown's father, was the lead plaintiff who was suing to have his daughter attend the school nearest her home, rather than having to ride a bus to attend a black elementary school. It was not contested that her assigned school was inferior, and in Topeka black and white students attended integrated high schools. Topeka's high school had been integrated for more than half a century, though not in athletics, as described in Jack Alexander's reminiscence quoted at the beginning of this chapter.

President Dwight D. Eisenhower, hero of World War II and a moderate Republican, had recently appointed former governor of California Earl Warren as chief justice of the Supreme Court. With his appointment and three Southerners on the Court, segregationists believed they had a good chance to win the case. Seeing the decision coming, many states that practiced segregation upgraded black educational facilities in order to make them more like white schools, but they were far from being equal, and it was too late.

Thurgood Marshall argued that at issue was the experience of segregation itself. In focusing on elementary school students, he was challenging the entire edifice of segregation in everyday life, not just the specialized settings of graduate and law schools. Segregation, he argued, might appear to restrict both the dominating group and the subordinated group to their own institutions, but the cost of segregation was born by the subordinate group. Racial segregation unconstitutionally privileged one race over another.[23]

---

[23]Martin Luther King, Jr., expresses the humiliation of segregation in his account of a train trip he took as a teenager. "I had to change to a Jim Crow car at the nation's capital in order to continue the trip to Atlanta. The first time that I was seated behind a curtain in a dining car, I felt as if the curtain had been dropped on my selfhood. I could never adjust to the separate waiting rooms, separate eating places, separate rest rooms, partly because the separate was always unequal, and partly because the very idea of separation did something to my sense of dignity and self-respect" (Carson 1997: 11–12).

Again, empirical facts were at the heart of the case, not merely ideas and concepts divorced from social reality. The use of social science research in the case was significant (Tsesis 2008: 252 ff.). Relying on the equal protection clause of the Fourteenth Amendment, the Court in a unanimous decision agreed with Marshall and struck down *Plessy v. Ferguson* as legal precedent. Segregation in any form was now against the law everywhere.

Voting Rights and Violence. For generations, violence and intimidation, as well as literacy tests, the poll tax, and denial of participation in party primaries had kept voter registration among Blacks extremely low and actual voting even lower. Across the South, most Blacks had lost the right to vote after the collapse of Reconstruction. Laws in the 1880s and 1890s codified this after violence and intimidation had made it a fact of life. In 1940, only 3 percent of southern Blacks were registered to vote. By 1947, the figure was 12 percent and by 1952, 20 percent (Payne 1995: 36). Voting by Blacks in Alabama increased from six persons to fifty thousand, and in Louisiana it went from ten to 100,000 between the late 1940s and early 1950s. This was a small fraction of people who could have been voting, however, and in other southern states black voting numbers were even lower.

Of the 39,000 Blacks living in Leflore County, Mississippi, in 1946, only twenty-nine were registered, and none voted. Humphrey County, also in the Delta region of Mississippi, was almost two-thirds Black. Not one African American had voted in the county since Reconstruction. Margaret Price gathered voter registration data for Mississippi in the mid-1950s and found that "as late as 1954, in the thirteen Mississippi counties that had majority Negro populations, a *total* of fourteen votes were cast by Blacks in that year's election" (Payne 1995: 26).

The 1944 Supreme Court decision against all-white primaries gave hope to Blacks, and the 1954 *Brown* decision seemed to be a turning point. Still, Blacks knew well that white supremacists would not go away quietly, and they didn't. Fearing the impending changes, violence by Whites was carried out throughout the Deep South.

Because it argued the most visible and successful cases, the NAACP was targeted. Its members were intimidated, and its leaders were beaten and murdered. Membership dropped precipitously (losing 48,000 members between 1955 and 1958), and voter rolls began to shrink. Civil rights participants were denied loans, threatened with bodily harm, and lost their jobs. Leaders were beaten, had their houses bombed, and were murdered. Geoffrey Payne (1995) documents the violence in Mississippi in the years following *Brown*, violence repeated in Alabama, Georgia, and other southern states.

Myrlie Evers, widow of Medgar Evers, writes that she remembers the fifties by the names of the victims—1956, Ed Duckworth shot to death, Milton Russell burned; 1957, Charles Brown killed in Yazoo City; 1958, George Love killed by a posse, Woodrow Wilson beaten to death by a sheriff; 1959, Jonas Causey killed in Clarksdale, William Roy Prather, a teenager, killed in a Halloween prank, Mack Charles Parker dragged from a jail cell and lynched. (Payne 2005, 53)

The Justice Department's Civil Rights Division played something of a dampening role by investigating crimes the southern states would not act on, and federal legislation expanded their authority. While Eisenhower used federal troops in the South only once, in 1957 in Little Rock, Arkansas,[24] the Justice Department was increasingly active. It investigated violence and relied on the interstate commerce clause to support integration of public transportation. Its actions were limited, however, and the persistence of segregation in interstate travel in the South prompted the Freedom Riders in 1961. Their insistence on their right to travel by bus evoked violent resistance by segregationists and became an indelible part of the civil rights legacy. The three 1964 Selma-to-Montgomery marches, known as Freedom Summer, resulted in six hundred activists being beaten by local police and Alabama state troopers and marked the pinnacle of the civil rights struggle for voting rights (McAdam 1988).

Over the years, black voter registration had steadily increased, but only with the final successes of the civil rights movement and passage of the Voting Rights Act of 1965 did the numbers begin to substantially reverse the denial of black political participation (Andrews 2004: 110–111) and bring political freedom to all of the country's citizens.

Moving Beyond *Brown.* When she didn't like something, Iris Summers would say, "There oughta be a law against it." Passing or affirming a law, however, doesn't ensure that things will change without a struggle. The Court's decision in *Brown* became the law. It was the basis for striking

---

[24]The plight of the Little Rock Nine provided a vivid picture to the entire nation of segregationist resistance, now kept abreast by television. Three years after *Brown*, Arkansas governor Orval Faubus, an opponent to school integration, reluctantly called in National Guard troops to protect nine young Blacks who sought to begin school in September at the all-white Central High School. Under public pressure, he withdrew the troops. An angry mob attacked the remaining local police and black bystanders, forcing President Eisenhower to send an Airborne Division and federalize Arkansas National Guard troops in order to stem the violence and ensure the safety of the students to attend school.

Selma-to-Montgomery "Freedom Summer" marchers. Photographer, Charles Moore.

Source: Black Star http://www.blackstar.com/

down thousands of statutes, but it was another ten years before federal civil rights legislation would firmly advance social change.

The Court's language in *Brown* had been imprecise in securing an end to segregation. During the second phase of the case, it agreed that the dismantling of segregated schools should proceed "with all deliberate speed." This was taken to mean "delay" by segregationists who vehemently opposed the *Brown* decision. "On the national scale, many a segregated school district exploited this ambiguity to stall implementation" (Tsesis 2008: 256). The "Massive Resistance Campaign" in Virginia, orchestrated by U.S. senator Harry Byrd, got the state to pass laws that denied funds for integrated schools, closing many schools for a year or more. Families were given vouchers to attend segregated private schools, a move that proved temporary and was overturned by the courts in suits often brought by white parents who objected to the closing of their children's schools.

In Mississippi, Citizens' Councils were formed a month after *Brown* and were soon supported by the new State Sovereignty Commission. State legislation was passed over the next few years to erect a barrier to integration. In

1956, state offices were "directed and required" to block implementation of federal civil rights decisions (Andrews 2004: 156). Eight years after *Brown*, in 1962, Governor Ross Barnet personally sought to prevent James Meredith from attending the University of Mississippi. Only in 1964, after a federal court ordered the integration of the state's schools, was the tide of social change recognized as the law of the land.

The hold southern Democrats had on Congress, often using the filibuster, began to weaken as some border-state legislators broke ranks and rule changes reduced southern Democrats' power in Congress. The 1948 Dixiecrats rebellion from the Democratic Party gave more power to liberals like Hubert Humphrey who pushed the Democrats to embrace a civil rights agenda. Illinois senator Adlai Stevenson, an advocate of integration, challenged Eisenhower in the 1956 presidential race. Until the mid-1960s, when southern white conservatives left the Democratic Party and began to exert power in the Republican Party (Phillips 1969), they were pursuing a rearguard action with less and less success.[25] It was an FDR-inspired Democrat from Texas, Lyndon Johnson, who as president championed and muscled through the passage of the 1960s civil rights legislation.

From the mid-1950s onward, the civil rights movement used legal challenges, legislation, and the enforcement powers of the federal government to force an end to segregation and the discriminatory treatment against persons of all ethnic groups, but especially Blacks. Known as the King Years (Branch 1988), the violence did not abate with the *Brown* decision. The story continued with the murder of civil rights workers, the bombing of churches and homes, and the politics of states' rights. Later debates over school busing and affirmative action and the use of coded language by political entrepreneurs, e.g., *law and order*, *welfare cheats*, and *reverse discrimination*, expressed racial prejudices or widespread anxiety over changing social relations that influenced and in some cases dictated politics for decades.

Undoubtedly, the federal government and courts could have acted sooner and more effectively to end the scourge of Jim Crow. The price in lives lost and human possibilities diminished was much greater than it should have been. A democracy owed more to its people than to tolerate as long as it did the crimes great and small of white supremacy. But the state did act, largely through decisions of the U.S. Supreme Court and civil rights legislation, to dismantle Jim Crow and consign it to the dustbin of history.

---

[25]The political calculation of losing the South to the Republican Party accounted for President Kennedy's reluctance to send in troops. As Kevin Phillips (1969) explained, the Democrats' support for civil rights provided the Republican Party with a "southern strategy" that has created a largely white Republican Party. There were no black Republicans in Congress between 2003 and 2010.

State-Driven Social Change in Modern China

*A new theory on the need for states at China's level of develop-
ment to adopt a "neo-authoritarian" form of government had
been approved by Deng Xiaoping. He believed that "the modernization process in a backward country needs strong-man
politics with authority rather than Western-style democracy as a
driving force."*

—John Gittings (2005: 225), quoting
Prime Minister Deng Xiaoping

Two Versions of Democracy

State-driven social change is a worldwide phenomenon, especially in
the twenty-first century. How the nearly two hundred nations of the
world define themselves as democracies distinguishes how the state seeks
to effect change. Capitalist democracies that emphasize private economic
activity and market forces see themselves as democratic in terms of the
process by which state power is used. The three cases discussed earlier—
public health, public projects, and civil rights—can be read this way, with
various constellations of power and interests vying for the attention and
resources of the state.

In comparison, socialist systems and other "command economies"
entrust economic decisions to the state, which directs economic operations
much the way corporate CEOs direct managers to carry out their decisions.
Their idea of democracy focuses on *outcomes*. That is, the society's resources
are developed and distributed in terms of the society's needs—as determined
by the state. In China's socialist[26] system of the Mao Zedong era (1949–
1976), the emphasis was on democratic (i.e., socialist) outcomes: using state
power to build a modern nation on behalf of the hundreds of millions of
people who needed education, health care, housing, and security.

---

[26]The concept of socialism is a very negative one for Americans who have been
encouraged to equate it with authoritarianism. Of course, capitalism can also be the
dominant economic system of a nation with an authoritarian political system, and
socialism can be the economic system of a nation that makes political decisions
democratically. A nondemocratic political system where the economy is in private
hands (capitalism) is often associated with fascism but can take the milder forms of
what Zakaria (1997) calls illiberal democracy and Yiftachel's (2010) characterization
of ethnocracy. Conversely, a democratic political system can exist in a nation where
there is widespread public ownership, i.e., a socialist democracy.

Process Democracy. Rural households in the United States had no electricity many decades after European farmers had electric lights. In Europe it was seen as a matter of fairness. If families in cities had electricity, so should families in the countryside. Progressives in the United States for years talked of the economic costs of out-migration of farm families and the many ways productivity was compromised due to a lack of modern technology and lights that would provide a longer workday. Electricity for rural families was framed as a solution to problems, not a matter of fairness or equity. Progressives offered solutions through the political system, but because commercial interests are powerful forces in what the state does, corporate power was needed to pass the legislation. Rural electrification became a reality only when manufacturers of electrical appliances and machinery, like GE and Westinghouse, got behind rural electrification. By then, the state had gone ahead with hydroelectric projects like the TVA and Hoover Dam and created electricity that could be distributed for a profit by private companies (Nye 1990).

In capitalist democracies, the *process* includes popular elections of individual representatives who occupy law-making positions in the government. To some degree, these individuals must remain accountable to the popular will, expressed as public opinion and voter preferences. Courts are guided by the rule of law, and though some judicial positions may be elected, they need not be so long as individuals are accountable for their actions. But the process of democracy is more than the composition and practices of the government.

Citizenship in democracies is available to all, without ethnic, national, religious, gender, or class qualifications. It is a right that carries responsibilities to uphold the law, be informed about political matters, and engage with the political system when needed. Democracies work best when a professional military is subservient to the civilian government and does not succumb to the temptation to overthrow the government when it performs poorly. A free press is also vital as a check on the inevitable desire of the state for secrecy and the ability to hide its mistakes. In a democracy, legitimate power rests with the office, not the person. It is a breach of the public's trust for individuals, families, or group of families to use political positions for personal gain or as family heirlooms to be passed from parents to their children.

As Alexis de Tocqueville (2000) recognized a hundred and eighty years ago, democracy also rests on a set of underlying beliefs, an ethos, about the rules of the political process. These are the social norms, attitudes, and values of political culture and include such things as the belief that electoral winners are rightfully the representatives in government, at least until the next election. Political culture sets informally agreed-upon limits on what is beyond fair play in electoral contests, how government representatives should balance leadership with their being a representative of the public, and a sense of common interest and purpose among all citizens.

Between the individual and the democratic state is civil society, that broad and varied band of activities and associations of people pursuing interests normally not the province of government (Alexander and Smith 1993). They enter the political domain to voice opposition or support, muster resources, align with interested parties, and exert whatever power they can to affect the actions of the state. In a capitalist democracy, civil society itself does many things that in other countries are done by the state. For example, organizations created by the corporations themselves may be charged with regulating some corporate activities. In addition to local, state, and federal governmental social services, the poor and homeless may be supported by food pantries, homeless shelters, churches, and others who have no ties to the state. Health and safety measures are often devised by university scientists and industries as part of their research agenda and then adopted and enforced by the state.

Democratic Outcomes. Just as societies with capitalist economies may be more and less democratic, socialist systems may or may not involve democratic process. The former Yugoslavia prided itself on a governmental structure that gave every person two ways of being represented, as a citizen of a geographic area and as a member of the economy. In addition, there was considerable democratic participation in the workplace (Singleton and Topham 1963), something rarely found in capitalist democracies.

More often, socialist states are dominated by a strong, even authoritarian, political center. Whether African, Asian, or Latin American, the state is ideologically committed to allocating the valued goods of the nation broadly but by a process that may involve little popular participation. The priority for solving problems is determined not by broad public participation but by a single party guided by the ideology of a socialist alternative to capitalism.[27]

---

[27]Capitalism is understood by socialists to be a chaotic system that trusts in positive social outcomes through a process that expects individuals to pursue their own selective interests. A famous quote attributed to John Maynard Keynes resonates with socialists: "Capitalism is the astonishing belief that the most wickedest of men will do the most wickedest of things for the greatest good of everyone" (quoted in Albert 2000: 128). For example, prices set by markets are irrational when this puts out of reach those things most needed by the least advantaged, e.g., health care, housing, and public safety. Many socialists deny the possibility of "free markets," believing the private economy of capitalism dominates the state and is able to set prices and obtain policies to benefit the corporate elite.

The historical lesson has been that administration "on behalf of the people" has usually, if not always, devolved into a monopoly of power that in time locks most of the people out of any significant role in governance. If everyone benefits, however, this centralization of state power is not recognized as a critical flaw or a violation of democratic ideals, based as they are on outcomes rather than process.

A socialist economy doesn't necessarily eliminate all private property but limits it to individual and family property for their immediate use rather than investment. Property is outside the political sphere, not a tool to leverage political influence. If this sounds idealistic or utopian, that may be why there are so few socialist systems operating in the world today. It is important to keep in mind the differing versions of democracy, however, to understand the transition of China from authoritarian socialism to one-party state capitalism or what some term market authoritarianism (Cassidy 2010: 97).

Mao's Revolutionary China, 1949–1976

The humiliation of colonial power domination that exposed the military and economic weaknesses of China more than a century ago thrust the country on a path of nation building, sweeping away much of its past traditions and social structure. A brutal and prolonged civil war early in the twentieth century, the Japanese invasion and World War II, and a thoroughgoing revolution that brought Mao Zedong and the Chinese Communist Party (CCP) to power marked China's tumultuous first five decades of the twentieth century.

Under the banner of socialism, the leaders of the Chinese Revolution entrusted the CCP, a single-party political apparatus, to build a powerful nation and quickly overcome the poverty and international weakness endemic to nonindustrialized nations. As the world's most populous country, this was a formidable task. It was made even more difficult by the Cold War rivalry between the USSR and the United States. Anticommunism on the part of the United States, a staple of the Cold War, threatened China's national security and limited its network of political and economic partnerships.

China's revolutionary leaders built a strong state that penetrated every aspect of life and charged the CCP with carrying out an ambitious agenda of economic and social change. Though calling itself communist, China pursued a path of economic growth that was first and foremost nationalist, i.e., committed to building China. Its leaders saw in the public control of the means of production—including industry, trade, and agriculture—the most rational and efficient way to both mobilize resources and harness the enthusiasm of the people for the hard work of nation building.

The "Correct Path" of Chinese Socialism. Can there be socialism without there first being capitalism? Capitalism is an engine for growth and development, a necessary historical stage. Marx and Engels were enthusiastic about the possibilities of science and technology in nineteenth-century Europe, developed and exploited in the entrepreneurial drive for private gain. The feudal social order of peasants and gentry was abolished by capitalism. Replacing it were social classes of bourgeoisie (owners) and proletariat (workers), a structure of inequality that, when it exploded, would make possible a classless society. Once industrial capitalism had taken its course, revolutionaries—with the enthusiastic support of those who had the most to gain, the masses—could socialize, for the benefit of all, what capitalism had made possible.

One question was frequently asked within the CCP leadership of the Mao era: is ours the "correct path"? China seemed to defy the formula of historical social transformation. It had never been capitalist. The Chinese Revolution seemed to be skipping over the capitalist stage. It would have to benefit from the technologies of industry, finance, and administration developed by capitalism in Europe and the United States as well as learning from the first communist revolution in Russia. Skipping the stage of capitalism seemed not only possible but necessary in order to "catch up" and resume China's rightful place as a powerful and advanced nation (Jacques 2009).

Democratic outcomes required solving the problems of hunger, inadequate education, health insecurity, poor housing, backbreaking drudgery, and thwarted life chances for women, the poor, and others in positions of long-standing subservience. When Mao Zedong died in 1976, no one in China believed that the nation, in one generation under Chairman Mao, had achieved the high ideals of human equality and the satisfaction of needs that characterizes the goals of socialism, but great progress had to be made. The state's obligation in the post-Mao era was to continue this progress by expanding economic growth and national sovereignty. It has done this in ways quite different from what Mao envisioned on the eve of the Chinese Revolution.

Great Leaps and Stumbles

> *A revolution is not a dinner party, or writing an essay, or painting a picture, or doing embroidery . . . A revolution is an insurrection, an act of violence by which one class overthrows another.*
>
> —Mao Zedong (1970)

Revolutionary change, as discussed in Chapter 6, offers hope to the victor and often unleashes popular enthusiasm by those convinced that a new day will be theirs. China was no exception. Following the 1949 establishment of the People's Republic of China, the peasants embraced the possibilities of a new society by organizing massive flood control projects, public health cleanup efforts, and literacy campaigns. Traditional rural inequality was attacked and replaced by a meritocracy based on commitment to the revolution and personal effort.[28]

In 1955, the CCP began to construct largely voluntary agricultural cooperatives and in 1958 replaced these with People's Communes that effectively required land to be collectively farmed. It organized peasants into groups of villages, using economies of scale and a more industrial approach to farming to increase agricultural production. Peasants owned their homes and kept fields and gardens for household consumption, but all other aspects of agriculture were guided by the party, including production quotas, market prices, and the allocation of profits back to the communes and villages.

In 1956, Mao announced the Great Leap Forward, and again the peasants especially—more than three-quarters of the country's population—responded enthusiastically. The projects to clear land, create irrigation systems, and till more aggressively were poorly designed, however. This, combined with poor weather conditions over the next several years, led to a "demoralizing disaster," in the words of John Gittings (2005: 35), in which millions of people starved to death. By 1961, the consequences of the Great Leap Forward were clear, and the party reassessed China's path of social change. It would be twenty years before rural collectivization was dismantled, largely leaving the peasants who today number approximately 450 million to the vagaries of the market.

Political Fissures I: Revolution Versus Bureaucracy. As is always the case in postrevolutionary situations, there is a tension between those who make the revolution and those who consolidate power and create the apparatus to exercise social control and, in the case of China, guidance for the new society. Isaiah Berlin's (1969) metaphor of foxes and hedgehogs is one way to think about this difference. Revolutionary activists (foxes) must improvise and boldly lead, sometimes unpredictably, into new realms. Those who take up the job of implementing the new society (hedgehogs) are more plodding,

---

[28]Mao believed the greatest barrier to China's success was not a legacy of colonialism or a history of capitalist exploitation but its own feudal inequality, i.e., the structures of power, wealth, and customary status that allowed a privileged minority to live in material comfort at the expense of everyone else. William Hinton's account (1966) of the Chinese Revolution as it took place in rural villages, *Fanshen* (Turning Over), provides a fascinating look at the Maoist efforts to transform class relations and the distribution of power.

circumspect, and determined leaders who understand bureaucracy and more patiently work through challenges by mustering the power of organizations. Mao Zedong was much more a fox than a hedgehog.

Mao was the undisputed leader of China, revered by the people and respected by the CCP leaders for the power he wielded. The state was dominated not only by Mao but by the upper echelons of the party. Its control extended deep into every province, city, town, and village through a maze of organizations arranged hierarchically, with directives and most communication going from the party center to the periphery. Mao gave direction to the party's work and could be ruthless when his ideas were not followed. It was left to others to enact the policies. These were discussed and disputed outside of public view by the core of those who had led the revolution alongside Mao, increasingly powerful military figures as the Red Army grew in importance, and CCP-appointed provincial leaders. Following the disaster of the Great Leap Forward, there was discussion about the character of socialist transformation in China, as China split with the Soviet Union and its Marxist-Leninist-Stalinist orthodoxy. Mao's ideas increasingly conflicted with those of men who were to become the future leaders of China, especially Deng Xiaoping, party general secretary from 1954 to 1966 and leader of China from 1978 until his death in 1997.[29]

Sensing a waning of popular revolutionary zeal after the disasterous Great Leap Forward, Mao was very sympathetic to the enthusiasm of young Chinese for the idea of "permanent revolution" and a continual sifting, through outbursts of dissatisfaction at the grassroots level, to find the best solution to problems. In particular, Mao's attack on the legacy of Chinese feudalism, especially status and material inequality, was taken up by young Chinese who resented the emerging bureaucratic elites and who chaffed at the routines that seemed an impediment to experimentation and personal initiative.

In 1966, spontaneous demonstrations and the seizure of local power by young Red Guards were endorsed by Mao and the CCP. Mao promoted an original ideology, that socialism in China would be a long process of building and rebuilding, trial and error, and half truths that gave way to greater understanding. Maoism stressed the resolution of contradictions. Any effort would be incomplete and in need of change. New efforts would themselves come to be rejected in an unfolding cycle that would, at a future date, deliver prosperity and the highest possible quality of life to everyone in China.

---

[29]Zhao Ziang became premier in 1980 and party secretary general in 1987, followed by Ziang Zemin as party secretary general in 1989. Until just before his death, however, Deng remained the most powerful figure in China.

The chaos of the Red Guard Movement, violent and intimidating, was short lived, though thousands of people lost their lives, and many priceless objects of traditional beauty and craftsmanship were destroyed. By 1969, more conservative elements in the party reasserted their power. Revolutionary Committees composed of workers, party cadres, and others replaced local governments in towns and workplaces, initially to root out corruption and elitism but, in time, to continue national planning from the top, under the guidance of a centralized CCP.

What usually are called the excesses of the Red Guard Movement were reigned in, but much of the tumult it created did not abate until Mao's death in 1976, marking the end of the Cultural Revolution. Workers were given more power in their workplaces, and students continued to demonstrate for greater freedom to speak out and criticize the government. Local initiatives were tolerated in ways they had not been earlier. In asserting party control, sixteen million youth—Red Guards, but many other urban students and their teachers—were "sent down" to the countryside where they lived and worked alongside peasants for up to a decade. They became the *shiluo de yidai*—lost generation.

The End of the Beginning. After almost thirty years of Mao Zedong's often erratic experiments in mobilizing popular enthusiasm and collectivist efforts guided by the revolutionary generation, China seemed to be mired

Mao Zedong, 1937 or 1938

Source: http://news.boxun.com/news/gb/china/2008/04/200804171803.shtml

Deng Xiaoping

Source: http://news.qq.com/a/20090729/001646.htm

in political conflict, chronic economic shortcomings, and an unpredictable future. Until 1972, China was internationally isolated by the United States and its allies. It had become an adversary to the USSR. It was in the unenviable position of having a fifth of the world's population that was growing so quickly as to become an impossible barrier to China's development goals. The powerful state seemed increasingly to be incapable of "catching up" with the fast-changing West.

The Chinese Revolution had accomplished tremendous progress in advancing democratic outcomes, e.g., public health, literacy, life expectancy, housing, and other quality-of-life features. Gender discrimination was banned, and educational and occupational opportunities were closely tied to merit: i.e., ability and hard work, as well as political allegiance, rather than ascriptive features like family wealth and connections. Nevertheless, inequality among China's provinces and the gap in incomes and lifestyles between urban and rural Chinese belied Mao's insistence on socialist equality. The nation's economy had failed to grow sufficiently in the preceding decades, industrial production was inefficient, and its products were shoddy and expensive (Walder 1989: 407–409). From 1956 to 1976, personal income actually may have declined while population increased by two hundred million.

The period from 1966 to 1976 saw considerable political intrigue, with ideological divisions and shifting alliances in the upper reaches of the party. Deng Xiaoping, a leader during the Chinese Revolution, was demoted and largely stripped of power in 1966. He retreated during the Cultural Revolution, only to emerge as deputy to the venerated revolutionary leader Zhou Enlai in 1974. At Mao's instruction, he was again stripped of his party offices and went to Canton under the protection of a powerful military figure, Xu Shiyou (Guthrie 2009: 37). As Mao's health declined, his hand on power also began to slip. Deng reemerged the next year, issuing a series of documents criticizing the direction the country had gone over the past decade. Those he attacked fought back, but unsuccessfully. The "Gang of Four" that included Mao's wife, Jiang Qing, was put on trial soon after Mao's death, convicted of crimes against the state, and imprisoned.

Post-Mao China: The Deng Xiaoping Era

China's economy was being left behind by those of South Korea, Taiwan, and Japan. The country ran a trade deficit (the value of its imported goods exceeded its exported goods) for the first time. This gave impetus to the conservatives who wanted to take a path less tied to Mao's idea of a socialist economy. They, and especially Deng Xiaoping, were much more willing to pragmatically explore the best means for economic growth. In Deng's famous phrase, the color of the cat doesn't matter, so long as it can catch mice. Deng was selected as one of several vice premiers in 1977 and in 1978 became premier and China's undisputed leader.

The state social welfare system that administers health care, educational opportunities, and jobs was gradually reduced and today increasingly is based not on where a person works but where he or she lives, with payment required for what had been a right. The "iron rice bowl" of protections and benefits from cradle to grave was replaced by a "plastic rice bowl" of far fewer guarantees and supports. Working away from one's village or town jeopardizes a person's ability to receive social services, yet a hundred and fifty million people are now migrant workers who have streamed to the cities. They are individually rather than collectively in search of a better life.[30]

Workers across the country protest regularly over layoffs, pay and benefits, workplace safety, price increases, and the lack of reemployment. To date, peoples' grievances have had little apparent impact on the control the state exercises over the conditions of work and the enterprises' pursuit of

---

[30]The film *Last Train Home* powerfully captures the experiences of migrant workers in China today.

greater productivity by trimming workforces. Low wages and long working hours in nonstate or joint (state and private) ownership firms have led to calls for unions to be independent of the state-controlled labor organizations. Near the end of the Cultural Revolution, in 1975, the Party Congress legalized workers' strikes if grievances were not addressed by their workplace managers. This right was removed from the Constitution, at Deng's insistence, in 1982 (Gittings 2005: 279). Workers today are not able to form independent unions or collective bargaining associations, regardless of whether their enterprises are state or privately owned.

Political Fissures II: Democracy Versus Economic Growth. The Hundred Flowers Movement in 1956–57 expressed students' idealism about participating in public life by speaking out and freely criticizing what they saw as social injustices or misguided policies. The insistence on freedom of expression was repeated periodically, notably in the run-up to the Cultural Revolution and in the April 1st Movement of 1976 that continued to the end of the decade. Deng Xiaoping's orchestration of the state's response to the democracy movement in 1989—what has come to be known as the Tiananmen Massacre—made certain, however, that democracy in China would be different from what the students and workers had in mind.

In 2008, 303 Chinese intellectuals signed "Charter 08" urging China's leaders to, among other things, adopt a multiparty political system and open the way for greater democracy of decision making, i.e., democratic process. Though the state censored all reference to "Charter 08" on the web, more than eight thousand people signed it. Many of the signers were questioned, but only one, Liu Xiaobo, was imprisoned, where he remains. Liu Xiaobo was awarded the 2010 Nobel Peace Prize, giving the party some difficult moments, but most Chinese did not know his name or that he was given the Nobel award. China's leaders have learned, perhaps from the Tiananmen Square demonstrations, how to beat back calls for a more liberal political system or to suppress China's democracy movement with less visible or violent means.

The main features of democratic process seem unimportant, cumbersome, and occasionally threatening to the Chinese leadership (Wines 2009). The few leaders who venture too far from this view are summarily shunted aside. For example, Zhao Ziang, former premier and secretary general of the CCP, showed sympathy with the students and other citizens in Tiananmen Square and the citizens/workers protesting throughout Beijing and China in May 1989. He was dismissed by the party executive committee just days before the troops attacked, confined to house arrest, and silenced until his death in 2005.

Today, popular unrest is widespread but rarely reaches the wider public's attention. Labor walkouts, farmers' protests, unrest among Uigurs in Xinjiang

Province and Buddhists in Tibet, the challenges posed by Falun Gong's millions of followers,[31] environmental damage and threats, and the Internet's opening to the wider world pose significant challenges to a one-party political system. Greater democratic process, if developing, is only very slowly taking shape.

"If capitalism has something good, then socialism should take it over and use it." Deng Xiaoping's message on a visit to the special economic zone in Shenzhen Province marked the end of any serious equivocation about the direction he had taken China's economy. Eighty percent of the Chinese people continued to make a living from the land as late as 1990, and a majority of workers were still employed in state-owned enterprises, but many things were changing.

The state under Deng Xiaoping set a course to develop China's manufacturing capacity through joint ownership between state and foreign or private Chinese capital, increasingly expanding the latitude for private ownership. State micromanagement of the economy was giving way to personal initiatives motivated by the prospects of profits. China's robust economy, second largest in the world, grew over the past two decades at a pace surpassing any other country. Average per capita income has doubled, tripled, and quadrupled, though two hundred million people remain desperately poor.

All of this would remain closely watched by the CCP. The party gave no ground in controlling economic activities that threatened the needs of the people, as determined by the state. John Gittings, a longtime China watcher, concludes that Deng saw the state's role "as that of building a strong and prosperous, not [necessarily] a socialist, China" (Gittings 2005: 289).

Maintaining a Powerful State. Chinese leadership remains in the hands of a small group of individuals, mostly men with an education in engineering, who are planning assiduously for China's future as a major world power. While consigning more political responsibility to regional and local officials, the top leaders of the CCP are firmly guiding the country's future. At

---

[31]Doug Guthrie (2009), discussing the trade-offs between a robust economy and democratic outcomes, observes, "As scores of workers are laid off from old-state-owned factories with no guaranteed alternative for employment, and as migrant workers . . . have no guarantee for education for their children, the trade-offs of the market economy become increasingly clear, and they are trade-offs that are experienced disproportionately, if not exclusively, by the poor. These trade-offs are important because they lay bare the challenges and contradictions that circumscribe the transition to a market economy. Indeed, it is in this context that the government's crackdown on the Falun Gong movement must be understood: this movement, which has the largest organizational membership in China (larger, even, than the Communist Party) . . . is filled with constituents who are left behind by the economic reforms" (Guthrie 2009: 25).

the national level, the Plenum of the Central Committee, i.e., the politburo, is the central organ of power of the Communist Party. Originally composed of two dozen wartime revolutionaries, it is now made up of about three hundred of the country's officials. They are the most powerful state authorities, initiating and interpreting laws and designating who will occupy leadership positions throughout the state apparatus.

Their decisions are routinely endorsed by the National People's Congress that meets annually to approve and codify decisions of the party leaders. Increasingly, the People's Congress hears disagreements in policy implementation and tries to mediate solutions (Tanner 1994), but in major matters it has little authority. The CCP continues to grow in membership.[32] Bringing more people into the party, the only official organization for discussion and debate of political process and state policies, is considered a measure of democratization.

In the past forty years, the rule of law has grown in China. Moving away from the sometimes capricious dictates handed down during the Mao era, governance in China is developing a sophisticated code of civil and criminal law, including laws governing businesses, personal and workers' rights, and family law, that is recognized as authoritative in settling disputes, making contracts, and gaining corrective actions.[33] The legal system that is continuing to emerge is part of a strong state's guidance system, setting the parameters for acceptable personal behavior and organizational practices while providing incentives for people and groups to pursue activities the state favors. As Michelson (2007) shows, the rule of law remains embedded in "enduring socialist institutions" but constitutes at least one element of democratic process in China.

"With no popular mandate, the government's legitimacy relies on its record in making China richer and stronger" (Economist 2009b: 12). As is often the case in democratic societies as well, political legitimacy of the state and quiescence of would-be dissenters can be secured by means of economic development. The lives of most Chinese have improved since

---

[32]In 1979, Communist Party membership was 35 million. It increased to 48 million in 1988 and today is about 79 million, a bit more than one in eight adults. For many, as is typical in one-party states, CCP membership is an implicit requirement for holding a state position and is useful for career mobility.

[33]Among the most significant legislation were the Labor Law (1994) that recognized the rights of workers in dealings with employers, the Company Law (1994) to establish a form of limited-liability corporation, the Prison Reform Law (1994), and the National Compensation Law (1995). Guthrie (2009) estimates that in the first decade after 1978 seven hundred new national laws and two thousand local laws were passed.

the early 1980s. Restrictions on personal and political freedoms—features of process democracy—are the price they paid for greater prosperity (Halper 2010).[34]

The CCP has evolved, as the revolutionary generation of leaders died and was replaced by well-educated individuals who carried out Deng Xiaoping's urging: "to get rich is glorious." Millions of ordinary people have left their jobs in the state sector and become private entrepreneurs, what is popularly called "jumping into the sea." Deng was referring to the country as a whole, and not necessarily to individual wealth accumulation, but today income and wealth inequality rivals that of the United States and far surpasses Japan, Britain, and European nations. Deng's famous observation that "some people can get rich first" has become official policy.

According to Deng and the other leaders of China in the post-Mao era, only a strong state that operates largely behind closed doors, sets policy unilaterally, and allocates public resources without serious public input could modernize China. As Wan Li, a leader of the National People's Congress said in 1992, "A ruling party that cannot develop the economy and improve the people's living standards is not qualified [to rule]" (Gittings 2005: 252). The party has proved itself capable of doing this, at least for now.

The disagreement among the Chinese leaders over revolutionary ideology versus pragmatically driven growth (Mao versus Deng, in very simplified form) largely took place among the elite of China. The fissure over democracy versus state-guided planning was between the party and a minority of the Chinese people. The outcome remains uncertain today. Recent scholarship shows the continued strength and control of the CCP in China, including controlling the direction of the economy and investment, education, the use of natural resources, trade, and ownership of industry (Ian Johnson 2010; McGregor 2010).

Socialism With Chinese Characters. Mao Zedong understood social change in dialectical terms, with inevitable conflict incubating innovation and

---

[34]Other features of mature democracies, such as freedom of the press and the freedom of assembly, are significantly compromised. As well, Wu Bangguo, chairman of Parliament and China's second-ranking Chinese Communist Party official, declared in 2009, "China would never adopt a . . . multiparty political system, separation of powers, a bicameral legislature or an independent judiciary" (Wines 2009: A5).

clarity of purpose. "Trouble," he liked to say, "will clear things up." On the other hand, Deng wanted social change to be carefully and forcefully guided by strong control and oversight of the state. As problems arise, solutions will be found, or as Deng is to have said, as China "gropes for stones to cross the river."

"To build socialism we need large numbers of path breakers who dare to think, explore new ways, and generate new ideas. Otherwise we won't be able to rid our country of poverty and backwardness or to catch up with—still less surpass—the advanced countries." These words express Deng's effort to transform China into what many call a nation of market socialism (Gittings 2005: 225). Deng called this "socialism with Chinese characters," and his successors have continued to support and develop it. Public ownership remains a dominant part of the economy, but foreign investment and private Chinese investment are widespread. State planning continues to drive much of what happens in China, especially for large and costly projects like the construction of universities, nuclear energy and hydroelectric power, and transportation.

In international efforts to bolster China's economy, and anticipate future economic needs, the state is a major force by making investments and devising state-to-state agreements and partnerships. The foreign policy of China is apparently untroubled by the policies of other states with regard to human rights, labor, and environmental issues. Its policy of "noninterference" sometimes puts China at odds with other major nations seeking international approbation or censure for countries with a record of human rights abuses. These include Myanmar's dictatorship, Sudan's complicity in the Darfur genocide, and countries that are thought to be developing or enhancing nuclear weapons capability such as Iran and North Korea. The state's focus is on economic ties and opportunities that serve China's interests as a growing world power.[35] Its goal is to catch up with and surpass nations of the West that for a brief time treated China as a second-rate nation.

---

[35]Commentators refer to a "Beijing Consensus" that emphasizes national interests in economic growth and stability. This is sometimes described as an amalgam of Confucian respect for hierarchy and order, capitalism's enthusiasm for economic growth, and communism's use of state power to direct the use of resources (e.g., Economist 2010a). Joshua Ramos (2004) offers the alternative Beijing Consensus that encourages countries to try less orthodox ways of helping their national economies. He argues that progress should be measured by indicators of sustainability and income equality and recommends greater economic sovereignty for small countries, i.e., greater self-determination.

THE THREE GORGES DAM

Now we have entered a different kind of battle. We have declared war on nature. . . . To sum things up, it's a new task of a new era. When class struggle is over, we declare war on nature.

—Mao Zedong (quoted in
Schoenhals 1986: 102)

In 1957, less than a decade after the Chinese Revolution, Mao expressed this sentiment—popular in the nineteenth century—disquieting today in the midst of global warming and a heightened awareness of the earth's fragile environment, resource depletion, and the long-term costs of toxic pollution. A year earlier he had written a poem that said, in part,

A bridge will fly to span the north and south,

Turning a deep chasm into a thoroughfare;

Walls of stone will stand upstream to the west,

To hold back Wushan's cloud and rain,

Till a smooth lake rises in the deep gorges (in Chetham 2002: 146)

The "gorges" Mao wrote about are the three gorges of the Yangtze (*Chang Jiang* or Long River), among China's greatest geological marvels. Its limestone peaks and spires rise three and four thousand feet above deep gorges that narrow the flow of the world's third-longest river in its more than 20,000 feet fall from its source to the sea. The Yangtze's 700,000 square miles of watershed are a fifth of China's landmass. The river essentially divides north and south China. Legend has it that a troop of dragons, handled by the folk hero Yu, carved the gorges, peaks, and valleys "to drain the land and make it habitable" (Zich 1997: 19). For more than two thousand years, China's emperors imagined projects to stem the massive flow of waters when the Yangtze floods, on average once every ten years. Floods in 1931 and 1935 had terrible consequences. Millions of acres of land were submerged, more than a million homes were destroyed, and approximately 150,000 people lost their lives in each flood.

The United States consulted with Chiang Kai-shek's prerevolutionary government on the project. Mao's new China began work in the 1950s, with the help of Soviet technicians. The power of the state, on which such a project was wholly dependent, would be used to its fullest. For example, in the early 1950s Mao assigned seventy of the one hundred graduates in engineering geology and hydrology to the Three Gorges project (Chetham 2002: 154). When the Great Leap Forward revealed that the country's other needs were greater than the massively expensive dam, combined with the 1960 Sino-Soviet split that caused the Soviet Union to cancel all projects with China, the project was abandoned.

Deng Xiaoping renewed the state's intention to build a Three Gorges Dam in the 1980s, to be a colossal cornerstone in his plan for modern China's rapid development. Four major goals were to be pursued: flood control, improved safety of river navigation, hydroelectricity production, and water conservancy for irrigation, including a huge south-to-north diversion channel taking the Yangtze's water to northern China. Ground was broken in 1994, and the project was completed in 2007. In the spring of 2010, a surging Yangtze River was rushing more than half a million cubic feet of water every second into the gorges behind the dam.

Now filled, the reservoir stretches 370 miles, having inundated 1,400 rural towns and villages and displaced between 1.13 million (the government's figure) and 1.9 million people (Zich 1997). A quarter of a million acres of farmland were submerged, along with nine hundred factories, many of whose soils are rich in chemical pollutants. Stone Age archeological sites and more than 1,200 historical sites were buried under the water, most without having been thoroughly researched and only a few being relocated.

The dam itself is one and a third miles long and 607 feet high, capable of housing twenty-six of the largest turbines ever built and generating as much power as eighteen nuclear power plants: 18.3 megawatts of electricity. An international airport is being built beside the gorges and the world's longest single-arch bridge will span the river. This is only one of China's projects to provide the hydroelectricity needed for economic growth. It has an ambitious plan to build three dozen nuclear power plants as it reduces energy dependence on coal-fired electric generation. These major emitters of greenhouse gasses are being shut down by government order. The state is able to do this and marshal billions of dollars in funds for other projects, about which the vast majority of the people have little say. China's powerful state is sending the message that "China is a nation in transition to another kind of society altogether" (Zich 1997: 17).

State Power and Social Change. The Chinese Communist Party was synonymous with the state in the early years of the People's Republic of China, during the decades of revolutionary zeal under the sway of Mao Zedong. While Mao met opposition from within the party, it was largely invisible to most Chinese. The course of social change in China was an outcome of decisions made by a small group of individuals, decisions that had both intended and unintended effects. Twenty-five years after the revolution, the disasters during the Great Leap Forward, and the excesses of the Cultural Revolution, power was slipping from Mao's hands. Opposition was more obvious, though it remained within the party.

The rise of Deng Xiaoping to the positions of general secretary of the party and prime minister altered the course of social change in China, with his diminution of ideological fervor and his focus on economic growth. Deng and his allies set China on a course of quasi-capitalist development, directed by the state. Gradually, the party became less the sole bearer of state power, as the rule of law, decentralization of authority, village and township enterprises, and the emergence of private ownership and wealth shifted day-to-day decision making away from the party elite. The party retained sole authority over strategic planning and policy formation that guided the evolution of the economy, but some elements of democracy outside the party began to emerge. The press was given wider latitude in what it reported. The National People's Congress became less a rubber stamp for decisions made by the party elite. Local and municipal elections were more openly competitive.

For Deng Xiaoping, China's socialism no longer meant collective ownership and control of the means of production. Socialism was increasingly associated with efforts to end poverty, spur economic growth and institutional modernization, and slowly incubate political reform. Only in this way could the nation regain the international stature and power most Chinese feel belongs to them. They are, after all, a nation of one-fifth of the world's population and the birthplace of much of the science and technology that made possible industrialization and modernization elsewhere.

Resistance to State-Directed Social Change

The state is a powerful force. Citizens throughout the world look to the state for leadership and solutions to problems. The case of preventable diseases and the threat of epidemics are typical.

When vaccinations were developed for many childhood diseases in state-supported research facilities, government immunization programs quickly proliferated. The polio vaccine in the mid-1950s was made available at schools through state health programs. Postnatal vaccination against diphtheria and whooping cough became standard with the passage of the Vaccination Assistant Act of 1962, and every state passed laws requiring immunizations prior to attending school. Fears of swine flu, avian flu, and other potential pandemics elicit calls for the state to act in protecting its citizens and improving their lives. Still, there is skepticism and resistance.

Using the State in Opposition

The political spectrum in democratic societies is not divided between liberals who favor and conservatives who oppose the use of state power to effect social change. Conservatives are not opposed to using a powerful state. They support the death penalty, laws restricting abortions, robust policing and surveillance, and a large military that can exert national interests worldwide. Rather, political divides are based much more on the way to use state power. The example of abortion illustrates the contestation over the role the state should play as a force for and against social change.

Throughout the world, with the exception of most Muslim nations, the termination of pregnancy is widely practiced. The question facing the state is whether to use its power in support of safe abortions or deny the use of medical practices to terminate pregnancies. In heavily Catholic countries in Latin America, it is often difficult to obtain contraceptives as well as a medical abortion, despite the fact that an estimated 40 percent of pregnancies are terminated early, most illegally. This results in widespread injury, disease, and death among women as a consequence of failed illegal abortions (Mollmann 2010).

In 1933, when Iris Summers was a young woman, as many as 20 percent of maternal deaths in the United States were the result of septic infection from illegally induced abortion, and deaths from illegal abortions were "12 times higher for non-white than white women" (Darroch 2007: 270). Thirty-five years later, before abortion was legal, for every one hundred deliveries at New York City municipal hospitals there were twenty-three admissions for complications from illegal abortions. Concern about women's health led many states to pass liberalized abortion laws immediately prior to the 1972 U.S. Supreme Court ruling legalizing contraceptive practice for unmarried persons and the Court's 1974 decision, *Roe v. Wade*. Its passage caused abortion mortality to fall to negligible levels thereafter, and a majority of the public supported this change, as discussed in Chapter 5.

Until the twentieth century and the expansion of modern medicine, there were few laws specifically targeting abortion. This was the status quo for the next seven decades. Legal abortion became a major political issue only after the mid-1970s, largely driven by the Catholic Church's theological interpretation of conception. Conservative opposition to abortion thereafter sought to dismantle or weaken the actions of the state that made abortion an option to unintended and unwanted pregnancy—e.g., legislation, regulations, court decisions, public funding, policy initiatives—by using the mechanics of government to reverse course and return to the situation of an earlier time.

Joined by a growing Christian conservative movement, the social changes brought on by readily available technologies—i.e., new pharmaceuticals, surgical techniques, contraceptive devices, and abortion services—have been opposed by mobilizing the state's power to pass and enforce legislation and restrictive regulations limiting the use of these technologies, ban public funding of family planning services, and sponsor only reproductive health programs that promote sexual abstinence until marriage.

Another example of strategies of resistance to social change by employing state power revolves around civil rights. It would be difficult to reverse the major civil rights legislation of the 1960s—the Civil Rights Act of 1964 with its prohibition of discrimination in voting, public facilities, places of commerce, and schools—and the 1968 Fair Housing Act. Conservatives opposing the changes brought about by this legislation have sought to employ the state to minimize the impact of the laws. They have urged the Court to restrict the scope of the laws, offering legal challenges to the laws' enactment that seem to tread on other legal claims and reducing the state's enforcement ability against the laws' violations.

Opposing the Power of the State

Much resistance to state-directed social change is based not on how to use state power but on beliefs about where the state should act and where it should defer to others to act. One might think this, too, would divide the political landscape between liberals in favor of using state power widely and conservatives opposed to using state power, but nothing is so simple.

Conservatives objecting to the state as an agent of change may oppose the state as the agent of change, even if it advances the public's well-being. Instead, nonstate organizations, including churches, civic groups, small businesses, and corporations, are the preferred agents of social betterment. This is especially the view when there is a market opportunity for doing what liberals might see as the state's responsibilities. Opportunities include such things as generating and selling electricity, building and running prisons, distributing mail, providing health care, operating highways, maintaining and supervising public lands and national parks, engaging in space exploration, and even fighting wars.

Liberals are often at odds with conservatives who emphasize the state's obligations to protect its citizens and their property. At the same time, conservatives tend to reject almost anything that compromises private property rights, while liberals favor land use planning and other measures that give the state some control over private property.

On the other hand, while today's liberals see a positive benefit in collective efforts to effect social change, they often line up against the state as an agent of change. Liberal democracies emerged 250 years ago as opponents of authoritarian and absolutist rule. As a check on state power, traditional liberalism conferred rights to individuals that even the most democratically configured state cannot abrogate, based on a belief in the sovereignty of individuals and the basic equality of all citizens. This liberal tradition limits social change that extends the reach of the state into personal affairs and private lives and protects unpopular sentiments and minority practices from what Tocqueville called the "tyranny of the majority."

Where conservatism takes issue with liberals' idea of the state/individual relationship is not necessarily in terms of the use of state power. Today's liberals tend to champion constitutional freedoms of speech, assembly, and belief, and contend there is a right of privacy the state cannot abrogate. They are likely to oppose the state's efforts to compromise these in the name of security or national interest. For conservatives, intrusions into private lives are selectively accepted. When threats increase, so must the state's protective functions. Conservatives (but not libertarians) are more likely to accept the state placing restrictions on personal freedoms, in defense of traditions and beliefs they contend are a fundamental part of the national identity. Liberals reject this and any state-sponsored privilege given to, or promotion of, a specific religion, language, or set of cultural practices.

Most democracies pride themselves on tolerating, even encouraging, a lively debate about the power of the state. Democracies make heroes of those who have resisted the state as well as those who worked successfully to improve life and save lives by means of the state. As an instrument of social change, the state is formidable, though its terrain is and will remain a major site of contestation.

Topics for Discussion and Activities for Further Study

Topics for Discussion

1. What state intrusions into personal behavior do you favor? What do you oppose? Discuss the reasons for this by looking at such things as daylight saving time, seat belts and laws mandating the use of head- and taillights on bikes, antipornography laws and local ordinances on adult book shops, pharmacists' recordkeeping of drug prescriptions, zoning laws, airport security measures, and other things that evoke some differences of points of view.

2. In effecting social change, what do states do better than corporations? What do corporations do better than states? Make two lists and try to have some items of consensus. Looking at the lists, discuss the reasons for one over the other.

3. A provision of civil rights legislation concerning education is Title IX. It is the basis for schools and colleges to develop sports programs and opportunities to participate for females comparable to those for males. What social changes have resulted from Title IX? How have the schools you have attended met this obligation? How have they fallen short, compromised, or circumvented Title IX enactment?

4. Imagine the state no longer authorizing marriage. In what respects would that be a withdrawal or reduction of state power? What social changes can you imagine if legally binding marriage ended? If "family" no longer had legal status based on marriage? No more legal status of wife or husband. Would this eliminate legal rules for who can and can't be partners in reproduction and childrearing? Would religiously sanctioned marriage suffice? Would people create other institutional arrangements?

Activities for Further Study

1. Despite civil rights legislation and being a more racially tolerant society, ethnic and racial segregation in schools is at an all-time high. Gather some data on this and read the research that has provided reasons for this phenomenon. What role has the state played, or not played, in this?

2. How does China's one-child policy typify its use of state power? What is this policy? How did it originate and how was it enforced? How has it changed in the past decade or two? (Caution: this is a highly charged issue, with a great deal of poor information and many uninformed opinions. This is a good chance to test your skills at critically evaluating information sources.)

3. Hortense Powdermaker's *After Freedom: A Cultural Study of the Deep South* includes the survey of questions she asked people in her research and the tabulation of their answers. It was not guided by hypotheses and is very long, so it is not a particularly good survey instrument. Still, much of the data are fascinating. Read the data and make a report on what you found most interesting. You might make your own survey instrument and give it to a group of people and then compare what you find to what Powdermaker found.

4. In most countries, the state has been the major actor in setting aside land not to be developed. In the United States, this began with Teddy Roosevelt's inspiration, gained from personal experience and reading people like John Muir, the national park system. Do some reading about this or the same phenomenon in another country. Why was this conservation done? How was it done? Most importantly, was it an effort to direct social change in one direction rather than another?

9

Making Social Change

Engaging a Desire for Social Change

"It is the confluence of biography and history, the personal and the public, private efforts and the social milieu, that allows us to understand social change." You read that at the beginning of Chapter 1. It could now be rewritten to say, " . . . that allows us to understand *and make* social change." Nothing is more frustrating for engaged students than to be drawn into a topic about which they want to become involved, only to be left asking, What can I do?

You should now have a good idea about the forces that impel the direction, speed, and scope of social change. As stressed throughout earlier chapters, there is no stopping and no end to social change. It is an ongoing process. It is not, however, arbitrary or random. In many respects, it is contested terrain with varied possibilities. Being able to think intelligently about social change is an important skill for everyone, activist or observer. It provides clarity and a sense of proportion that helps to modulate between hopes and realities, tempering expectations while suggesting modes of involvement. This chapter explores the ways you can influence the process, if you so choose. It provides an opportunity to see and consider the ways you can be involved as an agent of social change.

Using Human Agency, Now or Later

Some of you already consider yourselves activists who work with others to make change. Perhaps now, having learned more about social change than you previously knew, you feel better prepared to continue or redouble your efforts. Maybe some of you will now want to get started. For others of you, it will take a long time to fully appreciate your potential. Most people are like that. They do not pursue a career in social activism but will occasionally take the opportunity to become involved when they feel strongly about something. They do this through an existing organization or in a more immediate, ad hoc fashion when something sets off an alarm. Increasingly they do it through, or are aided by, the Internet. Who knows what avenues and opportunities may exist twenty or thirty years from now? That may be the first time some of you work for social change. Life is like that.

Whether as a vocation, an avocation, or something that just happens, being involved in social change can be hard work, but good work. It can be exhilarating, absorbing, and a source of lifelong friendships. It can also do the opposite. It can divide friends and family, challenge the fabric of a community, and create deep fissures between opponents and proponents. It may happen without your intending, through your job, or something you do that results in unexpected consequences. What you deliberately do to effect social change also may have consequences you didn't intend, and you may have to work to set things right. In all these respects you are an agent of social change. It's not for the faint of heart, but it's also not only for the showboat and rabble-rouser. Anyone can do it if they care enough to set their mind to it.

In a democratic society, the burden of responsibility for what is done "in the name of the people" falls more heavily on the citizenry than in nondemocratic societies. Along with this responsibility comes the invitation for involvement and engagement. No one is exempt from the burden of knowing what is going on, contributing to the dialogue, and sorting through possibilities, differing opinions, and the costs and benefits to various interests. This may be done around the kitchen table, at the watercooler, or in the locker room. It may be at a PTA meeting, the VFW post, professional conventions, or a gathering of corporate stockholders. It can be done face-to-face or by phone, e-mail, blogging, or texting. It might be a group of scientists and investors considering a start-up company, a public meeting of hunters, anglers, and ranchers debating the uses of a mountain watershed, or a campus symposium on the future of print journalism, nuclear weapons, the local food system, marriage, or intercollegiate athletics.

No matter what the drivers of social change may be, there is always the personal element—the engagement of human agency embedded in social structures, organizations, and historical processes—that plays a part in directing social change. In previous chapters, technology, social movements, war, corporations, and the state have been shown to be major forces of social change. None of these exists apart from individuals and groups of people working, pulling, and pushing together. People act in ways appropriate to networks of roles and relationships, cultural norms, bureaucratic procedures, and laws governing their behavior. But none are robots immune from original, creative, even idiosyncratic ideas and initiatives. Much of social life offers an open invitation to become involved and use the power of human agency to make things happen.

Vocations of Social Change

Veterans of just about any occupation can describe where they think their work is going. Chemists and other researchers recognize nanotechnology as central to much of what they will be doing in the next decade. Writers and publishers fear that a declining number of serious readers cuts into their audience, but ebooks and iPads are thought to be increasing readership, especially among young adults. Factory workers recognize that their skills are becoming obsolete, that those working in manufacturing will need to be able to read blueprints, do more high-level math, and understand computer programming in order to hold on to a job. People in these and other fields recognize social change, new opportunities, and the forces they will have to contend with. *Globalization* describes many of the social changes on peoples' minds, with jobs and investment capital flowing into and out of countries at an accelerating rate, markets shifting, expanding and contracting, and centers of research and innovation proliferating across the globe.

Corporate power and new technologies are significant factors in the work people will be doing and the products they will be using in the years ahead. Many avenues to becoming an agent of social change lie in the private sector. Despite the significance corporations play in effecting social change, however, they give less consideration than they might to the way their actions and operations affect social change. The first order of business for corporations is profits, and corporations are increasingly driven by financial decisions and the capacity of financial transactions to improve the bottom line. This takes them away from considerations of the social changes their actions bring about.

There have been at least two trends in corporate behavior that run counter to this, as described in Chapter 7, and offer interesting options for those working in corporations. The first, corporate social responsibility, when seriously embraced, is an effort to balance profits with citizenship, at least with regard to the environment, if not with regard to employment security and community loyalty. In recognizing that there are choices to be made, socially responsible firms are making decisions and following paths that use less energy, provide for recycling, are less polluting, and use materials that are legitimately and, where possible, sustainably acquired.

Another, more challenging, trend is what Paul Hawken and his colleagues call "natural capitalism" that follows four principles: 1) maximizing the efficient use of resources; 2) practices continuous cycles that mimic biological systems; 3) elevates the value of quality and service to meet real rather than manufactured needs; and 4) invests in restoration to expand the stock of natural capital (Hawken, Lovins, and Lovins 1999: 10–11). The power of corporations is enormous, and the future of the planet depends on their working in new ways. This work is designed and carried out by individuals who can play an active role in directing social change, if they will.

As discussed in Chapter 6, the nature of war and military action is changing in the twenty-first century. Barnett (2004) calls this "new maps" that involve rebuilding as much as destroying. Those in the military and those providing support services will need to use skills in community building and strengthening civil society as well as security. It is too much to hope that the terrible power of war will not be with us forever. Those working in the "management of violence" (Miliband 1969: 51), however, have an opportunity to be social change agents in new ways, with less destructive outcomes.

There is skepticism among some people in affluent societies that the state can ever be the change agent it once was, or that it should be. Public service and leadership in using the collective resources available to the state have always been, and probably always will be, both admired and distrusted in democracies. Polling data consistently score Congress low in the public's trust. Public service, however, is the main reason young people pursue political office and is a source of pride for employees who devote their careers to public work.

Social movements often press the state to take actions to solve problems, as discussed earlier. They are not alone. People working in government are the ones who often imagine and implement initiatives to solve public problems. How much the state hears the public, responds to special interests, or carries water for powerful elites is less the issue than the fact that the state has been and can be a significant force for change. Those who work in the state system are in a position to effect changes, small and large—to make a difference. They devote their careers to keeping the streets repaired, lit, clean, and safe; the parks groomed; commerce fairly conducted; air travel safe; support checks coming; food clean; and burglars at bay.

Their efforts to do the things that the market can't or won't, things that must be done, can be progressive or only a holding action to keep things from completely falling apart. Unlike authoritarian governments that can spend the collective resources of a nation at will, without consulting the citizenry, democracies are more cumbersome. That is the challenge of those in a democracy who see the state as a driver for effective change. They accept the responsibility of involvement in creating collective goods and work to make them as widely and equitably available as possible to those who need and can use them.

More than a few social activists have found themselves teaching as a way to further their agenda. Environmental activists, community organizers, political office seekers, innovative entrepreneurs, and founders of nongovernmental organizations can invariably point to a teacher who inspired them to pursue their career. More mundane topics, e.g., math, social studies, languages, literature, biology, engineering, and history, are taught by the best teachers with a sense of optimism that their work makes a difference. A teacher's personal measure of success is not a paycheck or a pat on the back. It is the idea that the world is somehow, in some measure, better because of their effort to work with young people and adults, to help them become better, more capable, more inspired individuals who can make a positive contribution.

Dominique Lee began teaching in the fall of 2010 at Brick Avon Academy. It's an elementary school with thirty-eight teachers who run the school. In fact, there is no one working at the school who does not teach, i.e., there is no full-time principal or administrative staff. A few years ago, Dominique and five others who founded the school taught, for the first time, as Teach for America volunteers. Teach for America is a highly competitive program of the federal government that puts young college graduates into the nation's most troubled public schools. Having taught for two years, Dominique and his colleagues wanted to keep teaching, but in a new way (Hu 2010).

No one doubts that the United States has to do a better job of educating its young people. It ranks dismally in math and other subjects relative to other affluent countries (Adamson 2010; PISA 2009). High school dropout rates in some urban schools exceed 50 percent. Dozens of school districts are trying new models of education, both in what happens in the classroom and in running the schools themselves. This is nothing new. John Dewey, a prominent philosopher in the first half of the twentieth century, began what might be called experiential learning. He and his followers created schools in a new mold, with gardens and workshops that gave students a means of integrating what they were learning—rarely sitting at desks—with practical applications.

Dominique and his fellow teachers are trying something new as well. They are more explicitly advancing a social change agenda than most teachers, but only as a matter of degree.

Nongovernmental Organizations and Gap Year Experiences

Programs like Teach for America, AmeriCorps, VISTA, and the Peace Corps are federal organizations that challenge individuals, especially young people, to stretch themselves and join others in solving problems and engaging in social change. Paul Blackhurst's *Alternatives to the Peace Corps* (2005) offers a long list of similar organizations that offer many other short- and long-term opportunities, depending on the skill set and experience applicants can bring to them.

The worldwide proliferation of nongovernmental organizations (NGOs) working for social change has shifted the entire playing field of international relations. As incorporated nonprofit organizations, they provide an organizational structure to raise money, employ staff, and address issues and problems of every shape and size. There were reported to be ten thousand NGOs in Haiti when the devastating earthquake struck the country in January 2010. Around the world, NGO projects in education, public health, job training, police training, forestry, human rights, farming, construction, and nearly every other aspect of life seek to solve problems and initiate change by giving people a chance to enlarge their skill sets and have more control over their circumstances.

Many students become involved with NGOs. They do internships while they are students, often in the summer months or for a semester. Others join NGOs as volunteers or interns after graduation, gaining experience and testing their tolerance and enthusiasm for the work NGOs do. Rothenberg's (2006: 598–601) sampling of NGOs working for social change includes the Basel Action Network that tries to prevent toxic chemical crises by publicizing trade in toxic products and waste. Public Services International is a federation of more that six hundred trade unions focused on increasing gender equality and workers' rights worldwide and improving the quality of public services. Refugee Women in Development works in the United States and overseas to improve the lives of refugees, focusing especially on human rights abuses and women's empowerment. Even a brief web search reveals hundreds more.

Thousands of students around the world are taking time to gain experiences that will help them decide on career goals and improve their chances of finding the kind of lifetime work they would like to do. A gap year between high school and college, college and graduate or professional school, or just taking time off from school can be enormously beneficial and life changing. Finding out what you are (and are not) good at doing, as well as realizing that something that looked interesting really isn't (at least to you) can help you make choices and the best use of your own resources. Many NGOs understand these motivations on the part of students and make a limited commitment to interns. They pay little or nothing, provide no funding for travel, and offer no promises about future employment.[1] For many students, that's okay. The experience is worth it.

Cities, states, and the federal government offer opportunities to be engaged in public service as an intern. Nonprofit and corporate foundations offer a variety of possibilities as well. These can be highly competitive, but they are good opportunities for young people to see from the inside how agencies, departments, firms, and foundations in their city and state operate.

Agency and Ethical Responsibility

Social change activity's first tenet of ethical behavior is to take responsibility for what you do. Participating in a social movement organization is neither anonymous nor free of personal responsibility. Even signing a petition can be a public display of a point of view that tells others what a person believes. On the other hand, being in a large state or corporate bureaucracy provides many ways to hide responsibility. Not the least of these is to invoke the contention that you are only doing your job, that you are working for the good of the organization, or that you don't know very much about what the organization is doing. These are excuses for not taking responsibility, even if true.

When mechanisms of social change require more than just taking responsibility as a participant, a second kind of ethical question arises. Social change activities can involve the deliberate effort to change the lives of those

---

[1] If possible, work with an academic advisor, internship coordinator, or other knowledgeable person before committing to an internship. In this day of unpaid labor pretending to be an internship, it is important to look closely at the organization and the terms for doing an internship. An internship should have a significant training component and should not displace someone who might otherwise have a paid job.

over whom a person exercises significant power, but in ways that will not affect that change agent's own life. Groups lobby Congress, boycott corporations, oppose stem cell research, develop military applications for scientific findings, and "brand" new lines of products for overseas sale. The expectation is that something will be changed; social change will take one direction rather than another, will be broadened or slowed down, or in some other way will be consequential—for others, but probably not for oneself.

A case in point is international development projects. Poor, even illiterate, peasants are not stupid. They may be ignorant of a lot of things, but they know their own situation. They have a vast store of local knowledge. To survive, even at a subsistence level, they have to be able to make complex calculations, use foresight, carefully allocate their meager resources most effectively, and meet their obligations to their family and community, what James Scott (1976) calls their moral economy.

For more than half a century, affluent nations have sought to contribute to solutions to problems plaguing poor countries. Though often state-sponsored and part of a nation's strategic planning, international development projects don't necessarily have the political motive of winning strategic allies and gaining influence with national elites. Countries like Sweden and Japan that have very active international development agendas keep the state's international development efforts carefully cordoned off from the states' international political affairs. The object is to solve problems of hunger and malnutrition, illiteracy, poor sanitation and ill health, overpopulation, group conflict, and so forth. If the problems can be solved, peoples' lives will be changed. Positive social change will have been deliberately created, and people will be better off.

The ethical problem in international development work is this. The change agents who are affecting others bear none of the costs of failure nor, if the efforts succeed, experience any of the pain of adjustment to the new circumstances. This dilemma has given rise to what is variously called participatory action research or community participation research (McIntyre 2008). Essentially, the way to address the ethical problem is to abandon top-down strategies and adopt a bottom-up approach. The term "research" is a bit deceptive. What is really happening is social change, guided by efforts to understand the problematic situation from the point of view of those who have problems to be solved. As the changes unfold and new information becomes available, adjustments are made at the instigation of both outside experts and the persons for whom the change is intended. This is a flexible approach that requires a great deal of ongoing discussion.

Very few of you will work in international development. The illustration is applicable to a wide range of activities, however, whenever a more powerful entity is working on behalf of social change for others. Parents of children who go to private schools may care a great deal about public education and want to help, but they will not experience the consequences of their social change efforts unless they switch their children to public schools. They are, like the top-down international development manager, asking other people to accept changes, while they are largely immune from the consequences.

What is the alternative? No one can expect a cardiologist to wait until he has a heart attack before telling patients with heart disease to change their diet, get some exercise, and stop smoking. Experts cannot be expected to be part of every milieu in which they work. They are bound to escape the consequences of their work simply by virtue of the way modern societies are structured, with a complex division of labor and a highly decentralized personal economy. How, then, can those who would help or try to change the lives of others best proceed in an ethical way? Participatory action research offers a way.

Some people, including public officials, find decaying, poverty-ridden urban neighborhoods highly objectionable. They enlist experts in real estate, urban finance, urban architecture, and business development to draw up plans to raze a neighborhood and replace it with new mixed-income housing, businesses, and shops. The design is a much more attractive neighborhood, complete with a park, senior center, and school. That all sounds great, but what about the people living there? They, too, need to be an integral part of the processes that will change their lives. They need the power to help direct the changes. This was rarely the case fifty years ago, and much disappointment and failure was the result. It is not only a practical matter, however. It is also a matter of ethics in social change.

Responsible social change efforts involve the participation—from the outset—of the people who will be affected. It takes their concerns seriously rather than imposing a set of problems on them. It enlists them in the practices that will change their lives rather than structuring their new activities from above. It consults with them and involves them in working out solutions to emergent problems. Most importantly, it gives them options, including the option to do nothing. This is hard for people who believe in social change and think they know how to make it happen in others' lives. It is, however, the only ethical way to do this kind of work.

Activism as a Part of Life

Iris Summers was never involved in a social movement organization. When she was older, she gave small donations to a few political candidates, and she sent a small check to a couple of organizations she liked. One was trying to stop the killing of baby seals and the other was to buy school clothes for poor girls in a far-off country. And when she saw pictures of people in Ethiopia dying of famine, she sent a large check. Frank was dead by then, and she didn't have to explain herself to anyone. Iris never thought she knew enough to take a strong public stand or argue about something so distant from her own life, and it wasn't the kind of thing her friends did or would understand her doing. Just living day to day and participating in family events was enough.

For probably a majority of people, life is to be lived. Their social relationships and some of their personal identity may be attached to their work, but work is largely a means to an end: paying the bills so they can pursue leisure activities that are more important to them than paid work. On the other hand, some people seek out a job, start a company, or develop skills that can turn into an occupation that absorbs them completely. They live for their work, and nothing else gives them the satisfaction or occupies their thoughts and emotions like the work they do. There are also a lot of people somewhere in the middle of these two poles. They try to have meaningful work and balance this with fulfilling lives away from work.

Not directly in the work they do but in their nonworking lives, many people find this balance through involvement with social change. As Chapter 5 described, social movement participation can be episodic and unexpected. It can happen because a friend shows interest and invites you to join her. It can look like a chance to make a statement or back up talk with action. It can grow out of an affiliation with your employer, union, or a group you belong to that mobilizes for action whenever an issue needs its voice and resources. As varied as the reasons for participating, the level of participation can vary greatly. Signing a petition or writing a letter to the paper barely count as participation, but getting others to sign a petition does. Donating money from a garage sale, opening your house for a work party or a fundraiser, and phoning others are ways to participate. Most people never carry a sign or parade in the street, stand in a crowd urging a boycott, or board a bus for a mass rally. But lots of people do.

It is always a good idea to study up on what you're supporting. It is also a good idea to understand the other sides of an issue. Tolerance for other opinions goes along with respecting others' points of views. If knowing how one's opponent thinks diminishes a person's enthusiasm, tempers his strongest emotions, or causes him to hesitate before summarily dismissing any compromise, that's probably a good thing. Thoughtfulness, however, doesn't need to be an barrier to participation in a social movement organization. Rather, it makes for more informed involvement that can have greater credibility when questions come up or things don't go exactly as planned.

Social Change Happens

Iris Summers' youngest granddaughter, Amelia, left college before graduating and moved in with some young people in a city far from where she'd grown up, but near enough to Frank's niece and her husband that they could visit her. On one summertime visit, Amelia took them to a park where hundreds of young people were having a sort of impromptu festival. There were bands and other musicians, booths with people selling food, clothes, and homemade jewelry, and lots of dogs. People were playing games, sitting around eating and talking, and there were lots of bicycles. It looked like some kind of official bicycle event. When Amelia's aunt asked her, "What's the occasion?" Amelia told her it was "just to have fun."

Her uncle wasn't so easily convinced, telling Amelia that this was pretty well organized, despite the very laid-back demeanor of just about everybody. There were bike contests and demonstrations, bike decorations, and bike innovations on display. He asked someone if it was a fundraiser or a way to show the city that bikes and bicyclists should get more public attention and services, like redesigned streets and traffic stops. "No," the guy told him. "It's just to have fun."

Anyone living in that city, or many cities today, recognizes the visible changes of thousands of people using their bicycles (Mapes 2009). David Byrne (2009), formerly of the band Talking Heads, wrote about the cities in which he's biked (and lived to write about). He finds New York City quite a good place to bike. Despite the traffic snarls, the city government cut into driving space by putting in 250 miles of biking lanes on city streets between 2006 and 2010 (Goodman 2010). Minneapolis brags about being a great biking city and so do Davis, California, and Portland, Oregon. There's some sort of competition going on. This is social change, but many of the young people who seem to live on their bicycles just want to have fun.

Not everyone who makes social change does it through politics, as is obvious from what you've been reading. Not everyone joins a social movement, develops a transformative technology, or reorients a business to be more sustainable or socially responsible. There are many avenues to being a force for social change. Artists, dramatists, comics, athletes, farmers and gardeners, and just about anyone who cares about making things better can be an agent of social change. They do this by changing the culture, usually in small ways, but sometimes in big ways.

Amelia and her friends aren't acting blindly, ignoring the threat of global climate change, the meaninglessness of consumer culture, and the transformation of the labor market that seems to have so few positive options for them. Just the opposite. They are moving—however unusual it may look to Amelia's older relatives—toward a different way of organizing their lives, using their talents, and changing the world around them, for themselves and, by example, for others.

The young people in the house next to Amelia's are urban pickers. A few years ago they decided that the pounds and pounds of apples, cherries, and pears rotting in peoples' yards, on the trees growing in the parking strips, and falling into the gutters were an affront to the homeless people who were camped out a few blocks away. Not only the homeless, but the thousands of individuals and families who rely on food pantries and soup kitchens could be eating the fruit instead of it going to waste. A few young people decided to change things.

They started picking the fruit, and they took most of the good fruit to the food pantries that distribute food to the poor, homeless, and people down on their luck. Some of them knew how to store fruit by canning it, and they did this with the bruised and buggy fruit that could be salvaged. Other people became interested, so they all formed an organization, the Willing Gleaners. The next year they went to houses where fruit trees weren't being harvested and offered to do it. They registered hundreds of trees and organized dozens of work parties.

The third year they applied for a small grant from a local foundation to buy ladders, baskets, and pruning sheers. They filed papers and formed a nonprofit organization, with one of Amelia's neighbors becoming the director. That winter they got with some "old-timers" who knew how to make cider and decided to team up the next fall for a cider press. They turned the occasion into a cider festival and raised money for more ladders, tools, baskets, jars, and a used van. The next winter they got some people from the college and county agricultural extension offices to give them instructions on caring for fruit trees and natural ways to have fewer bugs and worms in the fruit. They had a lot of fun, and they are still making a difference.

"We live in a world that could easily go about its business without us." That's as true today as when you first opened this book and read those words. It could do this, but it won't. You are here, doing things and having consequence—for others, the environment, and for yourself. You aren't a ghost, nor do you need to act as if you are. You may want to leave a small footprint or a big impression. You may want to concentrate on "ways of being," finding the best way for you to live life, to be with others, to accept the world around you. That's fine.

You can also make a difference. What you've learned in reading this book over the past weeks is that there are many ways social change comes about. All of them offer opportunities and challenges. Some of them are very obvious, direct, and immediate. Others are less obvious, slow in developing, and take you on a crooked path. As you go through life, you'll find that there are always avenues for making social change, offering new ways of change.

Topics for Discussion and Activities for Further Study

Topics for Discussion

1. Has anyone in the class been a Vista, Peace Corps, or Teach For America volunteer? If so, have them discuss their experience, with particular attention to their sense of having been (or not been) an agent of social change. If anyone has served in the military, have this discussion as well. If no one in the class has been in these positions, invite someone to class who has.

2. Nita Eliasoph (1998), in her *Avoiding Politics*, talks about volunteers and others who try to do things for their communities, including tackling environmental issues, addressing problems of teen drug use, and improving the schools. She describes how volunteers avoid talking about the political and social context of these problems and shy away from addressing the structures of control and influence that underlie the problems. Instead, volunteers focus on nonpolitical activities and solving problems one person at a time. What are your experiences in this regard? Have you also been in situations that "avoid politics"? Why this avoidance?

3. What is something on your campus or in the nearby neighborhood that you would like to see changed? Make a list on the board and discuss the pros and cons of each. Which of these invites a "top-down" approach that was criticized in this chapter? How could this be avoided or mitigated in order to proceed with efforts in a more participatory-action-research way?

4. In the 1960s, high school and college students wanted greater "relevance" to their studies. One part of this was a demand that schools be part of the process of social change. Can schools be more effective agents of social change? Should they be? What could be done to give students a stronger sense of their own agency, not only for learning but for taking what they learn outside the classroom and contributing to the changes going on around them? What might be the downside to schools doing these things?

Activities for Further Study

1. Do some research about the gap year. When did this term come about? What drew attention to it? Has it changed over the years? Is it seen the same way in the United States as it is in Europe? What are the pros and cons of a gap year? Is it a consequence of, or reflective of, social change?

2. Betty Friedan (1993), in *The Fountain of Age*, suggests that a new stage in life is emerging among older adults. It goes by many names (e.g., encore years) that designate a time after children are raised, a lifelong career is over, and retirement is opening up new options. Some people find new careers, focus on their communities, develop skills they'd neglected, or apply what they've learned to new endeavors. Many of them are becoming social activists. Do some research and find out what's going on here. Who are these people, what are they doing, and why are they doing this?

3. Interview someone who has devoted his or her career to making social change. Find out what motivated them, if they did this intentionally or as an unintended consequences of something else. Write a summary chronology of their efforts, including successes and failures. What can their experience tell you about social change as a main focus of a person's life?

4. Go online and find an NGO, foundation, public agency, or private firm that is offering an internship. Study the organization and find out what it does and how it does it. What does its internship application require? If it is a statement of intent, an essay about yourself, or something else that requires you to explain why you would like to get the internship, complete this. You aren't actually applying, but this is good practice and helps you think about your interests and intentions.

References

Adamson, Peter. 2010. "The Children Left Behind." Innocenti Report Card 9. UICEF Innocenti Research Center. Available at www.unicef-irc.org/publications/pdf/rc9_eng .pdf. Accessed December 10, 2010.

Albert, Daniel M. 2007. "The Nut Behind the Wheel: Shifting Responsibilities for Traffic Safety Since 1895." Pp. 363–378 in John W. Ward and Christian Warren (eds.) *Silent Victories*. Oxford University Press.

Albert, Michael. 2000. *Moving Forward*. AK Press.

Alexander, Jack. 2004. "Black/White & Brown." Available at http://brownvboard .org/video/ blackwhitebrown. Accessed July 23, 2010.

Alexander, Jeffrey and Philip Smith. 1993. "A Discourse on Civil Society: A New Proposal for Cultural Studies." *Theory and Society* 22: 151–207.

Allport, Gordon W. 1954. *The Nature of Prejudice*. Addison-Wesley.

Almeida, Paul D. 2003. "Opportunity Organizations and Threat-Induced Contention: Protest Waves in Authoritarian Settings." *American Journal of Sociology* 109: 345–400.

Alwin, Duane F. and Ryan J. McCammon. 2004. "Generations, Cohorts, and Social Change." Pp. 23–49 in Jeylan T. Mortimer and Michael J. Shanahan (eds.) *Handbook of the Life Course*. Springer.

Alwin, Duane F. and Jacqueline Scott. 1996. "Attitude Change: Its Measurement and Interpretation Using Longitudinal Data." Pp. 75–106 in Bridget Taylor and Katarina Thompson (eds.) *Understanding Change in Social Attitudes*. Dartmouth University Press.

Amenta, Edwin. 2006. *When Movements Matter: The Townsend Plan and the Rise of Social Security*. Princeton University Press.

Anderson, Benedict, R. O'G. 1983. *Imagined Communities*. Verso/NLB.

Anderson, Robert T. 1971. *Traditional Europe: A Study in Anthropology and History*. Wadsworth.

Anderson, Margo J. and Stephen E. Feinberg. 1999. *Who Counts? The Politics of Census-taking in Contemporary America*. Russell Sage Foundation.

Andrews, Kenneth T. 2004. *Freedom is a Constant Struggle*. University of Chicago Press.

Appleby, Joyce. 2009. *The Relentless Revolution: A History of Capitalism*. W. W. Norton.

Arreguin-Toft, Ivan. 2005. "How the Weak Win Wars: A Theory of Asymmetric Conflict." *International Security* 26: 93–128.

Bachrach, Peter and Morton S. Baratz. 1970. *Power and Poverty.* Oxford University Press.

Bailey, Carol A. 1996. *A Guide to Field Research.* Pine Forge.

Baker, Peter. 2010. "With Arms Treaty, a Challenge Remains." *New York Times* (April 8): A8.

Baker, Russell. 2009. "A Revolutionary President." *New York Review* (February 12): 4–7.

Banton, Michael. 1978. *The Idea of Race.* Westview.

Barnett, Thomas P. M. 2004. *The Pentagon's New Map.* G. P. Putnam's Sons.

Barry, Dan. 2011. "Baby Boomers Hit Another Milestone of Self-Absorption: Turning 65." *New York Times* (January 1) A1, A3.

Barry, Ellen. 2010. "Ex-Separatist Says He Planned Moscow Attacks and Vows More." *New York Times* (April 1): A1, A10.

Bartlett, Thomas. 2002. "Evaluating Student Attitudes Is More Difficult This Year." *Chronicle of Higher Education* 48 (February 1): 35–38.

Bayer, Israel. 2010. "I Feel Like I'm a Part of Something." *Street Roots.* Portland, OR. (January 22): 3.

Becker, Howard. 1951. "The Professional Jazz Musician and His Audience." *American Journal of Sociology* 57: 136–144.

Becker, Howard and Blanche Geer. 1969 [1957]. "Participant Observation and Interviewing: A Comparison." Pp. 322–331 in George J. McCall and J. L. Simmons (eds.) *Issues in Participant Observation.* Addison-Wesley. Reprinted from *Human Organization* 16 (1957), 28–32.

Bellah, Robert N., Richard Madsen, William M. Sullivan, Ann Swidler, and Steven M. Tipton. 1985. *Habits of the Heart.* Harper & Row.

Ben David, Joseph. 1960. "Roles and Innovation in Medicine." *American Sociological Review* 65: 557–568.

Bender, Bryan. 2010. "From the Pentagon to the Private Sector." *Boston Globe* (December 26). Available at www.boston.com/news/nation/washington/articles/2010/12/26/ defense_firms_lure_retired_generals/. Accessed December 29, 2010.

Benford, Robert D. 1993. "Frame Disputes Within the Nuclear Disarmament Movement." *Social Forces* 71: 677–701.

Berle, Adolf and Gardiner Means. 1932. *The Modern Corporation and Private Property.* Macmillan.

Berlin, Isaiah. 1969. *Four Essays on Liberty.* Oxford University Press.

Berman, Ronald. 1981. *Advertising and Social Change.* New York University Press.

Bernhardt, Annette, Martina Morris, Mark S. Handcock and Marc A. Scott. 2001. *Divergent Paths: Economic Mobility in the New American Labor Market.* Russell Sage Foundation.

Bernstein, Irving. 1966. *The Turbulent Years.* Penguin.

Bijker, Wiebe E. 1995. *Of Bicycles, Bakelite, and Bulbs.* MIT Press.

Blackhurst, Paul. 2005. *Alternatives to the Peace Corps.* 11th edition. Food First Books.

Blake, Judith. 1971. "Abortion and Public Opinion: The 1960–1970 Debate." *Science* 171 (February 12): 540–549.

BLS. 2009. "Men and Women in Management, Professional, and Related Occupations." Bureau of Labor Statistics, Washington, DC. Available at www .bls.gov/opub/ted/2009/ted_20090807.htm. Accessed January 10, 2011.

————. 2011. "Union Members Summary." Economic News Release, Bureau of Labor Statistics, Washington, DC. Available at http://www.bls.gov/news.release/union2.nr0.htm. Accessed January 23, 2011.

Bornstein, David. 2005. *How to Change the World*. Penguin Books.

Boudon, Raymond. 1986. *Theories of Social Change*. University of California Press.

Bourdieu, Pierre. 1990 [1963]. "Time Perspective of the Kabyle." Pp. 219–237 in John Hassard (ed.) *Sociology of Time*. St. Martin's.

Bowles, Samuel and Herbert Gintis. 1976. *Schooling in Capitalist America*. Basic Books.

Branch, Taylor. 1988. *Parting the Waters*. Simon and Schuster.

Brandt, Allan M. 2007. "The First Surgeon General's Report on Tobacco: Science and the State in the New Age of Chronic Disease." Pp. 437–456 in John W. Ward and Christian Warren (eds.) *Silent Victories*. Oxford University Press.

Braverman, Harry. 1974. *Labor and Monopoly Capital*. Monthly Review Press.

Brecher, Jeremy. 1977. *Strike!* South End.

Breyer, Stephen G. 2010. *Making Our Democracy Work*. Random House.

Brinton, Crane. 1965 [1938]. *The Anatomy of Revolution*. Revised and expanded edition. Random House.

Brooks, David. 2009. "Cell Phones, Tests and Lovers." *New York Times* (November 3): A25.

Brown, Clair. 1994. *American Standards of Living*. Blackwell.

Brown, David. 2004. *Social Blueprints*. Oxford University Press.

Bumiller, Elisabeth. 2010. "The War: A Trillion Can Be Cheap." *New York Times* (July 25): WK3.

Burbach, K. D. and Roger Burbach. 2000. *Globalize This!* Common Courage Press.

Burns, John F. 2010. "Britain Approves Tuition Increase as Violent Protests Erupt in London." *New York Times* (December 10): 12A.

Burnstein, Paul. 1999. "Social Movements and Public Policy." Pp. 3–21 in Marco Giugni, Doug McAdam and Charles Tilly (eds.) *How Social Movements Matter*. University of Minnesota Press.

Buroway, Michael. 1979. *Manufacturing Consent*. University of Chicago Press.

Buttel, Frederick H. and Kenneth A. Gould. 2006. "Environmentalism and the Trajectory of the Anti-Corporate Globalization Movement." Pp. 269–288 in Christopher Chase-Dunn and Salvatore J. Babones (eds.) *Global Social Change*. Johns Hopkins University Press.

Button, James W. 1978. *Black Violence: Political Impact of the 1960s Riots*. Princeton University Press.

Byrne, David. 2009. *Bicycle Diaries*. Viking.

Canetti, Elias. 1984 [1962]. *Crowds and Power*. Farrar Straus Giroux.

Carr, Patrick J. and Maria J. Kefalas. 2009. *Hollowing Out the Middle*. Beacon Press.

Carson, Clayborne, ed. 1997. *The Autobiography of Martin Luther King, Jr.* Warner Books.

Cassidy, John. 2010. "Enter the Dragon." *New Yorker* (December 13): 96–101.

Chan, Steve. 1985. "The Impact of Defense Spending on Economic Performance: A Survey of Evidence and Problems." *Orbis* 29: 403–434.

Chandler, Alfred D. Jr. 1965. *The Railroads*. Harcourt, Brace & World.

————. 1977. *Visible Hand*. Harvard University Press.

Chetham, Deidre. 2002. *Before the Deluge: The Vanishing World of the Yangtze's Three Gorges*. Palgrave.

Chirot, Daniel. 1994. *How Societies Change*. Pine Forge.

Chua, Amy. 2003. *World on Fire*. Doubleday.

Cohen, Stanley and Laurie Taylor. 1990. "Time and the Long-Term Prisoner." Pp. 170–187 in John Hassard (ed.) *The Sociology of Time*. St. Martin's.

Coleman, James S., Elihu Katz, and Herbert Menzel. 1966. *Medical Innovation: A Diffusion Study*. Bobbs-Merrill.

Collins, Randall. 1975. *Conflict Sociology: Toward an Explanatory Science*. With Joan Annett. Academic.

Coontz, Stephanie. 1992. *The Way We Never Were*. Basic Books.

Cortenraad, Wouter H. F. M. 2000. *The Corporate Paradox*. Kluwer Academic.

Coser, Lewis. 1957. "Social Conflict and the Theory of Social Change." *British Journal of Sociology* 8: 197–207.

———. 1967. "Greedy Organizations." *European Journal of Sociology* 8: 196–215.

———. 1971. *Masters of Sociological Thought*. Harcourt Brace Jovanovich.

Crane, Langdon T. et al. 1977. *Petroleum Industry Involvement in Alternative Sources of Energy*. Prepared at the Request of Frank Church for the Subcommittee on Energy Research and Development of the Committee on Energy and Natural Resources, United States Senate. U.S. Government Printing Office.

Cronne, H. A. 1939. "The Origins of Feudalism." *History* 24: 251–259.

Cross, Mary, ed. 2002. *A Century of American Icons*. Greenwood.

Crowther, Hal. 2010. "One Hundred Fears of Solitude." *Granta* (Summer): 97–117.

CRP. 2010. "Lobbying Database." Center for Responsive Politics. Available at http://www.opensecrets.org/lobby.index.php. Accessed November 2, 2010.

Cuzzort, R. P. and Edith W. King. 2002. *Twentieth Century Social Thought*. Harcourt Brace.

Dalfiume, Richard M. 1969. *Desegregation of the U.S. Armed Forces*. University of Missouri Press.

D'Arista, Jane. 2007. "Financial Concentration." Wall Street Watch, Working Paper No. 3. Political Economy Research Institute, University of Massachusetts, Amherst.

Darroch, Jacqueline E. 2007. "Family Planning: A Century of Change." Pp. 253–278 in John W. Ward and Christian Warren (eds.) *Silent Victories*. Oxford University Press.

Davey, Monica. 2010. "Doctor's Killer Puts Abortion on the Stand." *New York Times* (January 29): 1A, 11A.

Davies, James C. 1969. "The J-curve of Rising and Declining Satisfactions as a Cause of Some Great Evolutions and a Contained Rebellion." Pp. 690–730 in Hugh Davis Graham and Ted Robert Gurr (eds.) *The History of Violence in America*. Praeger.

Davis, James A. 1996. "Patterns of Attitude Change in the USA: 1972–1994." Pp. 151–183 in Bridget Taylor and Katarina Thomson (eds.) *Understanding Change in Social Attitudes*. Dartmouth University Press.

Deák, István. 2002. "The Crime of the Century." *New York Review* (September 26): 48–51.

Dellinger, Ann M., David A. Sleet and Bruce H. Jones. 2007. "Drivers, Wheels, and Roads: Motor Vehicle Safety in the Twentieth Century." Pp. 343–362 in John W. Ward and Christian Warren (eds.) *Silent Victories*. Oxford University Press.

Democracy Now. 2010. "On Front Lines of BP Oil Spill: Democracy Now! Travels Across Coastal Louisiana." Available at http://www.democracynow.org./2010/6/2/democracy_now_travels_across_the_bayous_and. Accessed July 1, 2010.

de Tocqueville, Alexis. 1945 [1839]. *Democracy in America.* A. A. Knopf.

Deutschmann, Paul J. and Orlando Fals Borda. 1962. "Communication and Adoption Patterns in an Andean Village." Report, Progtrama Interamericano de Informatión Popular, San Jose, Costa Rica.

Diamond, Jared. 1997. *Guns, Germs, and Steel.* W. W. Norton.

———. 2005. *Collapse.* Viking.

Domenico, Desirae M. and Karen H. Jones. 2006. "Career Aspirations of Women in the 20th Century." *Journal of Career and Technical Education* 2: 1–9.

Domhoff, G. William. 1967. *Who Rules America?* Prentice-Hall.

———. 1978. *The Powers That Be.* Random House.

———. 1990. *The Power Elite and the State.* Aldine De Gruyer.

Doukas, Dimitra. 2003. *Worked Over.* Cornell University Press.

Drew, Elizabeth. 1992. "Letter from Washington." *New Yorker* 68 (July 6): 70–75.

Durkheim, Emile. 1950 [1938]. *The Rules of Sociological Method.* The Free Press.

———. 1956 [1933]. *The Division of Labor in Society.* The Free Press.

Dworkin, Ronald G. 2008. *The Supreme Court Phalanx.* Random House.

Dyson, Freeman. 2010. "Strangest Man: The Hidden Life of Paul Dirac, Mystic of the Atom." *New York Review* (February 25): 20–23.

Earl, Jennifer. 2004. "The Cultural Consequences of Social Movements." Pp. 508–530 in David A. Snow, Sarah A. Soule, and Hanspeter Kriese (eds.) *Blackwell Companion to Social Movements.* Blackwell.

Easterlin, Richard A. 1987. *Birth and Fortune.* 2nd edition. University of Chicago Press.

Easterly, William. 2001. *The Elusive Quest for Growth.* MIT Press.

Eckholm, Erik. 2010. "Saying No to 'I Do,' Economy in Mind." *New York Times* (September 29): A14.

Economist. 2001. "The Right to Good Ideas." *The Economist,* 359 (June 23): 21–23.

———. 2009a. "Falling Fertility." *The Economist,* 393 (October 31): 15.

———. 2009b. "China's Place in the World." *The Economist,* 288 (October 3): 12.

———. 2010a. "The Beijing Consensus Is to Keep Quiet." *The Economist* (May 8): 41–42.

———. 2010b. "To the Last Drop." *The Economist* (May 22): Special Report Pp. 17–19.

Edwards, Richard. 1979. *Contested Terrain.* Basic Books.

Edwards, Paul and Judy Wajcman. 2005. *The Politics of Working Life.* Oxford University Press.

Eisenhower, Dwight D. 1960. "Farewell Speech" Available at www.americanrhetoric.com/speeches/dwightdeisenhowerfarwell.html. Accessed November 2, 2010.

Eisenstadt, S. N. 1964. "Social Change, Differentiation and Evolution." *American Sociological Review* 29: 375–386.

Elder, Glen H. 1974. *Children of the Great Depression.* University of Chicago Press.

Eliasoph, Nina. 1998. *Avoiding Politics.* Cambridge University Press.

Eriksen, Michael P., Lawrence W. Green, Corinne G. Husten, Linda L. Pederson, and Terry F. Pechanek. 2007. "Thank You for Not Smoking: The Public Health Response to Tobacco-Related Mortality in the United States." Pp. 423–436 in John W. Ward and Christian Warren (eds.) *Silent Victories.* Oxford University Press.

Evans, David S. 2002. "Who Owns Ideas?" *Foreign Affairs* 81 (November/December): 160–166.

Ewen, Stuart and Elizabeth Ewen. 1992. *Channels of Desire*. University of Minnesota Press.

Faludi, Susan. 1991. *Beauty Myth: How Images of Beauty are Used Against Women*. William Morrow.

Fathi, Nazila. 2010. "Iran: The Deadly Game." *New York Review* (February 25): 12–14.

Federal Writers' Project. 1939. *These Were Our Lives*. University of North Carolina Press.

Fischer, Claude S. and Michael Hout. 2006. *Century of Difference*. Russell Sage Foundation.

Fischer, David Hackett. 1970. *Historians' Fallacies*. Harper & Row.

Fogerty, Robert S. 1972. *American Utopianism*. F. E. Peacock.

Foner, Eric. 2010. *The Fiery Trail: Abraham Lincoln and American Slavery*. W. W. Norton.

Frank, Andre Gunder. 1966. "The Development of Underdevelopment." *Monthly Review* 18: 17–31.

Franklin, J. H. 1956. *From Slavery to Freedom: A History of American Negroes*. 2nd edition. Knopf.

Friedan, Betty. 1963. *The Feminine Mystique*. W. W. Norton.

———. 1993. *The Fountain of Age*. Simon and Schuster.

Friedman, Milton. 1962. *Capitalism and Freedom*. University of Chicago Press.

Furstenberg, Frank F., Sheela Kennedy, Vonnie C. McLoyd, Ruben G. Rumbaut, and Richard A. Settersten. 2004. "Growing Up Is Harder to Do." *Contexts* 3 (3): 33–41.

Fussell, Paul. 1989. *Wartime: Understanding and Behavior in the Second World War*. Oxford University Press.

Gage, Beverly. 2009. *The Day Wall Street Exploded*. Oxford University Press.

Galbraith, John Kenneth. 1956. *The Affluent Society*. Houghton Mifflin.

———. 1967. *The New Industrial State*. Houghton Mifflin.

Gamson, William A. 1968. *Power and Discontent*. Dorsey Press.

———. 1975. *The Strategy of Social Protest*. Dorsey Press.

Gartman, David. 1979. "Origins of the Assembly Line and Capitalist Control of Work at Ford." Pp. 193–205 in Andrew Zimbalist (ed.) *Case Studies on the Labor Process*. Monthly Review Press.

Gelbard, Alene, Carl Haub, and Mary M. Kent. 1999. "World Population beyond Six Billion." *Population Bulletin* 50 (1): 1–44.

Gelman, Andrew, Jeffrey Lax, and Justin Phillips. 2010. "Over Time, a Gay Marriage Groundswell." *New York Times*. (August 22): WK 3.

Genovese, Eugene. 1976. *Roll, Jordan, Roll*. Vintage.

Germino, Dante L. 1972. *Modern Western Political Thought: Machiavelli to Marx*. Rand McNally.

Gerstle, Gary. 2008. "The Immigrant as Threat to American Security." Pp. 217–245 in Elliott R. Barkan, Hasia Diner, and Alan M. Kraut (eds.) *From Arrival to Incorporation*. New York University Press.

Gitlin, Todd. 1980. *The Whole World Is Watching: Mass Media and the Making and Unmaking of the New Left*. University of California Press.

Gittings, John. 2005. *The Changing Face of China*. Oxford University Press.

Giugni, Marco G. 1998. "Was It Worth the Effort? The Outcomes and Consequences of Social Movements." *Annual Review of Sociology* 98: 371–393.

———. 2004. "Personal and Biographical Consequences." Pp. 497–507 in *Blackwell Companion to Social Movements*. David Snow et al. (eds.). Blackwell.

Glaeser, Bernhard, ed. 1987. *The Green Revolution Revisited*. Allen and Unwin.

Glenn, Norval D. 1987. "Social Trends in the United States (Evidence from Sample Surveys)." *Public Opinion Quarterly* 51: 109–126.

———. 2005. *Cohort Analysis*. 2nd edition. Sage.

Goldin, Claudia Dale. 1990. *Understanding the Gender Gap*. Oxford University Press.

Goldman, Robert and Stephen Papson. 1998. *Nike Culture*. Sage.

Goldstein, Melvyn C. 1976. "Fraternal Polyandry and Fertility in a High Himalayan Valley in Northwest Nepal." *Human Ecology* 4: 223–233.

Goldstone, Jack. 1986. "The Comparative and Historical Study of Revolutions." Pp. 1–20 of Jack Goldstone (ed.) *Revolutions*. Thompson.

Goldthorpe, John, David Lockwood, Frank Bechofer, and Jennifer Platt. 1969. *Affluent Worker in the Class Structure*. Cambridge University Press.

Goodman, J. David. 2010. "Bike Lanes Proliferate, and Protest Gets Louder." *New York Times* (November 23): A23–24.

Gordinier, Jeff. 2008. *X Saves the World*. Viking Press.

Gottschalk, Louis, Clyde Kluckholm, and Robert Angell. 1945. *The Uses of Personal Documents in History, Anthropology and Sociology*. Social Science Research Council.

Gould, Stephen Jay. 1980. "The Episodic Nature of Evolutionary Change." in his *The Panda's Thumb*. W. W. Norton.

———. 1981. *The Mismeasure of Man*. W. W. Norton.

Gourevitch, Philip. 1994. *We Wish to Inform You That Tomorrow We Will Be Killed with Our Families*. Farrah, Straus, and Giroux.

Granovetter, Mark. 1984. "Small Is Bountiful: Labor Markets and Establishment Size." *American Sociological Review* 49: 323–334.

Gravel, Mike, ed. 1971–72. *Pentagon Papers: The Defense Department History of the United States Decisionmaking in Vietnam*. Beacon Press.

Graybill, Lyn. 2002. *Truth and Reconciliation in South Africa: Miracle or Model?* Lynne Rienner.

Griffin, Keith. 1974. *The Political Economy of Agrarian Change*. Harvard University Press.

Griffin, Larry J. 2004. "'Generations and Collective Memory' Revisited: Race, Region, and Memory of Civil Rights." *American Sociological Review* 69: 544-557.

Griffin, Larry J., Michael E. Wallace, and Beth A. Rubin. 1986. "Capitalist Resistance to the Organization of Labor before the New Deal: Why? How? Success?" *American Sociological Review* 51: 147–167.

Gurivitch, George. 1964. "Varieties of Social-Time" in his *The Spectrum of Social Time*. D. Reidel.

Gurr, Ted Robert. 1970. *Why Men Rebel*. Princeton University Press.

Guthrie, Doug. 2009. *China and Globalization*. Revised edition. Routledge.

Habermas, Jurgen. 1970. *Toward a Rational Society*. Translated by Jeremy J. Shapiro. Beacon Press.

Hacker, Jacob S. and Paul Pierson. 2010. *Winner-Take-All Politics*. Simon and Schuster.

Halper, Stefan. 2010. *The Beijing Consensus: How China's Authoritarian Model Will Dominate the Twenty-First Century*. Basic Books.

Halperin, Sandra. 2004. *War and Social Change in Modern Europe*. Cambridge University Press.

Hampden-Turner, Charles. 1970. *Radical Man*. Schekman.

Hareven, Tamara K. 1991. "The History of the Family and the Complexity of Social Change." *American Historical Review* 96: 95–124.

Harff, Barbara and Ted Robert Gurr. 2004. *Ethnic Conflict in World Politics*. 2nd edition. Westview.

Harper, Charles L. and Kevin T. Leicht. 2007. *Exploring Social Change*. 5th edition. Pearson Prentice-Hall.

Harris, Marvin. 1979. *Cultural Materialism*. Vintage.

Hasso, Frances S. 2001. "Feminist Generations? The Long-term Impact of Social Movement Involvement on Palestinian Women's Lives." *American Journal of Sociology* 107: 586–611.

Hawken, Paul. 1993. *The Ecology of Commerce*. HarperCollins.

Hawken, Paul, Amory Lovins, and L. Hunter Lovins. 1999. *Natural Capitalism*. Little Brown and Company.

Hawking, Stephen. 1996. *The Illustrated A Brief History of Time*. Bantam Books.

Headrick, Daniel R. 1981. *The Tools of Empire*. Oxford University Press.

Herrera, Geoffrey L. 2006. *Technology and International Transformation*. State University of New York Press.

Hertz, Nora. 2010. *Silent Takeover*. New York University Press.

Hewstone, Miles and Rupert Brown (eds.). 1986. *Contact and Conflict in Intergroup Encounters*. Basil Blackwell.

Hinton, William. 1966. *Fanshen*. Vintage.

Hirschman, Albert O. 1970. *Exit, Voice, and Loyalty*. Harvard University Press.

Hochschild, Arlie Russell. 1989. *The Second Shift*. Viking.

———. 1997. *Time Bind*. Henry Hold and Company.

Hodson, Randy. 1999a. *Analyzing Documentary Accounts*. Sage.

———. 1999b. "Management Citizenship Behavior: A New Concept and an Empirical Test." *Social Problems* 46: 460–478.

———. 2001. *Dignity at Work*. Cambridge University.

Hodson, Randy, Duško Sekulić, and Garth Massey. 1994. "National Tolerance in the Former Yugoslavia." *American Journal of Sociology* 99: 1534–1558.

Hodson, Randy and Teresa Sullivan. 2008. *The Social Organization of Work*. 4th edition. Wadsworth.

Hoffer, Eric. 1951. *The True Believer*. Harper & Row.

Hogan, Michael J. 1987. *The Marshall Plan*. Cambridge University Press.

Hooks, Gregory and James Rice. 2005. "War, Militarism, and States: The Insights and Blind Spots of Political Sociology." Pp. 556–584 in Thomas Janoski, Robert R. Alford, Alexander M. Hicks, and Mildred A. Schwartz (eds.) *Handbook of Political Sociology*. Cambridge University Press.

Hooks, Gregory and Chad L. Smith. 2005. "Treadmills of Production and Destruction: Threats to the Environment Posed by Militarism." *Organizations and Environment* 18: 19–37.

Horton, John. 1964. "The Dehumanization of Anomie and Alienation." *British Journal of Sociology* 15: 283–300.

Hu, Winnie. 2010. "In a New Role, Teachers Move to Run Schools." *New York Times.* (September 7): A1, A20.

Hughes, Mary Elizabeth and Angela M. O'Rand. 2002. "The Lives and Times of the Baby Boomers." American People: Census 2000 Series. U.S. Government Printing Office.

Jackson, Toby. 1977. "Parsons' Theory of Social Evolution." Pp.1–23 in Talcott Parsons' *The Evolution of Societies*. Prentice-Hall.

Jackson, Wes. 2010. *Consulting the Genius of the Place.* Counterpoint.

Jacques, Martin. 2009. *When China Rules the World.* Penguin Books.

Johnson, Chalmers. 2000. *Blowback.* Owl Books.

———. 2004. *The Sorrows of Empire.* Metropolitan Books.

———. 2006. *Nemesis.* Metropolitan Books.

———. 2010. *Dismantling the Empire.* Metropolitan Books.

Johnson, Ian. 2010. "The Party: Impenetrable, All Powerful" *New York Review* (September 30): 69–72.

Johnson, Steven. 2006. *Ghost Maps.* Penguin.

Kagarlitsky, Boris. 2008. *Empire of the Periphery.* Pluto Press.

Kaplan, Robert. 1993. *Balkan Ghosts: A Journey through History.* Paparmac.

Kateb, George. 1972. *Utopia and Its Enemies.* Schocken Books.

Keynes, John Maynard. 1936. *The General Theory of Employment, Interest and Money.* Harcourt, Brace.

King, Brayden G. and Sarah A. Soule. 2007. "Social Movements as Extra-Institutional Entrepreneurs: The Effect of Protest on Stock Price Returns." *Administrative Science Quarterly* 52: 413–442.

Klare, Michael T. 2001. *Resource Wars.* Owl Books.

Kolb, Felix. 2007. *Protest and Opportunity.* Compus/Verlag.

Kornhauser, William. 1959. *The Politics of Mass Society.* The Free Press.

Korten, David C. 1999. *The Post-Corporate World.* Kumarian Press.

———. 2001. *When Corporations Rule the World.* 2nd edition. Kumarian Press.

Kotelchuck, Milton. 2007. "Safe Mothers, Healthy Babies: Reproductive Health in the Twentieth Century." Pp. 105–134 in John W. Ward and Christian Warren (eds.) *Silent Victories*. Oxford University Press.

Kotkin, Stephen. 2009. *Uncivil Society: 1989 and the Implosion of the Communist Establishment.* Modern Library.

Krolokke, Charlotte and Anne Scott Sorensen. 2006. *Gender Communication Theories and Analysis.* Sage.

Kryder, Daniel. 2000. *Divided Arsenal.* Cambridge University Press.

Kuhn, Thomas. 1970. *The Structure of Scientific Revolutions.* University of Chicago Press.

La Porta, Rafael, Florencio Lopez-De-Silanes, and Andrei Shleifer. 1999. "Corporate Ownership Around the World." *Journal of Finance* 54: 471–517.

Laird, Pamela Walker. 1998. *Advertising Progress.* Johns Hopkins University Press.

Lansing, Stephen J. 1987. "Balinese 'Water Temples' and the Management of Irrigation." *American Anthropologist* 89: 326–341.

Lauer, Robert H. 1991. *Perspectives on Social Change.* Allyn and Bacon.

Laumann, Edward O., Robert T. Michael, and John H. Gagnon. 1994. "A Political History of the National Sex Survey of Adults." *Family Planning Perspectives* 26: 34–38.

Lazarsfeld, P. F. 1970. "Sociology." Pp. 61–165 in *Main Trends of Research in the Social and Human Sciences*. Part I, *Social Sciences*. Mouton/UNESCO.

Le Bon, Gustav. 1903. *The Crowd*. T. F. Unwin.

Lee, Taeku. 2002. *Mobilizing Public Opinion*. University of Chicago Press.

Leibowitz, Arleen and Jacob Alex Klerman. 1995. "Explaining Changes in Married Mothers' Employment over Time." *Demography* 32: 365–378.

Lemann, Nicholas. 1991. *The Promised Land*. A. A. Knopf.

Lenski, Gerhard. 1986. *Power and Privilege*. University of North Carolina Press.

Levine, David N., ed. 1971. *George Simmel (On Individuality and Social Forms)*. Translated by David N. Levine. University of Chicago Press.

Light, Paul C. 2008. *The Search for Social Entrepreneurship*. Brookings Institution.

Lindbloom, Eric. 2010. "Toll of Tobacco in the United States of America" Available at www.tobaccofreekids.org/research/factsheets/pdf/0072.pdf. Accessed November 19, 2010.

Liptak, Adam. 2010. "At 89, Stevens Contemplates the Law, and How to Leave It." *New York Times* (April 4): A1, A3.

Lipton, Eric, Mike McIntire, and Don Van Natta Jr. 2010. "Large Donations Aid U.S. Chamber in Election Drive." *New York Times* (October 22): 1 Op.

Luo, Michael and Griff Palmer. 2010. "A Surge in Democratic Spending." *New York Times* (October 30): A11.

Madrick, Jeff. 2011. "The Wall Street Leviathan." *New York Review*. April 28. 70–73.

Mailer, Norman. 1957. *The White Negro*. City Light Books.

Mann, Michael. 1988. *States, War and Capitalism*. Basil Blackwell.

Mannheim, Karl. 1936. *Ideology and Utopia*. Harcourt, Brace & World.

Mapes, Jeff. 2009. *Pedaling Revolution: How Cyclists Are Changing America*. Oregon State University Press.

Marcuse, Herbert. 1964. *One-Dimensional Man*. Beacon Press.

Marwick, Arthur. 1974. *War and Social Change in the Twentieth Century*. St. Martin's Press.

Marx, Karl. 1956. *Karl Marx: Selected Writings on Sociology and Social Philosophy*. Edited and translated by T. B. Bottomore. McGraw-Hill.

———. 1964 [1852]. *The Eighteenth Brumaire of Louis Bonaparte*. International.

———. 1967. *Capital*. 3 Volumes. International.

———. 1967 [1844]. *Writings of the Young Marx on Philosophy and Society*. Edited and translated by Loyd D. Easton and Kurt H. Guddat. Anchor Books.

Marx, Karl and Frederich Engels. 1978 [1848]. "Manifesto of the Communist Party." Pp. 469–511 in Robert C. Tucker (ed.) *The Marx-Engels Reader*. 2nd edition. W. W. Norton.

Massey, Garth. 1986. *Subsistence and Change: Lessons of Agropastoralism in Somalia*. Westview.

Mayer, Jane. 2010. "Covert Operation." *New Yorker* (April 20): 44–55.

McAdam, Doug. 1988. *Freedom Summer*. Oxford University Press.

———. 1999a. *Political Process and the Development of Black Insurgency 1930–1970*. Revised edition. University of Chicago Press.

———. 1999b. "The Biographical Impact of Social Movements." Pp. 117–146 in Marco Giugni, Doug McAdam, and Charles Tilly (eds.) *How Social Movements Matter*. University of Minnesota Press.

McAdam, Doug and Yang Su. 2002. "The War at Home: Anti-War Protests and Congressional Voting, 1965–1973." *American Sociological Review* 67: 696–721.

McAdam, Doug, Sidney Tarrow, and Charles Tilly. 2001. *Dynamics of Contention.* Cambridge University Press.

McGregor, Richard. 2010. *The Party: The Secret World of China's Communist Rulers.* Harper.

McIntyre, Alice. 2008. *Participatory Action Research.* Sage.

McLuhan, Marshall. 1962. *Gutenberg's Galaxy: The Making of Telegraphic Man.* University of Toronto Press.

McMahon, Kevin J. 2004. *Reconsidering Roosevelt on Race.* University of Chicago Press.

McMichael, Philip. 2004. *Development and Social Change.* 3rd edition. Pine Forge.

McMillan, Neil. 1989. *Dark Journey.* University of Illinois Press.

McNeal, James U. 2007. *On Becoming a Consumer.* Elsevier.

McNeil, Donald G. Jr. 2009. "Brazil and India Join the Top Ranks of Governments Supporting Research." *New York Times* (December 22): D6.

McPhee, John. 1989. *Control of Nature.* Farrah, Straus, Giroux.

Meier, August and Elliott Rudwick. 1973. *CORE: A Study in the Civil Rights Movement 1942–1968.* Oxford University Press.

Melman, Seymour. 1985. *The Permanent War Economy.* Simon and Schuster.

Melosi, Martin V. 2005. *Garbage in the Cities.* Revised edition. University of Pittsburgh Press.

Merton, Robert K. 1968. *Social Theory and Social Structure.* 2nd edition. The Free Press.

Michelson, Ethan. 2007. "Lawyers, Political Embeddedness, and Institutional Continuity in China's Transition to Socialism." *American Journal of Sociology,* 113: 352–313.

Micklethwait, John and Adrian Wooldridge. 2005. *The Company.* Modern Library.

Miliband, Ralph. 1969. *The State in Capitalist Society.* Basic Books.

Miller, Claire Cain. 2009. "The Cell Refusniks, an Ever-Shrinking Club." *New York Times* (October 23): B1, B5.

Miller, Danny. 1990. *The Icarus Paradox.* HarperCollins.

Mills, C. Wright. 1940. "Situated Actions and Vocabularies of Motive." *American Sociological Review* 5: 905–913.

———. 1959. *The Sociological Imagination.* Oxford University Press.

Mintz, Alex and Randolph Stevenson. 1995. "Defense Expenditures, Economic Growth, and the Peace Dividend." *Journal of Conflict Resolution* 39: 283–305.

Mokyr, Joel. 1990. *The Lever of Riches.* Oxford University Press.

Mollmann, Marianne. 2010. "Illusion of Care." Human Rights Watch Report. Available at www.hrw.org/en/reports/2010/08/10/illusions-care-0. Accessed August 19, 2010.

Montgomery, David. 1979. *Workers' Control in America.* Cambridge University Press.

Monthly Labor Review. 2007. "Changes in Men's and Women's Labor Force Participation Rates." *Monthly Labor Review.* Available at http://www.bls.gov/opub/ted/2007/jan/wk2/art03.htm. Accessed November 26, 2010.

Moore, Barrington. 1966. *Social Origins of Dictatorship and Democracy.* Beacon Press.

Moore, Kelly. 2008. *Disrupting Science.* Princeton University Press.

Moore, Wilbert E., ed. 1972. *Technology and Social Change.* Quadrangle Books.

Morris, Alden. 1984. *Origin of the Civil Rights Movement.* The Free Press

Moskos, Charles C. 1986. "Success Story: Blacks in the Army." *The Atlantic* 257 (5): 64–72.

Mueller, John. 1973. *War, Presidents and Public Opinion.* Wiley.

Naimark, Norman M. 2001. *Fires of Hatred.* Harvard University Press.

Nair, Kusum. 1979. *In Defense of the Irrational Peasant*. University of Chicago Press.

NCDC. 2011. "State of the Climate Global Analysis Annual 2010." National Climate Data Center. Available at www.ncdc.noaa.gov/sotc/global/. Accessed January 17, 2011.

Nisbet, Robert A. 1969. *History and Social Change*. Oxford University Press.

Noble, David F. 1977. *America by Design*. Alfred A. Knopf.

Nolan, Patrick and Gerhard Lenski. 2009. *Human Societies*. Paradigm.

Nordhoff, Charles. 1966 [1875]. *Communistic Societies of the United States*. Dover.

NPR. 2010. "An Evangelical Crusade to go Green with God." Weekend Edition Sunday, National Public Radio (June 27). Available at http://www.npr.org/templates/rundowns/rundown.php?prgID=10&prgDate=6-27-10. Accessed November 16, 2010.

Nye, David E. 1990. *Electrifying America*. MIT Press.

NYT. 2010. "The Price of Broadband Politics." *New York Times* (June 30): A24.

Ohmann, Richard. 1996. *Selling Culture*. Verso.

Olson, Mancur. 1965. *The Logic of Collective Action*. Harvard University Press.

Olson, Philip G. 1963. *America as a Mass Society*. The Free Press.

Oreskes, Naomi. 2007. "The Scientific Consensus on Climate Change: How Do We Know We're Not Wrong?" Pp. 65–99 in Joseph F. C. Dimento and Pamela Doughman (eds.) *Climate Change*. MIT Press.

Oreskes, Naomi and Erik M. Conway. 2010. *Merchants of Doubt*. St. Martin's Press.

Orwell, George. 1958. *Road to Wigan Pier*. Harcourt Brace Jovanovich.

Panagopoulos, Costas and Peter L. Francia. 2008. "Labor Unions in the United States." *Public Opinion Quarterly* 72: 134–159.

Parsons, Talcott. 1951. *The Social System*. The Free Press.

Payne, Charles M. 1995. *I've Got the Light of Freedom*. University of California Press.

Pelto, Pertti J. 1973. *The Snowmobile Revolution: Technology and Social Change in the Arctic*. Cummings.

Perrin, Noel. 1979. *Giving up the Gun: Japan's Reversion to the Sword 1543–1979*. David R. Golding.

Pettegree, Andrew. 2010. *The Book of the Renaissance*. Yale University Press.

Phillips, Kevin. 1969. *The Emerging Republican Majority*. Arlington House.

Pinch, Trevor J. and Wiebe Bijker. 1987. "The Social Construction of Facts and Artifacts: Or How the Sociology of Science and the Sociology of Technology Might Benefit Each Other." Pp. 17–50 in Wiebe E. Bijker, Thomas P. Hughes, and Trevor J. Pinch (eds.) *The Social Construction of Technological Systems*. MIT Press.

PISA. 2009. "OECD, PISA 2009 Results: Executive Summary." Available at www.pisa.oecd.org/dataoecd/34/60/46619703.pdf. Accessed December 11, 2010.

Pisani, Donald J. 1984. *From Family Farm to Agribusiness*. University of California Press.

Piven, Frances Fox and Richard Cloward. 1979. *Poor People's Movements*. Vintage.

Polenberg, Richard. 1972. *War and Society*. J. B. Lippincott.

———. 1992. "The Good War? A Reappraisal of How World War II Affected American Society." *Virginia Magazine of History and Biography* 100 (July): 295–309.

Pollan, Michael. 2010. "The Food Movement, Rising." *New York Review* (June 10): 31–33.

Powdermaker, Hortense. 1968 [1939]. *After Freedom: A Cultural Study of the Deep South*. Atheneum.

Power, Samantha. 2002. *The Problem From Hell and the Age of Genocide*. HarperCollins.

Pugh, Allison J. 2009. *Longing and Belonging: Parents, Children, and Consumer Culture*. University of California Press.

Putnam, Robert D. 2000. *Bowling Alone*. Simon and Schuster.

Putnam, Robert D. and David C. Campbell. 2010. *American Grace*. Simon and Schuster.

Quadango, Jill S. 1984. "Welfare Capitalism and the Social Security Act of 1935." *American Sociological Review* 49: 632–647.

Ram, Rati. 1986. "Government Size and Economic Growth: A New Framework and Some Evidence from Cross-sectional Data." *American Economic Review* 76: 191–203.

Ramakrishnan, Venkitexh. 2005. "The Naxalite Challenge." *Frontline* 22: 13–19.

Ramet, Sabrina. 2004. "For a Charm of Powerful Trouble, Like a Hell-broth and Bubble: Theories About the Roots of the Yugoslav Troubles." *Nationalist Papers* 32: 731–763.

Ramos, Joshua Cooper. 2004. "The Beijing Consensus." Foreign Policy Center, London, UK. Available at www.fsblob/244.pdf. Accessed November 4, 2010.

Rapper, Arthur. 1969 [1933]. *The Tragedy of Lynching*. New American Library.

Rawick, George P. 1972–79. *The American Slave: A Composite Autobiography*. Greenwood.

Regan, Patrick M. 2002. "Third Party Intervention and the Duration of Intrastate Conflicts." *Journal of Conflict Resolution*. 46 (February): 55–73.

Reisner, Marc. 1986. *Cadillac Desert*. Viking Penguin.

Rex, John. 1961. *Key Problems of Sociological Theory*. Routledge and Kegan Paul.

Reynolds, Farley. 1996. *The New American Reality*. Russell Sage Foundation.

Ringdal, Gerd Inger and Kristen Ringdal. 2011. "War Experiences and War-Related Distress in Bosnia-Herzegovina, Croatia, and Kosovo." In Kristen Ringdal and Albert Simkus (eds.) *Aftermath of War*. Central European University Press (forthcoming).

Robbins, Anthony and Philip J. Landrigan. 2007. "Safer, Healthier Workers: Advances in Occupational Disease and Injury Prevention." Pp. 209–229 in John W. Ward and Christian Warren (eds.) *Silent Victories*. Oxford University Press.

Robinson, John P. and Geoffrey Godbey. 1997. *Time for Life*. Penn State University Press.

Robinson, Michael C. 1979. *Water for the West. The Bureau of Reclamation 1902–1977*. Chicago: Public Works Historical Society.

Rogers, Everett M. 2003. *Diffusion of Innovations*. 5th edition. The Free Press.

Rogers, Heather. 2005. *Gone Tomorrow: The Hidden Life of Garbage*. The New Press.

Rojas-Burke, Joe. 2009. "OHSU Team Makes Brain Breakthrough." *Oregonian* (December 2): C2.

Rosen, Stanley. 2000. "Forward." Pp. xi–xxviii in *Mao's Children in the New China*. Yarong Jiang and David Ashley. Routledge.

Rothenberg, Paula S., ed. 2006. *Beyond Borders*. Worth.

Roy, William G. 1997. *Socializing Capital*. Princeton University Press.

———. 2001. *Making Societies*. Pine Forge.

Rubin, Beth. 1996. *Shifts in the Social Contract*. Pine Forge.

Rudin, Ken and Alex Chadwick. 2005. "William Proxmire: Crusader Against Government Waste." Broadcast on "Day to Day," National Public Radio (December 15).

Russell, D. E. H. 1974. *Rebellion, Revolution, and Armed Forces*. Academic Press.

Ryan, Bryce and Neal C. Gross. 1943. "The Diffusion of Hybrid Seed Corn in Two Iowa Communities." *Rural Sociology* 8: 15–24.

Ryder, Norman B. 1965. "The Cohort as a Concept in the Study of Social Change." *American Sociological Review* 30: 843–861.

Samuelsson, Kurt. 1961. *Religion and Economic Action.* William Heinemann Ltd.

Schoenhals, Michael. 1986. "Original Contradictions—on the Unrevised Text of Mao Zedong's 'On the Correct Handling of Contradictions Among the People.'" *Australian Journal of Chinese Affairs,* 16: 99–112.

Schor, Juliet. 1991. *The Overworked American.* Basic Books.

———. 2004. *Time to Buy.* Scribner.

Schumacher, E. F. 1973. *Small Is Beautiful.* Harper & Row.

Schumaker, Paul D. 1975. "Policy Responsiveness to Protest-group Demands." *Journal of Politics* 37: 488–521.

Schuman, Howard and Jacqueline Scott. 1989. "Generations and Collective Memory." *American Sociological Review* 54: 359–381.

Schutt, Russell K. 2009. *Investigating the Social World.* 6th edition. Pine Forge.

Scott, James C. 1976. *The Moral Economy of the Peasant.* Yale University Press.

———. 1985. *Weapons of the Weak.* Yale University Press.

Seeman, Melvin. 1959. "On the Meaning of Alienation." *American Sociological Review* 24: 783–791.

Segal, Howard P. 1985. *Technological Utopianism in American Culture.* University of Chicago Press.

Sekulić, Duško. 2011. "Ethnic Intolerance as a Product Rather Than a Cause of War: Revisiting the State of the Art." Chapter 3 in Dario Spini, Dinka Corkalo Biruski, and Guy Elcheroth (eds.) *War and Community* (forthcoming).

Sekulić, Duško, Garth Massey, and Randy Hodson. 2006. "Ethnic Tolerance and Ethnic Conflict in the Dissolution of Yugoslavia." *Ethnic and Racial Studies* 29: 797–827.

Sellers, Christopher. 2007. "A Prejudice Which May Cloud the Mentality: The Making of Objectivity in Early Twentieth-Century Occupational Health." Pp. 230–250 in John W. Ward and Christian Warren (eds.) *Silent Victories.* Oxford University Press.

Sennett, Richard. 1998. *The Corrosion of Character.* W. W. Norton.

Shanahan, Michael J. and Ross Macmillan. 2008. *Biography and the Sociological Imagination.* W. W. Norton.

Sharp, Lauriston. 1952. "Steel Axes for Stone Age Australians." *Human Organization* 11: 17–22.

Shaw, Greg. 2003. "The Polls-Trends: Abortion." *Public Opinion Quarterly* 67: 407–429.

Shelley, Mary Wollstonecraft. 1969 [1818]. *Frankenstein; or, The Modern Prometheus.* Oxford University Press.

Sibley, D. 1988. "Purification of Space." *Environment and Planning D: Society and Space* 6: 409–421.

Simmel, George. 1950. *The Sociology of George Simmel.* Edited and translated by Kurt H. Wolff. The Free Press.

Singleton, Fred and Anthony Topham. 1963. *Workers Control in Yugoslavia.* Fabian Society.

Sivard, Ruth Leger. 1988. *World Military and Social Expenditures 1988.* World Priorities.

———. 1996. *World Military and Social Expenditures 1996.* World Priorities.

Sivulka, Juliann. 1998. *Soap, Sex, and Cigarettes.* Wadsworth.

Skocpol, Theda. 1979. *States and Social Revolutions.* Cambridge University Press.

———. 1992. *Protecting Soldiers and Mothers.* Belknap.

Small, Melvin and J. David Singer. 1982. *Resort to Arms.* Sage.

Smith, Anthony D. 1981. "War and Ethnicity: The Role of Warfare in the Formation, Self-Images and Cohesion of Ethnic Communities." *Ethnic and Racial Studies* 4: 375–397.

———. 1992. "Chosen People." *Ethnic and Racial Studies* 15: 440–449.

Smith, Douglas K. and Robert C. Alexander. 1988. *Fumbling the Future: How Xerox Invented, and Then Ignored, the First Personal Computer.* Morrow.

Snow, David, Daniel Cress, Liam Downey, and Andrew Jones. 1998. "Disrupting the Quotidian: Reconceptualizing the Relationship between Breakdown and the Emergence of Collective Action." *Mobilization: An International Journal* 3: 1–22.

Snow, David and Sarah Soule. 2010. *A Primer on Social Movements.* W. W. Norton.

Solomon, Stephen. 1981. "The Controversy over Infant Formula." *New York Times* (December 6): Science Digest, section 6, 92.

Solow, Robert M. 1957. "Technological Change and the Aggregate Production Function." *Review of Economics and Statistics* 39: 312–320.

Sorokin, Pitirim. 1937–41. *Social and Cultural Dynamics.* 4 volumes. American Book Company.

Soule, Sarah A. 2009. *Contention and Corporate Social Responsibility.* Cambridge University Press.

Soule, Sarah A., Doug McAdam, John McCarthy, and Yang Su. 1999. "Protest Events: Cause or Consequence of State Action? The U.S. Women's Movement and Federal Congressional Activities, 1956–1979." *Mobilization* 4: 239–250.

Staub, Ervin, Laurie Anne Pearlman, and Vachel Miller. 2003. "Healing the Roots in Rwanda." *Peace Review* 15: 287–294.

Stein, Rob and Donna St. George. 2009. "Number of Babies Born Out of Wedlock Jumps Sharply in the U.S." *Washington Post* (May 14): A6.

Stelter, Brian and Brad Stone. 2009. "Web Pries Lid off Iranian Censorship. *New York Times* (June 23): A1.

Stevens, John Paul. 2010. "*Citizens United v. Federal Election Commission.*" Available at www.supremecourt.gov/opinions/09pdf/08–205.pdf. Accessed May 11, 2010.

Stevens, Joseph E. 1988. *Hoover Dam.* University of Oklahoma Press.

Stiglitz, Joseph L. 2002. *Globalization and Its Discontents.* W. W. Norton.

Stiglitz, Joseph L. and Linda J. Bilmes. 2008. *The Three Trillion Dollar War.* W. W. Norton.

Stinchcombe, Arthur. 1978. *Theoretical Methods in Social History.* Academic Press.

Sullivan, Laura. 2010. "Prison Economics Help Drive Arizona Immigration Law." National Public Radio, October 28. Available at www.npr.org/templates/transcript/transcript.php?storyID=130833741. Accessed October 30, 2010.

Sumner, William Graham. 1954. "The Absurd Effort to Make the World Over" in Perry Miller (ed.) *American Social Thought: Civil War to World War I.* Rinehart.

Tainter, Joseph A. 1988. *The Collapse of Complex Societies.* Cambridge University Press.

Taniellian, Terri and Lisa H. Jaycox. 2008. *Invisible Wounds of War.* Rand Corporation.

Tanner, Murray Scot. 1994. "The Erosion of Communist Party Control Over Lawmaking in China." *China Quarterly* 138: 381–403.

Tawney, R. H. 1954 [1926]. *Religion and the Rise of Capitalism*. Harcourt Brace Jovanovich.

———. 1966 [1932]. *Land and Labor in China*. M. E. Sharpe.

Taylor, Verta and Nancy Whittier. 1992. "Identity Politics as High-Risk Activism: Career Consequences for Lesbian, Gay, and Bisexual Sociologists." *Social Problems* 42: 252–273.

Tedrow, Lucky M. and E. R. Mahoney. 1979. "Trends in Attitudes Toward Abortion: 1972–1976." *Public Opinion Quarterly* 43: 181–189.

Terkel, Studs. 1970. *Hard Times: An Oral History of the Great Depression*. Pantheon Books.

———. 1984. *"The Good War": An Oral History of World War Two*. Pantheon Books.

Thein, Alan. 1992. *How Much Is Enough? The Consumer Society and the Future of the Earth*. W. W. Norton.

Thomas, W. I. and Florian Znaniecki. 1958 [1918]. *The Polish Peasant in Poland and America*. Dover.

Thompson, E. P. 1963. *The Making of the English Working Class*. Random House.

Tilahun, Nebiyou Y., David M. Levinson, and Kevin J. Krizek. 2007. "Valuing Bicycle Facilities with an Adaptive Stated Preference Survey." *Transportation Research Part A: Policy & Practice* 41: 287–301.

Tilly, Charles. 1984. *Big Structures, Large Processes, Huge Comparisons*. Russell Sage Foundation.

———. 1992. *Coercion, Capital, and European States, AD 990–1992*. Blackwell.

———. 1999. "From Interactions to Outcomes in Social Movements." Pp. 253–270 in Marco Giugni, Doug McAdam, and Charles Tilly (eds.) *How Social Movements Matter*. University of Minnesota Press.

Trimberger, Kay. 1978. *Revolutions from Above*. Transaction Books.

Tsesis, Alexander. 2008. *We Shall Overcome*. Yale University Press.

Tsurumi, Kazuko. 1970. *Social Change and the Individual*. Princeton University Press.

Tufte, Edward R. 1997. *Visual Explanations*. Graphics Press.

Tuitt, Wesley B. 2006. *The Corporation*. Greenwood Press.

Turkel, Sherry. 2011. *Alone Together*. Basic Books.

Twenge, Jean. 2006. *Generation Me*. The Free Press.

UN. 2010. *World Fertility Report: 2007*. United Nations, Department of Economic and Social Affairs, Population Division. United Nations.

UNHCR. 2009. "UNHCR Annual Report Shows 42 Million People Uprooted Worldwide." (June 16). Available at www.unhcr.org/ra2fd52412d.html. Accessed April 29, 2010.

———. 2010. "UNHCR Statistical Online Population Database." United Nations High Commissioner for Refugees. Available at www.unhcr.org/statistics/populationdatabase. Accessed April 18, 2010.

USCRI. 2010. *World Refugee Survey 2008*. United States Committee for Refugees and Immigrants.

Useem, Michael. 1984. *The Inner Circle*. Oxford University Press.

van den Bulte, Christophe and Gary L. Lilien. 2001. "*Medical Innovation*" Revisited: Social Contagion versus Marketing Effort." *American Journal of Sociology* 106: 1409–1435.

Veblen, Thorstein. 1899. *Theory of the Leisure Class*. Macmillan Company.

————. 1967 [1923]. *Absentee Ownership and Business Enterprise in Recent Times.* Beacon Press.

Virgil. 2006. *The Aeneid.* Translated by Robert Fagles. Penguin Books.

Vogel, David. 1978. *Lobbying the Corporation.* Basic Books.

————. 1996. *Kindred Strangers.* Princeton University Press.

————. 2005. *The Market for Virtue.* Basic Books.

Volti, Rudi. 2001. *Society and Technological Change.* 4th edition. Worth.

Walder, Andrew G. 1989. "Social Change in Post-Revolution China." *Annual Review of Sociology* 15: 405–424.

Walker, Edward. 2009. "Privatizing Participation: Civic Change and the Organizational Dynamics of Grassroots Lobbying Firms." *American Sociological Review* 74: 83–105.

————. 2010. "Industry-Driven Activism." *Contexts* 9 (2): 44–49.

Wallerstein, Immanuel. 1974. *The Modern World System.* Academic Press.

Ward, John W. and Christian Warren, eds. 2007. *Silent Victories.* Oxford University Press.

Watson, Brad. 2010. *Aliens in the Prime of Their Lives.* W. W. Norton.

Watson, James L. 1997. "McDonald's in Hong Kong: Consumerism, Dietary Change, and the Rise of Children's Culture." Pp. 77–109 in James L. Watson (ed.) *Golden Arches East: McDonald's in East Asia.* Stanford University Press.

Weber, Max. 1946. "Politics as a Vocation." Pp. 77–128 in *From Max Weber: Essays in Sociology.* Translated, edited, and with an introduction by H. H. Gerth and C. Wright Mills. Oxford University Press.

————. 1949 [1904]. "'Objectivity' in Social Science and Social Policy." Pp. 49–112 in Edward A. Shills and Henry A. Finch (eds.) *Max Weber on the Methodology of the Social Sciences.* The Free Press.

————. 1958 [1920]. *The Protestant Ethic and the Spirit of Capitalism.* Charles Scribner's Sons.

————. 1964. *The Theory of Social and Economic Organization.* Edited and translated by Talcott Parson. The Free Press.

Weinstein, James. 1968. *The Corporate Ideal in the Liberal State 1900–1918.* Beacon Press.

Werum, Regina and Bill Winders. 2001. "Who's 'In' and Who's 'Out': State Fragmentation and the Struggle over Gay Rights, 1974–1990." *Social Problems* 48: 386–410.

White, Lynn. 1962. *Medieval Technology and Social Change.* Oxford University Press.

Whittier, Nancy. 1997. "Political Generations, Micro-Cohorts, and the Transformation of Social Movements." *American Sociological Review* 62: 760–778.

Wilder, Thornton. 1953. "The Silent Generation." *Harper's Magazine* 20 (April): 34–36.

Wilford, John Noble. 2008. "Plague: How Cholera Helped Shape New York." *New York Times* (April 15): F4.

Wilkerson, Isabel. 2010. *The Warmth of Other Suns.* Random House.

Williams, Raymond. 1980. "Advertising: The Magic System." Pp. 170–195 in his *Problems in Materialism and Culture.* New Left Books.

Williams, Timothy. 2010. "Geraldine Doyle, 86, Iconic Face of World War II." *New York Times* (December 30): A18.

Williamson, John. 1993. "Democracy and the 'Washington Consensus.'" *World Development* 21: 1329–1393.

Wilson, Duff. 2007. "Teenage Smoking Rates Spur Calls to Renew Anti-Tobacco Campaigns." *New York Times* (July 8): B2.

———. 2010. "Cigarette Giants in Global Fight on Tighter Rules." *New York Times* (November 13): A1, A22.

Wines, Michael. 2009. "In China, No Plans to Emulate West's Way." *New York Times* (March 10): A5, A8.

Winner, Langdon. 1977. *Autonomous Technology*. MIT Press.

Wolf, Eric R. 1982. *Europe and the People Without History*. University of California Press.

Wolf, Jacqueline H. 2007. "Saving Babies and Mothers: Pioneering Efforts to Decrease Infant and Maternal Mortality." Pp. 135–160 in John W. Ward and Christian Warren (eds.) *Silent Victories*. Oxford University Press.

Wong, Edward and Jonathan Ansfield. 2010. "China Replaces Leader of the Restive Xinjiang Region." *New York Times* (April 25): A10.

Wright, Wynne and Gerad Middendorf, eds. 2008. *The Fight Over Food*. Pennsylvania State University Press.

Yergin, Daniel. 1991. *The Prize*. The Free Press.

Yiftachel, Oren. 2010. "'Ethnocracy': The Politics of Judaizing Israel/Palestine." Pp. 269–306 in Ilan Pappé and Jamil Hilal (eds.) *Across the Wall*. I. B. Tauris.

Yunis, Muhammed. 2009. *Creating a World Without Poverty*. With Karl Weber. Public Affairs.

Zakaria, Fareed. 1997. "The Rise of Illiberal Democracy." *Foreign Affairs* 76/6: 22–43.

Zedong, Mao. 1970. "Report on an Investigation of the Peasant Movement in Hunan." Pp. 110–123 in *Selected Works of Mao Tse-Tung*. Abridged by Bruno Shaw. Harper & Row.

Zerk, Jennifer A. 2006. *Multinationals and Corporate Social Responsibility*. Cambridge University Press.

Zernike, Kate. 2007. "Why Are There So Many Single Americans?" *New York Times* (January 21): Section 4: 1, 4.

———. 2010. "In Power Push, Movement Sees Base in G.O.P." *New York Times* (January 16): 1A, 15A.

Zich, Arthur. 1997. "Before the Flood." *National Geographic* 192 (3): 2–22, 28–30.

Name Index

Subject Index

About the Author

Garth Massey (PhD, Indiana University-Bloomington) is professor emeritus of international studies at the University of Wyoming where he was director of international studies and professor of international studies and sociology. He authored *Subsistence and Change: Lessons of Agropastoralism in Somalia* (Westview) and edits *Readings for Sociology* (W. W. Norton). He currently lives in Portland, Oregon.

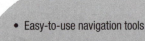